D0053317

# EARL WARREN

# EARL WARREN
## A PUBLIC LIFE

## G. EDWARD WHITE

New York     •     Oxford
OXFORD UNIVERSITY PRESS
1982

Copyright © 1982 by Oxford University Press, Inc.

Library of Congress Cataloging in Publication Data

White, G. Edward.
  Earl Warren, a public life.

  Includes index.
  1. Warren, Earl, 1891–1974. 2. Judges—United
States—Biography.  I. Title.
KF8745.W3W45      347.73'2634  [B]      82-2105
ISBN 0-19-503121-0      347.3073534 [B]    AACR2

Printing (last digit): 9 8 7 6 5 4 3 2 1

Printed in the United States of America

For

Elisabeth McCafferty Davis White

# Preface

This book has had a number of "working titles" as its shape and focus have changed and its purposes have become refined. At one point I thought of calling it an "interpretive biography", but the fact that all biographies are interpretive has deterred me from retaining that nomenclature. I had wanted the qualifier "interpretive" to emphasize that this has been a selective treatment of Earl Warren's life, and I had wanted "biography" to emphasize that the treatment here does follow a chronological narrative.

But perhaps it was better to abandon that title altogether, because it might suggest a stance of unintrusiveness in the author that is commonly identified with the biography genre. I have tried to be unintrusive in places throughout the narrative, but have not always succeeded, and sometimes I have not tried at all. The result is that this book is one man's portrait of Earl Warren and one man's reactions to his public career. I hope it will provide some enlightenment to those that are curious about Warren's life, and I even hope that it might cause some prospective Warren biographers to put off their efforts for a while. I have been pleased to discover a good many things about Warren's career that have not previously found their way into print, and I have tried to dispel some myths, correct what I take to be some errors, and advance some explanations of my own. But a person as complex and interesting as Warren can never completely be captured in one setting.

I have had a good deal of help on this book from colleagues, friends, acquaintances and supportive staff. Louis Auchincloss, John F. Davis, William H. Harbaugh, A. E. Dick Howard, J. C. Levenson, and J. Harvie Wilkinson III have read the entire manuscript, sometimes in embarrassingly early stages. Their readings have been responsible for changing my focus and improving my thinking. Amelia Fry, Douglas Leslie, Gabrielle Morris, George Rutherglen, Sara Sharp, and Miriam Feingold Stein have read portions of various drafts and given me the benefit of extensive commentary. Ms. Fry, Morris, and Sharp, and Ms. Willa Baum are also responsible for enabling me to take full advantage of the invaluable collection of documents at the Regional Oral History Office in the Bancroft Library at Berkeley. I am also indebted to the manuscript staff at the Library of Congress, the William and Mary Law Library, the California State Archives, The Harvard Law School Library, and the Butler Library at Columbia University. Terry R. Richards and Mimi Stein have assisted in the checking of West Coast sources. Gary L. Francione and Linda Ray have helped with the editorial process, and Ms. Ray and William Wells have assisted in the preparation of the bibliography of Warren's Supreme Court opinions. Marsha Rogers lent her expertise to the process of archival research. Linda C. Williams assisted with various editorial stages. Paula Price, Donna Walker and Diane Moss of the University of Virginia School of Law have typed more manuscript drafts than they would care to remember. Linda Woodson helped with the proofreading. Carol Franz has prepared the index. Susan Rabiner of Oxford Press has rendered her usual excellent editorial assistance. Portions of the book first appeared, in somewhat altered form, in the *Virginia Quarterly* and the *Virginia Law Review*.

Susan Davis White, Alexandra V. White, and Elisabeth McC. D. White have been companions in this enterprise. That is a role to which the first two persons are accustomed; this book is dedicated to Elisabeth in the hope that she will allow her father to write some more books in the years ahead.

*Charlottesville*
*August 1981*

# *Contents*

# EARL WARREN

# Introduction

Little more than a decade after his resignation as Chief Justice of the United States Supreme Court, Earl Warren is, curiously, no longer a figure of widespread public interest. Even his death in 1974 did not generate a surge of scholarly or popular retrospective commentary. Warren served as Chief Justice of the United States Supreme Court through one of the few periods in American history during which the Court became the center of intense public controversy, and was himself a highly visible and controversial figure. But a general consensus of opinion seems to have emerged that for all Warren's achievements, and for all the controversy surrounding his career, there is really not much one can say about him. He was, according to this opinion, a relatively simple man who happened to occupy a number of visible positions; a person of ordinary intellectual talents, of relatively bland disposition, and of limited horizons whose success, paradoxically, may be attributable to no more than the fact that he was "average."

Warren himself contributed to this stereotype. His public persona was that of the conventional politician; he appeared to strangers and to acquaintances as genial, hearty, affable, and perhaps a bit uninteresting. His public speeches, throughout his career, tended to be ponderous and wooden; neither humor nor pithiness was his forte. Nor was his conversation vivid: He was neither a raconteur nor a wit, he did not enjoy pointed analysis or gossip. He was devoted to the preservation of his privacy and his family's autonomy, so much so that few of the details of his intimate life became a matter of public knowledge. He was not a prolific letter writer and did not tend to leave public records of his actions, either as a California public official or as Chief Justice. He seemed perfectly content to have others think of him as an unassuming, easygoing fellow, who preferred conversations about sports and pleasantries to intellectual discourse.

The result of these impressions of Warren is that a certain pat explanation of his public life has become entrenched. In brief compass, the explanation goes as follows. Earl Warren was a conservative, law-and-order California politician who, through the vagaries of politics and accumulated political debts, was appointed Chief Justice of the United States by Dwight D. Eisenhower. Once on the Court something seemed to happen to Warren, for he began reversing a number of positions he had taken earlier, supporting the rights of criminals, forcing legislative reapportionment, and demanding that persons accused of being Communists be given full procedural protections. By the time Warren retired he was regarded as a thoroughgoing liberal judge, and his Court was considered one of the most liberal in history. The basis of this astonishing metamorphosis in Warren remains largely mysterious; but one likely supposition is that Warren, who was himself not much of a legal technician, was influenced by the positions of Justices Hugo Black and William O. Douglas, themselves outspoken liberals and experienced jurists.

There are three assumptions in this interpretation of events that this book seeks to challenge. The first is that Warren was a conservative California politician; I shall suggest that neither the term "conservative" nor the term "politician" accurately describes Warren's career as a California public official. The second is that Warren underwent a marked change in his attitudes once on the Supreme Court. I shall argue that his public life can be seen as of a piece and that the surface contradictions in his thought can be seen as manifestations of a deep commitment to a general set of principles that were consistent in themselves. The third is that Warren was not a legal technician and that his jurisprudential views were largely derivative. I shall contend that Warren was merely a different kind of legal technician, unorthodox rather than inept, and that his theory of judging, while uniquely his, was not without its own theoretical integrity.

In the process of seeking to revise the conventional portrait of Warren, I shall also address an apparent paradox left unresolved by the standard interpretation of Warren's career. The paradox emerged from a contrasting of Warren's apparently modest abilities with the length and breadth of his achievements. Such a contrast inevitably invites observers to speculate on how a person who was so average in his talents could have exercised such influence. For one cannot gainsay Warren his accomplishments. From 1920 to 1952 he was elected to every office in California he sought: He captured the nominations of both major political parties as governor, was elected to the governorship for an unprecedented third term, and achieved all that without any permanent political organization. Then, on retiring from the governorship, Warren was named Chief Justice of the United States, a position he held for sixteen years. After his retirement two polls placed him among the greatest justices ever to sit on the Supreme Court; the second of the polls ranked him the fourth leading justice in American history. In the face of these accomplishments, and given the apparent evidence of Warren's modest abilities, students of his career have resorted to words such as "luck," "magic," and "mystery."

My own efforts to investigate this paradox have led me to the very opposite conclusion; namely, that there was much more to Earl Warren than previous scholars have given him credit for. In a sense this conclusion squares with and perhaps was anticipated by my own personal reactions to Warren, for whom I served as a law clerk in the 1971 term. On going to work for Warren, I had a surface impression of him as a justice and as a person; over the year that impression was first reinforced and then disabused, leaving me with the feeling that there were dimensions to Warren's character I had not yet fathomed. In writing this book, that feeling has stayed with me. My subject, moreover, has not yielded up information easily; indeed he has laid many a false trail.

I originally envisioned this book as an extended essay on Warren's career, one in which I might address some of the themes suggested by the conventional stereotype. In the course of research I abandoned that plan and shifted to a format more resembling that of the conventional biography. I make no claims, however, for exhaustiveness of treatment; I have left out a good deal of information and much remains for other scholars to address. My decision to shift the focus of my work was made when I came to the realizations that much of the explanation for the success of Warren's career lay in his years in California and that there were incidents in those years, such as the *Point Lobos* case, the Max Radin affair, and the loyalty oath controversy, that needed to be reconsidered. As detail was added to some sections of my treatment, it seemed logical to add it to others, such as those on the Warren Commission and Warren's years in retirement, so that the result has been a longer and fuller account than I (or perhaps the publisher) originally thought necessary.

The thematic organization of this book rests on a juxtaposition of Warren the person against Warren the public figure. Individual chapters offset Warren's attitudes at a given point in his life with the offices he held at that point in his life. In a sense this book follows others in examining "the fit between an individual's style and the requirements of his political roles,"[1] although I have not used the term "style" in the technical sense in which it has been employed by social scientists. Put simply, I have been repeatedly interested in searching for common attitudes and themes in Warren's life and in considering the extent to which the public roles that he occupied contributed to those attitudes.

I find Warren to have been one of the significant figures in twentieth-century American history. Warren's significance is addressed in three of the book's later chapters, the conclusion of which can be stated here. As a California public official Warren represented a style of decision making and an attitude toward public service, which I have labeled Progressive, that transcended the political context of its origins and can be seen as a symbol of educated reformist thought in early twentieth-century America. As a judge Warren developed a theory of judging that combined an ethical gloss on the Constitution with an activist theory of judicial review. That theory has been a major, if misunderstood, influence on late twentieth-century American jurisprudence. And as a reformer whose ethos eventually abandoned Progressivism for liberalism, Warren was representative of many Americans whose lives spanned the first seventy years of the twentieth century. Warren and his contemporaries conceived of America in those years as a nation growing more and more hospitable to the ideas of affirmative governmental action and protection for civil liberties. Warren's vision, I conclude, is both a symbolic vision of his times and a vision that is passing into history.

# ONE

## THE CALIFORNIA PUBLIC OFFICIAL

# I

# Preparation

Earl Warren was born on what was then called "Dingy Turner Street" in Los Angeles, California, on March 19, 1891.[1] Seventeen years later, in the middle of August, 1908, Warren boarded a train for a "long, miserable trip [from Bakersfield, where his family had moved in 1894] . . . through the valley . . . to the San Francisco Bay area."[2] The journey was also a symbolic one. Warren was leaving "an arid landscape," with "long, hot Bakersfield summers" for "cool sea breeze(s)," which he found "exhilarating beyond description." His train arrived in Oakland, and there he boarded a ferry for San Francisco, where he would stay the night before embarking for Berkeley. He was to spend the next six years, from 1908 to 1914, as an undergraduate and law student at the University of California. "As I stood on the bow of the ferryboat," Warren recalled, "I filled my lungs with refreshing air and said to myself, 'I never want to live anywhere else the rest of my life.' "[3]

The journey from Bakersfield to the Bay Area was predictive of much of Warren's public life. Warren would make a number of journeys in his lifetime, some involving relocation in the literal sense and some in the figurative. On several such journeys, he left not only a place but also the provincialism its inhabitants represented. Each of the journeys, until the very end of his life, would be in some way exhilarating: an opportunity for him to grow, a new challenge, an accomplishment in which he could take pride. On each of these trips,

one could also say, he left in the "dust" of their limited expectations
for him those who had been his critics and opponents.

If the themes of contrast and change were present in that initial
journey in 1908, so too was the theme of continuity. Warren was
beginning a sixty-six-year love affair with northern California, an
affair that did not cool down despite a hiatus of nearly two decades
outside the state. In the early 1970s, after he had been living in
Washington and serving as Chief Justice of the United States since
the fall of 1953, Warren, recalling the feelings that were first evoked
on the 1908 Oakland-San Francisco ferry, concluded that "I have
never really changed my mind" about living in the Bay Area.

Warren's roots were deeply Californian. Until 1948, when he ran
as Thomas Dewey's vice-presidential candidate, he had neither held
a job outside the state nor traveled outside its borders for any extended period. His previous campaigns had all been for statewide
office; he had been elected governor of California twice and would
be elected, in 1950, for an unprecedented third term; he understood
California's political culture as well as anyone else in the state. The
experience of being a Californian was the principal constant of Warren's public life from 1919, when he obtained his first position in
public service, until 1953. His preparation for the chief justiceship
centered in California.

This chapter considers a dimension of that preparation; one might
say that it focuses on the "formative years" of Earl Warren. That
conventional phrase, however, conceals as much as it reveals when
applied to Warren's life. The phrase has a psychological tone, and
devotees of psychological biography will find, I fear, that Warren
has not been a very cooperative subject. As noted, he was not inclined to put his thoughts on paper and was careful not to preserve
intimate records; his rhetoric as a public official was often bland
and ponderous; his relationships with family members were self-
consciously private matters; he chose as intimates persons of marked
loyalty and discretion. There is scarcely enough data for the professionally trained biographer, let alone for lay analysts. "Formative" in
this chapter is thus used in a modest sense.

&

Earl Warren was fourteen pages into his memoirs when he decided
that he had reached "an appropriate place . . . to tell something
about my father."[4] The context for this decision was Warren's dis-

cussion of the Southern Pacific Railroad, a dominant economic and political force throughout California in the 1890s. Methias ("Matt") Warren, Earl's father, had, shortly after his arrival in Los Angeles in 1889, found work as a car repairman and car inspector for the Southern Pacific. Matt Warren was to remain, with one significant exception, an employee of the Southern Pacific until 1929, when he retired on a pension and began to invest in real estate. Nine years later, at the age of seventy-four, he was murdered by an unknown assailant in Bakersfield.

For his only son a significant dimension of the figure of Matt Warren was his status as employee of the Southern Pacific. In the 1890s Southern Pacific employees such as Matt Warren received comparatively low pay, worked long hours, and could hold only fragile expectations of job security. An 1894 strike by Southern Pacific employees, part of the nationwide Pullman strike, dramatized those features of railroad employment. Matt Warren joined the strike, was "blacklisted," and could not find work in Los Angeles. He first went to San Bernadino, ninety miles east, and secured a job with the Santa Fe Railroad. Some months later a repairman position with the Southern Pacific became available in Sumner, two miles east of Bakersfield, where the Southern Pacific had moved its Los Angeles repair shops.

Matt Warren took the position and located his family "in a little row house across the street from the shopyards," and in that house Earl Warren spent his childhood. Matt first worked as a railroad car inspector on the night shift, then "as foreman of the Bakersfield car repair shops," and finally as "master car repairer for the San Joaquin division" of the Southern Pacific, which stretched from Fresno to Los Angeles.[5] As the railroad prospered, so did Bakersfield: After Earl Warren left for Berkeley the city of Bakersfield annexed the town of Sumner, where oil had been discovered in 1899.

Matt Warren's identification with and dependence upon the Southern Pacific made a vivid impression on his son. Earl himself worked as a "call boy" for the railroad, notifying crew members who were scheduled to service an incoming train. That job, coupled with his father's experience, generated a memory that was to endure in sharply etched form. In his memoirs Warren talked about the Southern Pacific:

> I was dealing with people as they worked for a giant corporation that dominated the economic and political life of the community. I

saw that power exercised and the hardships that followed in its wake. I saw every man on the railroad not essential for the operation of the trains laid off without pay and without warning for weeks before the end of a fiscal year in order that the corporate stock might pay a higher dividend. I saw minority groups brought into the country for cheap labor paid a dollar a day for ten hours of work only to be fleeced out of much of that at the company store where they were obliged to trade. I helped carry men to the little room called the emergency hospital for amputation of an arm or a leg that had been crushed because there were no safety appliances in the shops or yards to prevent such injuries. I knew of men who were fired for even considering a suit against the railroad for the injuries they sustained. There was no compensation for them, and they went through life as cripples. I witnessed crime and vice of all kinds countenanced by corrupt government. . . . As a train caller, I saw men rush from the pay car to the gambling houses and never leave until they had lost every cent of their month's laborious earnings. . . . I saw conditions in many of the houses where the breadwinner had lost his earnings at the gaming tables. . . . The things I learned about monopolistic power, political dominance, corruption in government, and their effect on the people of a community were valuable lessons that would tend to shape my career throughout life. . . .[6]

Several of the themes of that paragraph—the vulnerability of people to large corporate enterprises, the tendency of such enterprises to become political as well as economic units, the Southern Pacific's indifference to minorities and to the health of its employees, the tendency of persons who worked under conditions of economic dependence to squander their wages in pleasures and vices—were to become starting premises for Earl Warren the public official. They transcended his relations with his father, but they originated there.

According to his son's memory, Matt Warren dealt with economic dependency in three ways. He stood up to the physical demands of his work: He was remembered as "wiry and tireless." He shunned the temptations of a nomadic life: He was not a "boomer," moving "from one town or one railroad to another," but a self-contained, parsimonious, temperate family man, who, along with Earl's mother Christine (called Chrystal), was a "total abstainer" from cigarettes and liquor "as a matter of personal forbearance." Finally, Matt pressed his children to escape economic dependency through education. He "made it a life passion," Earl Warren recalled, "to see

that my sister, Ethel [who was four years older than Earl] and I
should have . . . a good education. . . . It was a great sorrow to
him that she was not able to go to college." [7]

In his father, then, Earl Warren saw the costs of economic dependency; he also saw the benefits of "personal forbearance." Matt
Warren husbanded his money carefully, planned for the future, and
encouraged his son to do the same. He once told Earl that " 'saving
is a habit like drinking, smoking, or spending,' " and that Earl should
" 'always save some part of what you earn.' " [8] Saving money became Methias Warren's "guiding principle." Educational advancement, as noted, was another major concern. Matt did "everything
possible to keep me interested in learning," Earl Warren said. Until
"he was well into middle age" Matt took correspondence courses in
accounting and mechanics; he cajoled the Summer grammar school
principal into admitting Earl at the age of five; he taught Earl to read
and to write by that age; he "always had [Earl] doing a little homework and mathematics in advance of [his] class at school." [9] Through
forbearance and education one escaped the thralldom of the Southern Pacific.

In later life Earl Warren was to retain many of the values of his
father. In succeeding chapters we shall note his strong commitment
to family responsibilities, his concern for parsimony and "saving for
the future," his belief in "personal forbearance" in his habits, and his
abiding belief in education, which he once called, in an unpublished
draft of a Supreme Court opinion, a "fundamental" constitutional
right. [10] But there were sad memories of Matt Warren as well, memories that Warren understandably did not recall in print. Toward
the end of his life Matt Warren grew increasingly irascible and eccentric, and by the time he was found murdered, on the morning of
May 15, 1938, Chrystal Warren had already left him. On arriving at
his father's home the morning after the murder District Attorney
Earl Warren, then in the midst of an ultimately successful campaign
for attorney general of California, unabashedly wept.

The thematic importance of Chrystal Warren in Earl Warren's
life is more difficult to discern. Christine Hernlund Warren was born
in the province of Holsingland, Sweden, in 1867. The Hernlunds
emigrated to the United States when Chrystal was an infant, settling
first in Chicago, where their home was burned in the Great Fire of
1871, and subsequently in Minneapolis. There she met Methias
Warren, who was himself of Scandinavian extraction, having been

born in Havgesund, Norway, and who had likewise been brought to America as an infant. The Warren family's original name had been Varran, but at the time of Methias' birth in 1864 it had been changed to Warren.[11] Matt and Chrystal were married on February 14, 1886, in Minneapolis, and left for California in 1889.[12]

Earl Warren's published recollections of his mother were not particularly revealing. She was described as having a "happy . . . home life as a girl," and "affectionate . . . relations with her sisters and her brothers," few of whom Earl had ever met. She was called "gentle" and "dear" at points in the narrative. An occasional anecdote was related, such as Chrystal's reluctance "to agree to cutting the little ringlet curls [on Earl] that she had nurtured since [his] earliest childhood," or her insistence on Earl's wearing "knee pants" to school, which made him feel "humiliated."[13] Chrystal Warren had a series of illnesses in her later life, including eye and abdominal operations, and died in 1941, three years after Matt's murder. For the last several years of her life she lived in Oakland in an apartment near Earl's sister, Ethel Warren Plank. While correspondence indicates Chrystal was warmly regarded in the Earl Warren family circle, she does not appear to have been as symbolic a figure for her son as was Matt Warren.

In his memoirs Warren described his childhood and youth as a series of "lessons."[14] Matt Warren had taught, consciously or otherwise, the lessons of economic self-sufficiency, personal forbearance, and educational achievement. He had also introduced Earl Warren to another, less pointed lesson. That lesson was a strategy for coping with the presence of authority. Matt Warren was both a person with authoritarian tendencies and a person forced, for most of his life, to be subservient to the authority of others. While Earl Warren was to exhibit the former tendencies in his public career, he rarely, and then uncomfortably, endured the latter state. Through his father's experience Earl was exposed to the struggles of a man seeking to survive, even to flourish, in an authoritarian atmosphere. Matt Warren's techniques for escaping oppression—frugality, sobriety, and self-education—were to appear as parts of the composite of his son's adult values. But the theme of coping with authority took on an added dimension for Earl Warren; it became a theme invested with moral overtones.

"Neither of my parents," Warren said in his memoirs, "was inclined to moralize greatly with me."[15] Nonetheless the example of a parsimonious, abstemious, achievement-oriented household in the midst of a profligate and drifting subculture of railroad "boomers" was one from which a young person might draw moral lessons. Earl Warren, in addition, had demonstrated a tendency to see the question of obedience to authority as one with moral dimensions. Two examples are illustrative.

On arriving in Berkeley, Earl Warren immediately went to the La Junta Club, a type of fraternity to which a friend of his family had belonged. In the years that Warren was an undergraduate the Berkeley fraternities accepted new freshman members shortly after fall registration; it was not uncommon for prospective recruits to live in the fraternities while being considered for membership. Warren stayed at the La Junta Club for several weeks before being formally initiated, and at "the first Sunday dinner at which the full membership was present" he told a story about Kern County High School, from which he had just graduated.[16]

He and his friends, Warren said, had had a " 'cat and mouse' relationship" with the Kern County High School principal. That individual had "had a suspicious and accusative turn of mind, and often charged an entire class with wrongdoing when no more than one or two could have been responsible." Late in Earl's senior year the principal had temporarily expelled Earl and two of his friends for being late to a play rehearsal; he "would spy on classes by various means"; he "even penalized classes for *suspicion* of cheating." This sort of behavior fostered among Earl and others "a game of who could outwit whom": They regarded the principal as "fair game" for retaliatory cheating, and they cheated with defiance and pride.[17]

Shortly after the dinner at which he had told the story, Earl was taken to dinner by a senior at Berkeley, Herbert Whiting, who was a member of the La Junta Club. Whiting brought up the "cat and mouse" story and told Earl "in a fatherly way" that "this was not the way things were done at the university," which administered its examinations through an honor system and where students "were expected to be honorable in taking examinations as in all other things." Warren was "greatly embarrassed" and "deeply touched" by Whiting's conversation and "promised him I would never violate the honor system."[18]

The second example comes from Warren's years as a law student at Boalt Hall, where he matriculated in 1912. At that time Boalt Hall had a two-year program for those with Bachelor of Arts degrees from

Berkeley; such students could get an LL.B. degree in one year and a J.D. in two. The classroom methodology at Boalt was socratic; the students read cases from casebooks and the professors engaged students in dialogue in class. "[Answering questions on casebook material] did not appear to me to be a practical approach to becoming a lawyer,"[19] Warren later said, and accordingly he took two steps. First, he refused to speak in class, and at the end of his first year was reprimanded by Dean William Carey Jones, who suggested that he might not ever receive a J.D. Warren, according to his recollection forty years later, had argued at the time that there was no requirement that one speak in class in order to qualify for the degree and that he had no intention of failing his written examinations.[20]

In addition, Warren took a job in a local law office, in direct violation of Boalt Hall's rules on outside employment. Such rules were a product of a period in which the process of becoming a lawyer was still significantly tied to reading law in law offices. Very few states required attendance at a law school as a prerequisite to admission to the bar, and California was not one of them. Warren would likely have been dismissed from Boalt Hall had he been discovered, but he then would still have been eligible to qualify for the California bar, the entrance requirements of which were not severe. (Warren remembered "a crowded courtroom of applicants" who "would all be tested [by the bar examiners] in a couple of hours."[21]) Warren himself, after eventually receiving his J.D. from Boalt Hall, was admitted to practice on the motion of a member of the law school faculty.

The examples from Warren's years in law school suggest that the lessons taught by Matt Warren had begun to blend in with insights Earl Warren was gleaning for himself. Matt had told Earl, when the latter set out from Bakersfield to Berkeley in 1908, "you are a man now, and I am sure you are going to act like one," and Herbert Whiting and the other "fatherly" La Junta Club residents had reinforced that notion: "Childish and irresponsible behavior" was to be put aside at Berkeley.[22] But not only did high-jinks persist during Warren's college years ("occasionally we stopped traffic on the streets, put streetcars out of commission for hours, overrode the police by disproportionate weight of numbers, made loud tumultuous noises until all hours of the night, and . . . even broke up political meetings"[23]), Matt Warren's ambivalence toward authority had persisted as well. Earl Warren, however, was to transform that legacy from his father, with its narrow economic origins, into a basic stance toward the world. He sought out positions of authority, but at the same time he scrutinized the directives of those in authority and reserved judgment on whether he would comply with them.

Warren began to show an interest in political matters during college. A classmate remembered him as being "very active politically" and "always backing somebody for some kind of office."[24] In his memoirs Warren downplayed his participation, speaking of being "only mildly interested in campus politics" and of "occasionally helping a friend who aspired to office."[25] The truth probably lies somewhere in between. Warren's friends were regularly active in politics, but he, personally, "sought no leadership and achieved none,"[26] either as an undergraduate or in law school. Originally, Warren seemed to link politics less with leadership than with "companionship," which he later called "the greatest thing I found at the University."[27] He sought out such companionship by joining organizations ranging from Skull and Keys and an "unnamed" university society to The Young Lawyers Club of Alameda County, which met to hear speakers on a biweekly basis.[28]

Perhaps more predictive of Warren's later career than his political activity at college was his early involvement with reform politics in California. Warren's years at Berkeley were years in which Progressivism became a major force in California politics. At the turn of the century, according to George E. Mowry, the principal historian of Progressivism in California, the state "had only the shadow of representative government, while the real substance of power resided largely in the Southern Pacific Railroad . . . ." Through "means usual and unusual," Mowry wrote, "the Southern Pacific eventually organized a government within a government." The Southern Pacific successfully ran hand-picked candidates for governor and United States senator, exercised a "stronghold on municipal politics," and controlled appointments to the state judiciary.[29] "From the village constable to the governor of the state," one contemporary critic of the railroad declared, "the final selection of the peoples' officials lay with . . . the railroad machine."[30]

As late as 1910 Frank Norris, in his novel *The Octopus*, was bemoaning the established power of the Southern Pacific, but by then the railroad was unmistakably on the defensive. A new political coalition was emerging in California, made up of recruits from both major parties. Originally known as the Lincoln-Roosevelt League, this coalition emphasized "good government" in municipalities, opposition to railroad machine politics throughout the state, mild support for organized labor, a broadening of the bases of political participation, and a restoration of morality in public life.

In 1909, while in his second year at Berkeley, Warren offered his services to the vice-president of the league, Hiram Johnson, a former Sacramento city attorney and the principal prosecutor in a notorious series of graft trials in San Francisco that had begun in 1906. On the night of the San Francisco municipal elections in 1909 Warren was asked to go down to a waterfront precinct in San Francisco to "detect fraudulent counting," an experience he recounted in his memoirs. Two months later the league took the name "Progressives."[31] On February 20, 1910, the league announced the candidacy of Hiram Johnson for governor. Warren was not old enough to vote in the 1910 election, but he enthusiastically supported Johnson, who won the Republican nomination in a state-wide primary and then the election. With Johnson's election, Progressivism was to become the reference point for Earl Warren's future politics.

Progressivism in California can be seen as a process by which a different set of political values were injected into public affairs. Three such values were especially prominent: the value of honesty and openness in government; the value of affirmative governmental action in the "public interest," as opposed to "special" economic or class interests; and the value of fidelity to public opinion, which was taken to be enlightened, forward-looking, and honorable.[32] The reforms of Hiram Johnson's administrations can be seen as efforts to foster some, or all, of these values. The democratization of political participation effectuated by the initiative, referendum, and recall and by cross-filing and direct primaries fostered the values of open government and fidelity to public opinion. The creation and revitalization of regulatory governmental agencies fostered the values of affirmative governmental action and, given the Progressives' assumption that governmental officials would be moral and honorable persons, the value of honesty and openness in government. The revival of the state's civil service apparatus fostered all three values: Civil servants would ostensibly be less susceptible to corruption, more active in their pursuit of the public interest, and more responsive to public opinion.

Even the Heney-Webb Bill of 1913, which provided that aliens not eligible for American citizenship could not own or possess land unless a treaty between the United States and their home nation permitted it, could have been said by its proponents to be consistent with public morality and the "public interest." That bill, which became known as California's "Alien Land Law," demonstrated that Orientalist racism and "reform" were compatible values for California Progressives. Or as an organ of reform, the *California Weekly*, put the case for the Alien Land Law in 1909, the legislature "should limit Mongolian ownership of soil to a space four feet by six," because "a white population and a brown population, regardless of nationality or ideals, can never occupy the same soil together with advantage to either."[33] The two should "dwell apart and in amity."[34]

Earl Warren was to espouse such Progressive political values in his career as a California public official. While the values did not wholly define the framework from which he would participate in politics or hold office, they were significant components of it. Indeed one sees, in the coming to maturity of Earl Warren at the same time that Progressivism reoriented the political culture of California, a historical convergence that is perhaps overfraught with symbolism and perhaps deceptive in its suggestiveness, but surely worth pondering.

A new type of person entered California politics with the triumph of Progressivism. That person was "new" not only in the sense of holding political values that differed from those of established political figures, but new in the sense of not having been tainted with the older "Southern Pacific" model of California politics that suddenly and dramatically had been repudiated by Hiram Johnson and his supporters. Appeals to honesty, to nonpartisanship, to decisive, "forward-looking" government, and to the "white light" of public opinion were, after 1910, appeals to the mainstream of California political rhetoric. Earl Warren was to make those appeals his own.

And Earl Warren was remarkably well suited to this role. He was a political unknown, with no strong ties to either major party and no apparent political ambitions. His interest in state political issues had been spawned by the dramatic success of Hiram Johnson and his fellow reformers, who had completely changed the rules of California politics within a decade. He was not only ideologically committed to Johnson's reforms of the process of political participation, he directly benefited from them. Most important, Warren *believed* in the political values of California Progressivism. His own childhood furnished dramatic evidence of the power of the Southern Pacific. He had been brought up in an atmosphere where one's moral values were taken with great seriousness. He had developed a tendency to scrutinize the actions of those in authority, holding them to his own moral or practical standards: He was ideally suited to be a watchdog for the "public interest."

Finally, Warren had shown in his youth a marked ambivalence towards issues of individuality and group cohesiveness. He had been simultaneously gregarious and stubborn, a clubman and a person who chafed in too hierarchical an office setting. Out of this ambivalence would come an approach toward officeholding that emphasized the powers and freedoms of a public-minded but autocratic executive. This approach was consistent with an idealized Progressive political role, that of the "white knight" reformer who led crusades on behalf of enlightened public opinion.

The congruence of Warren's evolving character and the growing prominence of Progressivism in California was thus a striking one. It is surely too late in the history of ideas to resurrect a "man for his times" interpretation of history; moreover, we shall subsequently see that Warren's approach toward officeholding was adaptable to political cultures in which Progressive values were no longer paramount. At this juncture, however, it seems worth underscoring the remarkable "fit" between the values Earl Warren extrapolated from his personal experiences as a youth and the values Progressives sought to inject into California politics.

George Mowry's portrait of Hiram Johnson might just as well have served as a portrait of Earl Warren, first as a California public official and later as Chief Justice of the United States. Mowry wrote of Johnson:

> When not on public view his manner was pleasant rather than urbane. He despised formality, whether in public or private life, and he was direct and simple with all of his associates. . . .
>
> Instead of digging for facts he preferred to analyze a situation in conversation with his advisors. He was not quick to act and was often extremely cautious. He preferred the advice of "practical men" and tended to distrust too much intelligence. Aside from the moralities that peppered his speeches, Johnson rarely indulged in speculative thought. . . .
>
> Hiram Johnson had a curious mental tenacity. Once he adopted a position he held to it with all his strength, often irrespective of later evidence. It is doubtful whether he ever really forgave what he considered a personal slight, whether real or fancied. Viewed

most kindly, Hiram Johnson's inflexibility, if not downright narrow
stubbornness, might be considered consistency and fixedness of
purpose. . . . In a struggle, his outward attitude belied all doubt.
He stood for the forces of light, and the opposition was obviously
allied with the devil.[35]

Johnson was, in fact, Warren's political hero, the person to whom
Warren felt indirectly indebted for his subsequent political success.
It was Johnson's "dynamic 1910 campaign" for governor, Warren
wrote in his memoirs, where "he promised to, and after his election
did, put an end to Southern Pacific dominance of the political life of
[California]," that first attracted Warren to state politics. The "reform measures" of Johnson's gubernatorial administrations, Warren
wrote, were "never equaled in California or probably in any other
state before or since." Warren "believed in the progressivism" of
Johnson: "I was committed to the nonpartisanship he brought into
city and county government. . . . I believed implicitly in the direct
primary for all state and local elections, and in the cross-filing procedure for partisan state offices. . . . I believed in the civil service he
established." Even Warren's participation in elective politics was
identified with Johnson. The "enlightened procedures" that Johnson
helped introduce, wrote Warren, "enabled a poor man to seek public
office without the necessity of mortgaging himself to a political boss"
and "enabled a man to maintain his independence in the performance
of his duties."[36]

᚛

As Earl Warren approached his law school graduation in his twenty-
fourth year, his career prospects did not seem particularly bright.
He was to graduate "in routine fashion" from Boalt Hall, not having
distinguished himself by his academic performance. Matt Warren had
supported him through college and law school, but family resources
were limited and the Warren family had no professional contacts that
might be useful to Earl's career. With the help of a few friends, he
first secured a position in the law department of the Associated Oil
Company in San Francisco. That office consisted of Edmund Tauske,
"an irascible old man" who was counsel to the company, and one
secretary. Warren and Tauske did not hit it off. Warren was personally offended by Tauske's "peremptory" attitude toward him and
professionally indignant at Tauske's habit of using Warren's research

without giving the latter credit. Nor did Tauske endear himself to Warren by using him as a messenger to get Tauske his cigars.[37]

A year later Warren resigned, taking pleasure in telling Tauske that "there was no human dignity recognized in the office." A few days after resigning, with his "morale . . . extremely low," he landed a position with the Oakland firm of Robinson and Robinson. There Warren had a much more pleasant experience, although his position, which consisted of keeping the court calendar, doing research, and serving as "leg man to the courthouse on minor matters," was "subordinate and not particularly important" and his salary "meager." A year and a half later, in the spring of 1917, Warren "decided it was time to build a practice for myself," and, teaming up with two of his Boalt Hall classmates and Peter J. Crosby, an established Alameda County trial lawyer, made plans to establish a new law firm.[38] His ambition was to do trial litigation. The firm never formally came into being, however, because the United States declared war on Germany on April 17, and Warren immediately applied for admission as an officer in the Army.

Warren's wartime service had its comic aspects. His application for admission as an officer was twice rejected, the first time because of the large number of applicants and the second because of a hemorrhoid operation that led to pneumonia. By the time Warren was again eligible, officer enlistments were closed, and Warren "volunteered" as a draftee. He was put in charge of a group of draftees, who were sent to Camp Lewis, Washington, in September, 1917. Until January, 1918, Warren, acting in the capacity of first sergeant of this company, supervised the barracks and mess of a group of "soldiers" who had no uniforms, no guns, and little bedclothing. The group drilled with sticks, dug trenches, held track meets, and "improvised in many other ways to give the impression of soldiers."[39] It was regularly quarantined because of outbreaks of measles and spinal meningitis.

Eventually Warren was offered an opportunity to become an officer, which he accepted. He then went with his original company to Camp Lee in Petersburg, Virginia, where he was given the job of training replacement troops for the European theater. The trainees would come for thirty-day intervals. Warren found that "it was impossible to do more than scratch the surface," so that when the troops departed for Europe "to say that they were unprepared for combat service would be a gross understatement."[40] After a miserably hot summer, swine flu broke out in September, and thousands of soldiers died. Warren's company, which at first had no cases, was designated a "model company" and was asked to take special hygienic precautions to ensure that the disease did not spread among them. Immediately after the precautions were instituted, cases of influenza appeared. Twelve soldiers from Warren's company died; he escaped the disease altogether. In November, 1918, Warren was transferred to Camp McArthur, in Waco, Texas, to serve as a bayonet instructor for embarking officers. Two days after his arrival, the Armistice was signed.

Although more American soldiers died from swine flu than from European combat, Warren's war experience was for him, at least, more boring than dangerous. Nonetheless, it left its mark. Chrystal Warren had written to her son during the war that while she "[felt] badly . . . about the necessity of your being in the Army," she would have felt "worse had you been unwilling to be there."[41] For the rest of his life, her son would share that strong sense of patriotism. Twenty years later, when the Second World War broke out, Warren, then in his fifties and attorney general of California, planned to accept a commission in the army, abandoning that decision only when persuaded to run for governor. Even later, in 1963, Warren would agree to take the chairmanship of the Warren Commission only when persuaded to do so by Lyndon Johnson, who asked him once more to "put a rifle to your shoulder."[42] In his last years Warren made provisions in his will to be buried with full military honors in Arlington Cemetery.

On December 9, 1918, Warren was discharged from the army; he returned to Bakersfield. The three men with whom he had earlier made plans for a law partnership were now scattered, two of the original partners still being in the service. A common decision was made that each would go his own way. Warren, "determined not to return to a subordinate position,"[43] accepted a temporary job as clerk to the judiciary committee of the 1919 California legislature, which he secured with the help of Leon Gray, a former associate in the firm of Robinson and Robinson, who had been elected to the legislature that session, and Charles Kasin, a La Junta club contemporary who was also a member of the assembly. With this job, Warren began a fifty-year career in public service.

While the job with the state legislature was certainly a step up for an unemployed lawyer who was now twenty-eight, Warren's ca-

reer prospects still seemed unpromising. Changes in the political culture of California, along with the fortuitous congruity of Warren's
ambition to do trial litigation and the availability of a position in the
district attorney's office of Alameda County, however, were soon to
launch a public career that would be capped by the chief justiceship
of the United States. The convergence, in the early 1920s, of Earl
Warren's special character traits, the political climate of California,
and the role of a law enforcement agency in that climate constituted
a remarkable historical coincidence. It was the first of a series of
coincidences—observers were later to call them evidence of Warren's
luck—that marked Warren's public life.

In the spring of 1919, while Warren was serving as clerk of the
judiciary committee of the California assembly, Ezra Decoto, the
district attorney of Alameda County, which lies adjacent to Oakland, "came to Sacramento to ask the Legislature for an additional
deputy." Warren had previously indicated to friends in the legislature
that he would be interested in joining a district attorney's office to
get some trial experience, and Decoto's visit seemed happily fortuitous to Warren's supporters in the Alameda County delegation. They
"seized upon" Decoto's request and "told [Decoto] that they would
be happy to create the position if he would apppoint [Warren] to
it." [44] Decoto demurred, however, saying that he had promised the
position to another and citing Warren's inexperience.

On hearing of the results of the discussion with Decoto, Warren
"sought Mr. Decoto out and told him that while I would very much
like to be in his office, I would not accept the position if he were
under such duress." [45] This was the first of a number of artful conversations Warren was to have about positions he coveted. As with
Decoto, who left Sacramento with a positive impression of Warren,
these conversations often paid off later on.

Decoto gratefully accepted Warren's withdrawal from consideration for the deputyship and appointed the person to whom he had
promised the position, Charles W. Snook, whom Governor Warren
would later appoint, along with Decoto, to state superior court
judgeships. A year after the conversation, however, when Decoto
was once again in need of a deputy district attorney, Warren's strategy with Decoto proved to have been successful. Having left the
legislature at the close of the 1919 session, Warren had taken a position as deputy city attorney for the city of Oakland, where he had
"a good general experience . . . advising city boards and officers,

writing legal opinions and having a fair amount of contact with the courts." While serving in the city attorney's office Warren received a call from Decoto, who now offered him a deputy position, effective on May 1, 1920. "I asked him," Warren said, "if he was offering it because of pressure from my friends in the legislature, and when he said he was not, I accepted with enthusiasm . . . For the first time in my legal career, I had a sense of liberation such as that which possessed me when I entered the University of California." [46]

The journey to the Alameda County District Attorney's Office was the last symbolic journey of Warren's youth. It was not the end of his preparation—growth and change were constant themes of Warren's public life—but it was the end of the scattered wanderings that had begun with the train ride from Bakersfield. After a relatively unprepossessing early legal career—at the age of twenty-eight he had held two jobs, had served in the army, and had landed an appointment as a legislative clerk making ten dollars a week—Earl Warren found an occupational role that suited him. He was to flourish in that role: Eleven years later, Warren, who by then had succeeded Decoto as district attorney of Alameda County, would be called by the political theorist and presidential adviser Raymond Moley "the most intelligent and politically independent district attorney in the United States." [47]

# 2

# *Law Enforcement:*
# *The District Attorney*

"I became," Earl Warren said of his first years with the Alameda County District Attorney's Office, "a sort of Jack-of-all trades on both the criminal and civil sides." Warren tried criminal cases, defended civil and criminal suits against county officers, advised county boards of education about their prospective civil liability, and wrote legal opinions on zoning and other real estate matters for the county board of supervisors. "It was exciting for me, every day of it," he recalled, "and I made quick progress."[1] Not only was Ezra Decoto's office a baptism in general law practice for Warren, it was a tonic for his self-esteem. Warren "found himself" in the role of law enforcement agent, which became a means of acquiring status, prestige, power, and visibility, all prerequisites for a political career. Law enforcement was Earl Warren's first love in public service.

Warren represented a new type member on the Alameda County district attorney's staff. Ezra Decoto, he felt, tended to see the office "merely [as] a conduit between the activities of the [county] sheriff and the chiefs of police at one end and the [county] Court on the other." Most of Decoto's deputies consisted of "old-timers who . . . engage[d] in private practice extensively," were paid low salaries, did little investigative work, and often tried cases "without adequate preparation." Warren, by contrast, "paid no attention to private legal activity but was ready to work with anyone, night or day, if he wanted help on an office case."[2] As law business expanded in the

Oakland area in the 1920s, several deputies left the office, and Warren absorbed their caseloads. He handled a variety of situations, ranging from prosecutions of radicals under the 1919 Criminal Syndicalism Act to advising governmental bodies about the scope of their powers. After about three years, Warren became one of Decoto's two chief deputies, as well as the adviser to the Alameda County Board of Supervisors.

Warren's experience with the board of supervisors "proved to be of inestimable value."[3] In 1925, Decoto, whose elected term was to expire the following year, resigned as district attorney to take a position on the State Railroad Commission. The board of supervisors had two candidates to choose from to fill the remainder of Decoto's tenure as district attorney—Frank Shaw, the other chief deputy in Decoto's office, and Earl Warren. The story of Warren's appointment has often been cited as an example of the luck that allegedly marked his career in public life.[4]

The Alameda County Board of Supervisors had three members representing urban constituencies and two from rural areas in Alameda County. The three urban votes, which represented the interests of the cities of Oakland and Emeryville, were considered to be under the domination of a machine boss named Mike Kelly. Kelly opposed Warren's appointment, believing the other candidate, Shaw, would be more complaisant about disturbing Kelly's interests, which included "every kind of vice, including gambling, prostitution, and bootlegging,"[5] as well as machine politics. The two supervisors representing rural constituencies were clearly for Warren, but without a crossover vote from one of the three urban representatives, Warren would not get the appointment. The crossover vote came, but it had less to do with luck than with one supervisor's belief in Warren's integrity. Although Emeryville was "absolutely controlled by Kelly interests,"[6] John Mullins, the supervisor from Emeryville, believed that the Kelly machine was planning to "do tricks with the [district attorney's] office" and decided to vote for Warren, even though such an act was certain to result in Mullins' losing his seat on the board of supervisors. Under extreme pressure, Mullins stayed with his decision, did lose his seat, and reportedly told Warren, after the latter's appointment, that "you owe allegiance to no one except your God and your conscience." Warren later called Mullins "not only my original sponsor but also my most ardent supporter during twenty-five years of California politics."[7]

In 1926 Warren sought a four-year elected term as district attorney and "won overwhelmingly, carrying every precinct except the two representing the two county hospitals," whose patients, he believed, invariably "vote against incumbent candidates."[8] He was now the head of the district attorney's office, his own boss, setting his own standards. He was never again to hold a position in the public service where he was not the head of an office, although as attorney general of California he felt "subordinate" to the governor. A position of executive leadership had become the vantage point from which Earl Warren approached public issues; it was as an executive, an administrator, a head of an office, a chief policymaker that he honed his skills and planned for his future. After several years in which he had been uninspired by or irritated with subordinate positions, Warren was granted the opportunity to provide executive leadership. He took full advantage of that opportunity.

In addition to the mandate of independence he received with his election, two other factors helped shape the character of Warren's approach to the office of district attorney. In 1927 Warren's office discovered irregularities in paving contracts involving the Greater Oakland Construction Company. William H. Parker, the deputy sheriff of Alameda County, was allegedly taking bribes in connection with the paving of Oakland streets, and Warren proposed to bring Parker and others to trial. While preparing his case against Parker and his confederates, Warren ran across Franklin Hichborn, a Progressive political observer and free-lance writer, whom Warren identified as "an old Hiram Johnson crusader."[9] On hearing that Warren was preparing his case against Parker, Hichborn referred Warren to an account he had written of the San Francisco graft trial of 1908 and 1909 entitled *The System*.

Warren "received it and read it, not once but a number of times." Later in his life he would comment that no book "did me as much good in my public career as that one did."[10] The thesis of *The System* was that political reform movements were dependent upon public sentiment and consequently could be frustrated simply by delaying tactics. The San Francisco graft trials were an example: Hichborn pointed out that through various delays attorneys for the defendants had been able to keep most of the cases from going to trial while public sentiment was still strongly aroused. In fact, two years after the original indictments had been handed down, only one conviction had been obtained.

Warren was well prepared not only to receive Hichborn's message but to apply it. As early as his days with Robinson and Robinson, a particular interest of his had been court dockets and office calendars. Now, as district attorney, he was to make opposition to delay a principle of law enforcement. First, in the prosecution against Parker, Warren "made up my mind that [delay] would not happen . . . if it was possible for me to prevent it." When counsel for Parker and the others attempted delaying tactics, Warren's office hastily prepared objections and eventually "proceeded to try the . . . cases expeditiously." [11] Thereafter, when the Alameda County District Attorney's Office made its indictments public, Warren and his staff were generally prepared to go to trial immediately and to file opposition motions to requests for continuances, even on a telephone call's notice.

But most important, Warren never lost sight thereafter of the fact that efficient and timely control of an office's calendar or a court's docket was not a substitute for effectively harnessing public opinion. While Warren had clear goals for his office and emphasized thorough preparation in order to further those goals, he also kept before him Hichborn's dictum that reform measures were dependent on the public mood for their success. Where possible, Warren sought to control the timing of his office's actions; later in his career, as he became more overtly political, he often held off action until he found public opinion behind him. For this same reason, once he had declared himself, he fiercely fought delaying tactics by the opposition, lest the moment of favorable public opinion pass before a case could come to trial.

Finally, the Oakland paving prosecution sensitized Warren early in his career to the meaning of Lord Action's observation that "power tends to corrupt and absolute power corrupts absolutely." In 1925 Sheriff Frank Barnett of Alameda County had been implicated in an unsolved murder case; in 1927 Parker and "a half dozen [other] deputies" were indicted for bribery and conspiracy; in 1930 the sheriff of Alameda County, Burton F. Becker, was convicted of collecting bribes from illegal vice operations in Emeryville and Oakland. Warren came to see these events as part of a pattern of "corrupt practices by law enforcement officers." [12] He had earlier noted, in his brief service with the California legislature in 1919, the presence of "downright corruption," a "lobby-ridden" atmosphere, and other "sinister" phenomena. The experience, he felt, had been "far better than a college course in the legislative process."[13] As district attorney, Warren would hold no false notions about the morality of popularly elected officials.

Warren the district attorney was thus committed to efficient, honest, and active executive leadership, and was distrustful of other lawmaking or law enforcing agencies, especially if they had ties with "machine" politics or special interests. The crusades of Warren's office were not only forays against vice, corruption, and crime, they were efforts to professionalize California law enforcement by making it efficient, insuring its independence from partisan politics or pressure groups, and demonstrating its sensitivity to public opinion. Warren was to run the Alameda County District Attorney's Office in the style of a Progressive reformer.[14]

In order to establish this new tone for his office, Warren stressed hard work, nonpartisanship, and a full-time commitment to public service as office virtues. He divided the office into five units and met with each unit once a week for an hour. The meetings, he later said, were "really rugged":[15] Warren kept very close tabs on his personnel. "We kept their feet to the fire," Warren said of his staff; "we didn't have any forty-hour week." He insisted that staff members be loyal and discreet as well as energetic; as a result he had a fairly high turnover of personnel. He tended to hire his deputies from two groups—veterans of World War I and recent law school graduates. This resulted in the district attorney's office being staffed with "young lawyers who were interested in acquiring experience," sometimes at very low or no pay. Warren was parsimonious with raises and slow with praise. His interest was in finding "a personal organization impervious to outside influences."[16]

Alongside efficiency, loyalty, industriousness, and careful office organization in Warren's office went a concern for civic morality and symbolic gestures of nonpartisanship. Warren remembered his office as being referred to as "a 'Boy Scout' organization."[17] The law enforcement issues on which his office was active and visible—prosecutions of corrupt bail bond brokers, bunco artists, corrupt law enforcement officials; the Oakland paving graft scandals; bootlegging, hijacking, and other Prohibition-related offenses—were moral issues. The prosecutions undertaken by Warren's office were repeatedly backed by newspaper stories; Warren regularly gave the press information about his office's activities. He portrayed his task as that of a spokesman against corruption, vice and lawlessness, and had himself photographed at the scene of graft trials and smashing bootleggers' stills.

At the same time Warren gave indications that he intended to preserve his independence from partisan politics as much as possible. He refused campaign contributions in his 1926 campaign for election as district attorney, investing a year of his salary in the campaign. In the absence of partisan support, Warren cultivated the support of civic groups and the press. He joined "several lodges,"[18] including the Masons, the Elks, and the anti-Orientalist Native Sons of the Golden West, and participated in University of California alumni affairs. He also sought out the counsel of prominent newspaper executives. One California newspaperman believed that Joseph Knowland's *Oakland Tribune* "was responsible for Warren's becoming District Attorney";[19] an Oakland reporter thought that Warren "[made] it a practice to cultivate every important person" on her paper.[20] J. Frank Coakley, who succeeded Warren as district attorney, remembered him as "a very aggressive district attorney, what you could call a crusading district attorney," who "had the Knowland people behind him."[21] In his second two campaigns as district attorney, in 1930 and 1934, Warren, who refused to let his staff participate on his behalf, had such a broad base of popular support that he "did not have an opponent who necessitated any campaign expenditure on my part."[22]

Between 1930, when his last major graft case was concluded, and 1938, when he ran successfully for attorney general, Warren turned his attention to the professionalization of law enforcement. His skepticism about the integrity and the effectiveness of local law enforcement authorities merged with his expanding sense of the success of his own methods; he began to widen his professional constituency. Warren conceived a plan for "coordinating the law enforcement agencies of the entire state"[23] as a means of preventing corruption or incompetence and of resisting the influx of organized crime, which threatened to become established in California in the wake of Prohibition. He took several steps to facilitate this plan.

Warren was chairman of the Board of Managers of the California Bureau of Criminal Identification, whose purpose was to maintain a central information bureau, a crime laboratory, and communications between local sheriffs in outlying districts. The board conceived the idea of establishing schools to train police and proposed the establishment of a police training program to a number of state universities, without success. Finally San Jose State College agreed to give a summer program, which was an immediate success, and criminology courses began to appear at other California educational institutions. The efforts to establish the police training school had the additional purpose of bringing together "three law enforcement agencies—the city police, the county sheriff, and the district attorney"—which, Warren noted in his memoirs, "previously had been disparate and envious of each other to the point of disunity."[24]

Warren felt strongly that the California law enforcement agencies should cooperate but remain autonomous and independent, and he conceived of "coordination" as an alternative to centralization. "I was always against a state police or any centralization of all the police power in the state government," he said in 1971,[25] but he was also eager to "spread goodwill among the law enforcement officers of the state and . . . to show that we had a common cause." Eventually he set up a law enforcement lobbying organization, consisting of "a joint committee of sheriffs, district attorneys, and chiefs of police,"[26] with himself as chairman. The committee had as its principal objective "withstand[ing] centralization by state agencies." Warren "kept track of every bill in the legislature affecting law enforcement" and "kept one of my principal assistants at Sacramento through the legislative session."[27] When a bill in which Warren's committee had an interest came before the judiciary committees of the legislature, Warren would make a presentation, flanked by the district attorney, the sheriff, and the chief of police from the districts of all the committee members. "I was probably the biggest lobbyist [law enforcement] had in Sacramento,"[28] Warren said.

"As I worked on this phase of my job," Warren recalled in his memoirs, "I became sufficiently interested in governmental affairs to want to carve out a continuing public career." The logical position to which Warren's techniques of officeholding could next be applied was that of attorney general of California. Warren "aspire[d] to and plan[ned] for the office of attorney general"; although in the early 1930s the position "was considered almost a sinecure," Warren "could see a great opportunity for developing it into an extremely important arm of state government."[29] The incumbent attorney general, Ulysses S. Webb, had held the position since 1902, at a salary frozen by the California Constitution of five thousand dollars a year. Like county district attorneys, Webb could and did practice law privately, as did his deputies, whose state salaries averaged three thousand dollars a year. The office had locations in San Francisco, Sacramento, and Los Angeles, lacked a central filing system, had no clearly defined law enforcement powers, and generally resembled the Alameda County District Attorney's Office as it had been run by Ezra Decoto.

In the early 1930s Warren, taking advantage of his visibility as an advocate for professional law enforcement, began to lay plans for expanding the powers of the attorney general's office. One of the themes of Warren's approach to officeholding was that of expansive jurisdiction. The Alameda County District Attorney's Office under Warren had involved itself "in any case in the county more serious than traffic violations and minor disturbances of the peace,"[30] even though technically it could use only the evidence provided by local law enforcement officials. Through this expansive interpretation of his jurisdictional powers Warren had built his reputation and consolidated his power as district attorney; now, thinking ahead to the next elective office he had his eyes on, he laid plans to expand the jurisdictional reach of the attorney general's office. His goal was to ensure that the attorney general "could displace the elected district attorney and the elected sheriff in any proceeding that he felt called for such action."[31] Amendments to the California Constitution were necessary to achieve this goal, and Warren, in his capacities as secretary and past president of the District Attorney's Association, chairman of the joint lobbying committee, and vice-chairman of the State Bar Committee on the Administration of Justice, was well suited to conceive and to lobby for these amendments.

Warren made his preparations and waited for an opportunity to galvanize public sentiment: That opportunity came in 1934. Prior to that year trains carrying the payroll of the Columbia Steel Works were twice robbed in northern California. On both occasions local police from Alameda and Contra Costa counties failed to cooperate with each other and the bandits were not caught. The press pointed out this lack of cooperation. Then in 1934 two confessed kidnappers, who had murdered their victim, were removed by a mob from the Santa Clara County jail and lynched while local police looked the other way. Warren took advantage of these events, and the spectre of organized crime in California, to call for the adoption of four constitutional amendments in 1934. Of the amendments, which were all passed by popular initiative, three were significant for his purposes: the creation of a State Department of Justice, an idea that was then novel; the expansion of the jurisdictional powers of the attorney general, a proposal that Warren later called "revolutionary at the time,"[32] and an increase in the attorney general's salary to eleven thousand dollars a year; and a license to prosecutors and judges to comment on a defendant's failure to take the stand in a criminal case.[33] In lobbying for passage of the amendments Warren said that "the manner in which the Dillingers, the 'Baby Face' Nelsons, the Machine Gun Kellys, . . . and numerous other criminal gangs have been playing hide-and-seek with the public authorities has truly become a national disgrace," and the "fault has largely been in the lack of organization of our law-enforcement agencies." The law enforcement "business of California," he maintained, "is a gigantic business costing the people of the State thirty million dollars a year, and it is being run in a most unbusinesslike manner."[34]

In addition to expanding the powers and raising the salary of the attorney general's office, the 1934 amendments had the effect of eliminating the unassailable incumbent, Webb, as a future candidate against Warren. Webb declared his candidacy for the 1934 election, and Warren, according to his memoirs, told Webb that "I . . . want your job, but not while you want it" and that "I will never run against you." He also asked Webb to inform him if Webb "should ever decide to retire." Three years later Webb told Warren "that he intended to retire at the end of his term in 1938" because "he did not believe at his age . . . he should undertake the expanded duties of the office."[35] Warren then filed for attorney general on all three major party tickets, eventually receiving the Republican, Democratic, and Progressive nominations. "At that time," Warren recalled, "I had . . . the good wishes of practically all the law enforcement officers of California. . . . They supported me actively in all parts of the state."[36]

&

In 1936, while he was laying groundwork to run for the attorney generalship, Warren encountered the most controversial case of his career as district attorney of Alameda County—the "*Point Lobos*" case, so called because it involved a murder on the freighter *Point Lobos* on the Oakland waterfront. As one of the episodes in Warren's career that commentators have relished for its ironies, the *Point Lobos* case has been widely discussed. A recent reviewer of Warren's memoirs has rearticulated what has become the most established interpretation of Warren's actions in *Point Lobos*. "Many people," he argues, "thought the defendants [in the *Point Lobos* case] were innocent, despite the verdict." He goes on to claim that the investigators from Warren's office engaged in "gross fourth amendment violations" in securing evidence and that "there were also allegations that confessions were coerced both physically and psychologically."[37] This view was also expressed at the time of the *Point Lobos* murder: Warren, who successfully prosecuted four of the five defendants, was accused of a "frame up."[38] The contrast between these claims and Warren's reputation as an outspoken critic of oppressive police tactics during his tenure as Chief Justice is apparent. In his memoirs Warren discussed the *Point Lobos* case in some detail, his account exhibiting a certain defensiveness. The confession of one of the actual murderers, Warren maintained, "was well corroborated by other witnesses"; in Warren's view, "there could have been no doubt about the guilt of the persons on trial."[39]

The *Point Lobos* case deserves extended attention[40] for two reasons. First, it introduces us to another significant theme of Earl Warren's law enforcement years. That theme, which will present itself again in Warren's opposition to the nomination of a Berkeley law professor, Max Radin, to the California Supreme Court (an episode which will be treated subsequently), was anticommunism. Second, the case deserves extensive examination because new research shows that previously published works have inadequately treated the entire *Point Lobos* affair.

In the 1930s the rise of organized labor, combined with depressed economic conditions and sudden public awareness of massive population growth, threatened an older nativist conception of California as an island paradise. Geographically and culturally, California had been isolated from most of the rest of the nation; its economy, climate, political culture and racial tensions had been perceived of by its residents as distinctive and unique in America. Warren, for example, had grown up in an atmosphere where, as far as blacks and whites were concerned, "there was no accepted policy of school segregation,"[41] but where the city of Bakersfield's telephone directory, in the years of Warren's adolescence, did not list the names of residents of Oriental extraction, referring to them only as "Oriental" at a given address.[42] Warren's ideological attitudes had, of course, been influenced by this form of regional provincialism.

In addition, Warren's patriotism was strongly held, his self-image, in the 1930s, was that of a "professional" law enforcement official, and he had begun to become involved in Republican party politics, supporting candidates, such as Frank Merriam, the successful Republican nominee for governor in 1934, who were not sympathetic to organized labor. Warren's views on unionization were not firmly formed at the time. He had an instinctive affection for unions that stemmed from his early experiences with the Southern Pacific and his own membership in a musician's union. But he was as concerned as the "average" nativist Californian was about the unsavory association, in the 1930s, of "foreign" ideologies, especially communism, with organized labor.

The *Point Lobos* case was saturated, from the outset, with the theme of alleged Communist activity in maritime unions on the Oakland waterfront. The murdered man, George Alberts, chief engineer of the *Point Lobos*, had been a vocal anti-Communist. The King-Ramsay-Conner Defense Committee, an organization formed to defend three of Alberts' accused murderers, Earl King, Ernest Ramsay, and Frank Conner, declared itself dedicated to the cause of "labor and progressive people"[43] and was believed by Warren's office to have Communist leanings. In the 1930s the Alameda County District Attorney's Office, Warren recalled, "made it our business to know what was going on in the Communist movement around the Bay and used undercover agents for that."[44] After the *Point Lobos* case broke, Warren's office began to keep files on all organizations that expressed any support for the *Point Lobos* defendants.

From the outset, the *Point Lobos* investigation was a difficult one for Warren's office. No witnesses came forward. One suspect, Ben Sakovitz, had apparently fled the area; other suspects had apparent alibis. The first break for the district attorney's office came after some time: A fireman on the waterfront, who was an informant of the district attorney's office, told Lloyd Jester, one of Warren's deputies, that rumors on the waterfront had linked Sakovitz and a man named George Wallace to Alberts' murder. Sakovitz and Wallace, the informant said, were part of a "beef squad" that beat up recalcitrant union members. The beef squad was closely affiliated with the Marine Firemen, Oilers, Watertenders and Wipers Union (MFOW), which Warren's office believed to be a locus of Communist sympathizers. Jester pursued the informant's lead and secured the cooperation of

Matthew Guidera, another informant, who was seeking to reduce the MFOW's power on the waterfront.

Guidera led Warren's office to Albert Murphy, the assistant secretary-treasurer of the MFOW, who was, unlike most of the high officials in that union, an anti-Communist. Murphy and Guidera shared rooms at the Terminal Hotel on the Oakland waterfront and also shared a concern for the growing power of Communists in waterfront unions. Murphy told Guidera that he had helped George Wallace leave Oakland shortly after Alberts's murder, and that Wallace had, shortly thereafter, sent Murphy a letter from Del Rio, Texas. The letter implicated Earl King, the vice-president of the MFOW, in the Alberts murder, but not explicitly. Murphy showed the letter to Guidera, who immediately telephoned Warren's office. Guidera, Warren's deputy Oscar Jahnsen, and Warren's chief assistant Ralph Hoyt conceived a plan whereby the district attorney's office would install an eavesdropping device in Guidera's and Murphy's rooms, and Guidera would engage Murphy in conversation about his connection with Wallace. This was accomplished, and three members of Warren's staff and a stenographer installed themselves in an adjoining room in the Terminal Hotel.

The eavesdropping was repeated for the next two days, with Guidera talking to Murphy and to Earl King, who said nothing incriminating. Eventually Guidera persuaded Murphy to talk to representatives of the district attorney's office, who, he revealed, were stationed next door. Jahnsen then interrogated Murphy, showing him the eavesdropping devices, and urged him to give a full account to Hoyt in the Oakland office. Murphy eventually did, and his statement implicated King and Ernest Ramsay, another MFOW member, as well as Wallace and Sakovitz. Murphy claimed that Wallace and Sakovitz, after being paid by King, were driven by Ramsay to the waterfront, where they were instructed to beat up Alberts. Sakovitz inadvertently killed Alberts, and subsequently both men left Oakland for Cleveland. They eventually took separate paths; Wallace was attempting to get over the border into Mexico when he wrote Murphy from Del Rio.

Meanwhile Warren dispatched two of his men to Brownsville, Texas, where Wallace had written to Murphy he was heading. Eventually Wallace was apprehended in Harlingen, Texas, questioned, and surreptitiously brought back to Oakland. The last portion of his

journey involved a covert train ride on the Santa Fe Railroad, with
the deputies and Wallace disembarking at Barstow, California, in the
Mojave Desert. An Alameda County patrol car met them in Barstow
and drove them to Oakland.

Wallace arrived in Oakland late on the afternoon of August 30,
1936 and was questioned in a hotel room until 8:20 that night, when
he was booked in the Alameda County jail. He freely talked to Warren's deputies, but because he was nearly illiterate and inarticulate,[45]
his story was difficult to piece together. Eventually he confirmed
Murphy's story, identified Sakovitz as the person who had actually
beaten Alberts to death, and implicated a further MFOW member,
Frank Conner, as a participant in the assault on Alberts. Wallace
said that after Sakovitz emerged from Alberts' cabin, with "blood on
one of his hands,"[46] both men, with King's help, decided to get out
of Oakland.

With Wallace's confession, Warren's office moved to arrest the
remaining members of the conspiracy. King and Ramsay, still in the
Bay Area, were easily apprehended. Both were interrogated; King
asked for a lawyer but was denied access to one. Neither revealed
any incriminating information. Warren had previously held a press
conference in which he named Wallace as one of the murderers and
King and Ramsay as conspirators. He stated that the murder was
tied to labor unrest and to "a campaign of terrorism and sabotage for
communists to gain complete control of the waterfront unions."[47]
On August 31, after a grand jury indictment of all five men, Warren
released another statement to the press. He called Alberts's murder
"a paid assassin's job," in which "the basis of the plot was communistic." He said that the *Point Lobos* case was "not a case against
union labor, but rather one against certain murderous and racketeering individuals." He claimed that "tens of thousands of union labor
men in this state will rejoice that this case has been cleared up."[48]

Of the remaining two suspects, Frank Conner was shortly arrested in Seattle and expedited to California, where he was questioned in a hotel room throughout the night, eventually being hand-
cuffed before being permitted to sleep. The next day Warren's
deputies read Conner a story in the *Alameda Times-Star* that his Seattle attorney, whose counsel he had requested, had not yet agreed
to represent him. Conner, "angry and provoked," then decided to
confess[49] and was led into Warren's office. He had been in the custody of Warren's deputies for twenty-one hours, during all but four

and a half of which he had been questioned. After securing Connor's confession Warren released yet another statement to the press, in which he stated that he hoped to make Conner one of his chief witnesses.

As for Sakovitz, he was never brought to trial. After winding his way to New York City with Wallace, he disappeared, reportedly either "aboard an unknown steamship"[50] or into hiding in the American Communist Party headquarters in New York.[51] He was subsequently discovered by federal authorities, six years after the original *Point Lobos* trial, in the French Foreign Legion in North Africa, and was returned to American Army Headquarters, where he was placed on a ship back to New York in the custody of the army provost marshal. Sometime during the ship's passage or after its landing in New York Sakovitz again escaped, "presumably," Warren felt, "through the assistance of members of the same union which had brought about the killing."[52] The FBI still lists Sakovitz as a fugitive from justice.

King, Ramsay, Conner, and Wallace were brought to trial, convicted of second degree murder, and sentenced to indeterminate terms of five years to life. The trial took eight weeks, ending in January, 1937 and regularly receiving front-page coverage in the Bay Area press. The presiding judge, Frank M. Ogden, had been an assistant to Warren in the district attorney's office from 1925 to 1930, and had received his judgeship that year in part because of Warren's strong recommendations. Attorneys for the defendants immediately moved to disqualify Ogden for bias, and a hearing was held in superior court in Alameda County, in which Ogden was held not to be disqualified.

Warren fenced off the charge of Ogden's bias, clashed with defense lawyers, one of whom referred to him as "somewhat of an orator,"[53] was cross-examined several times about why Conner had been incarcerated without assistance for such a long period, had his home in Oakland picketed by members of several waterfront unions,[54] and said, in his closing argument, that his encounter "with some of [the defense lawyers]" was "one of the most unpleasant experiences that I ever had in my legal career."[55] While he denied that he was motivated by antilabor biases, stating that "there is no one in this courtroom that believes any more firmly in the principles and the general aims of Union Labor than I do,"[56] he would be vigorously opposed in his 1938 campaign for attorney general by most representatives of organized labor. He believed that the newspaper coverage of the *Point Lobos* case was the worst treatment a case of his had ever received in the press.[57]

Warren's troubles with the *Point Lobos* case persisted into his years as attorney general, 1938 to 1942. In 1941, four years after the initial verdict, evidence surfaced that Charles Wehr, Warren's chief prosecutor, had had a close relationship with Julia Vickerson, one of the jurors. Warren said at the time that he had not known of the relationship, and he seems truly not to have known at the time of the trial. Indeed, Warren's office had listed Mrs. Vickerson as an "undependable and peculiar character" at the original trial,[58] and Warren later stated that he had had very strong reservations about accepting her as a juror.[59]

But the incident was embarrassing nonetheless, particularly because the new governor of California, Culbert Olson, who was sympathetic to organized labor and eager to forge a "New Deal" coalition in California, was not disinclined to embarrass Warren, who, he thought, was a potential political opponent. Early on in Olson's administration effective pressure from the King-Ramsay-Conner Defense Committee had resulted in a trial balloon being sent up regarding a gubernatorial pardon for three of the *Point Lobos* defendants. Warren's estrangement from Olson was nearly total by the 1940s, and in October, 1941, he made his first open break with Olson over the pardon issue. "[S]ilence on my part in this matter would be cowardice," he said. "Everyone who is loyal to our country in its present crisis should fight to prevent [the pardon]." A pardon, in Warren's view, would "appease the revolutionary radicals."[60] Olson countered by calling Warren's remarks "hasty and ill-considered."[61] Although no pardon was granted, the California State Parole Board, under strong pressure from Olson, paroled King, Ramsay, and Conner on November 27, 1941. Warren immediately gave a statement to the press. "The murderers are free today," he said, "not because they are rehabilitated criminals but because they are politically powerful communistic radicals. Their parole is the culmination of a sinister program of subversive politics."[62]

But even then Earl Warren was not finished with the *Point Lobos* affair. For a time it looked as if Wallace, the one remaining incarcerated defendant, would get the benefit of a mistrial on account of the Vickerson incident. While Wallace's motion for a retrial was moving through the California judiciary, the California legislature's Joint Fact-Finding Committee on Un-American Activities in California, just then established, resolved to hold hearings on the parole of the defendants, with Warren as their principal witness. Hearings began only four days after the announcement of the parole. Warren testified for two hours, attempting to document Communist infiltration into the waterfront unions and the participation of Communists in the *Point Lobos* case. He claimed that the paroles made California "attractive for the worst radicals in America."[63] The committee (popularly known as the Tenney Committee for its chairman, Assemblyman Jack Tenney) had its hearings interrupted by Pearl Harbor, and eventually postponed them indefinitely. Later, in December, 1941, the California legislature passed a resolution urging Olson to revoke the paroles, a resolution Olson ignored.

Eventually, in 1943, the Tenney Committee reported formally to the legislature on "un-American activities" in California, asserting that the *Point Lobos* murder and paroles could be linked to the activities and lobby efforts of Communists.[64] Finally, in 1944, Wallace, still incarcerated, had his case reach the California Supreme Court. Although on examination at subsequent hearings Mrs. Vickerson turned out to be a chronically unreliable witness, the Supreme Court of California ruled that Mrs. Vickerson had not shown, in any of her remarkably confused and contradictory testimony, that she had been unqualified as a juror.[65] Wallace was eventually released from prison in 1949.

The prickliness of Warren's tone in remembering the *Point Lobos* case can be traced to several factors. *Point Lobos* was the first major public incident in Warren's career in law enforcement where he had significant opposition from a constituency whose support he coveted and with whose interests he was largely sympathetic. Grafters, bootleggers, racketeers, gamblers, and corrupt sheriffs were obviously evil persons; members of labor unions, on the other hand, were persons with whom Warren could identify. Warren was distressed by the labor pickets outside his house; by his being charged, by supporters of unionization, with a "frame-up"; by his being called, by those same supporters, a "servant of the Hearst-Bank of America . . . gang which owns all of California."[66] Those charges stung not only be-cause of their excessiveness but also because they were leveled by a segment of society whose friendship and respect Warren coveted.

Warren attempted to deal with his ambivalence about the labor issues in *Point Lobos* by making a sharp distinction between Com-munist and non-Communist maritime workers. In his memoirs he stressed that Alberts was "known for his resistance on his ship to Communist influences,"[67] and Warren's rhetoric during the investi-gation and trial, as noted, had been filled with references to subver-sives, Communists, revolutionary radicals, and the like. This stance was not just good politics: Warren was convinced, at the time, that "subversives" of all types were a threat to the security of California and that Communists, Fascists, and other "enemies of democracy" were attempting to prey upon the fears and aspirations of the "aver-age working man."[68]

Warren's distinction may have had some applicability to the *Point Lobos* case. Some of the leadership of the MFOW did have close ties to the Communist Party's labor affiliate, the Trade Union Unity League; Earl King was politically radical, although he denied being a Communist; Wallace and Sakovitz were close allies of King. In 1934 and 1936 maritime workers went on strike on the San Francisco waterfront. Unions affiliated with the Communist Party, such as the Marine Workers' Industrial Union, were prominent in the strikes, and communism on the Bay Area waterfronts became a volatile is-sue, sparking intraunion disputes and contributing to the presence of "beef-squads."[69]

On the other hand, the *Point Lobos* affair demonstrated the pres-ence in Warren of a provincial version of patriotism that caused him to lump together indiscriminately groups he considered dangerous to America. An example was his statement in 1934 that "aliens . . . point with loving pride to the joys of communism and fascism" and "advocate the overthrow of the government by violence."[70] Subse-quently he may have regretted his earlier expression of this attitude, as he did his statements about the Japanese in California during World War II; his vigorous reprosecution and conviction of the *Point Lobos* defendants in his memoirs may have been an aggressive response to some anxiety and perhaps even some embarrassment.[71]

The principal feature of the *Point Lobos* case that has remained controversial has been the alleged illegality of Warren's treatment of the defendants. Here some of Warren's critics have overreacted. It is essential to distinguish in time the practices Warren's office engaged in during the *Point Lobos* affair, which took place in the 1930s, from the subsequent response to the legality of such practices by the Supreme Court of the United States during Warren's tenure as Chief Justice from 1953 to 1969.

In the *Point Lobos* case Warren's office engaged in three practices, the constitutionality of which were subsequently considered by the Warren Court, and two other practices that were indirectly affected by Warren Court decisions. The practices, taken together, were the use in court of a confession secured in the absence of counsel and through intimidation by law enforcement officials; the use in court of evidence secured by surreptitious electronic eavesdropping; the use by a prosecutor's office of adverse pretrial publicity; the employment of a California procedure that permitted juries to decide both the voluntariness of a confession and its authenticity; and the prosecution's use of an extensive delay between the original arrest of one of the *Point Lobos* defendants, Conner, and his formal arraignment.

The Warren Court declared the first of these practices squarely unconstitutional, with Warren joining the majority opinions. The Court also invalidated analogous procedures in sixteen states concerning the submission of evidence to a jury about the voluntariness of a confession, although the procedures were not precisely those of California. Finally, the Court struck down, in the federal courts, "unreasonable delays" between the arrest and the arraignment of criminal suspects. Chief Justice Warren said, in a case in which the facts resembled the circumstances of Frank Conner's confession, that "life and liberty can be as much endangered from illegal methods used to convict those thought to be criminals as from the criminals themselves." [72]

Under Warren Court standards of criminal procedure, King, Ramsay, Conner, and possibly Wallace would never have been convicted. Conner's confession, which later he repudiated, was "coerced" according to the rules laid down by Warren in *Miranda v. Arizona*. [73] King's and Ramsay's convictions were obtained by "eavesdropping" techniques that *Berger v. New York* [74] and *Katz v. United States* [75] invalidated. The principles pertaining to adverse pretrial publicity announced in *Sheppard v. Maxwell* [76] would have tainted Warren's release of the grand jury transcript to the press and his office's subsequent statement that the case was now "clinched," since Wallace was mentioned in the statement. Indeed there was an ironic echo of the Warren office's comment that the *Point Lobos* case was "clinched" in the 1963 statement on national television by Dallas Police Chief Jesse E. Curry that the case against Lee Harvey Oswald, the alleged assassin of President John F. Kennedy, was "cinched." Warren, of course, was to chair the presidential commission that investigated John Kennedy's murder.

*Jackson v. Denno*[77] invalidated state procedures similar to those that permitted jurors in the *Point Lobos* case to assess both the voluntariness of a confession and its authenticity. California passed a statute a year after *Jackson* altering its procedure, requiring that only judges initially consider evidence on the voluntariness of a confession.[78] And after *Mallory v. United States*[79] declared that federal authorities had violated the Sixth Amendment when they held an arrested person overnight without having him formally arraigned, California passed a statute requiring that arrested persons be arraigned within forty-eight hours.[80] Thus under modern standards of criminal procedure, which Chief Justice Earl Warren pioneered in formulating, District Attorney Earl Warren would probably have been prevented from convicting all the *Point Lobos* defendants.

But under the California and Supreme Court criminal procedure standards of 1936, none of the practices authorized by Warren were illegal. The area of state confessions was controlled by principles, reaffirmed in the decisions of *McNabb v. United States*[81] and *Crooker v. California*,[82] which tolerated intimidation of suspects by state authorities such as that which occurred with Frank Conner. Warren, well aware of such practices, stated in *Miranda* that "the modern practice of in-custody interrogation is psychologically rather than physically oriented" and was "created for no purpose other than to subjugate the individual to the will of his examiner."[83] Eavesdropping was controlled by the standards of *Olmstead v. United States*,[84] a 1928 decision, which held that telephone wiretaps did not violate the Fourth Amendment. The constitutionality of "adverse" pretrial publicity was not tested until the 1961 case of *Irwin v. Dowd*,[85] which *Sheppard* expanded upon. And, as noted, while *Jackson v. Denno* and *Mallory v. United States* were to affect California practices, the California statues conforming to their holdings were not in effect in 1936.

Thus Warren's memories of *Point Lobos* may likely have contained elements of both embarrassment and anger. He had engaged in criminal procedure practices that he later condemned as Chief Justice. He had made the kinds of glib associations between political radicalism and communism that were to provoke him, when made by Congress and the states, in the legislative investigation cases of the Warren Court. He had been ridiculed by organized labor, whose interests he strongly supported during his tenure on the Court. He had also had his home picketed, his integrity questioned, his efforts to bring murderers to justice opposed, even his courtroom style mocked. He had broken no laws, he had secured the convictions of "paid assassin[s]," [86] and he firmly believed in the guilt of the *Point Lobos* defendants. He hated communism as much as he hated fascism, and he genuinely believed that patriotism and anticommunism went hand in hand. Finally, he had done what his occupational role called for—convicting and incarcerating murderers. He was still very far from a belief that a tainted confession meant that a criminal suspect had not been "brought to justice."

On February 17, 1938, Warren had declared his candidacy for the attorney generalship.[87] His discussion with Ulysses S. Webb had taken place earlier that year; he had secured the endorsements of such prominent nonpartisans as William H. Waste, Chief Justice of the California Supreme Court. In his formal announcement he referred to the "great problems of law enforcement" that "now challenge the attention of all public officers and citizens of this state," and cited his thirteen years as district attorney of Alameda County, which had "not only acquainted me with these problems but [had] also given me the sincere desire to strive for their solution."[88] He followed up that announcement with a campaign white paper in which he stressed his nonpartisan approach to law enforcement, his experience, and the new responsibilities of the attorney general's office.[89]

The California system of cross-filing in primaries so deemphasized partisan affiliation that during the years Warren ran for elective office, candidates did not need to inform the voters of their party affiliations.[90] Warren's technique, which Hiram Johnson had also practiced, was to file for the nominations of all the major parties, including the Progressive party. All of the parties held statewide primaries, and, as noted, Warren won all three nominations. His campaign expenditures amounted to thirty-five thousand dollars, money which was raised by "friends in the San Francisco Bay area." He was opposed in the general election by Karl Kegley, whose sole campaign issue was support for "a radical pension plan," an idea then current in California.[91] Kegley received approximately five hundred thousand votes; Warren over two million.

Warren found his election "a hollow anticlimax,"[92] however, because of the murder of his father in May while he was preparing for the state primaries. More than thirty years later Warren remembered the incident:

> Early in the evening on Saturday, the fourteenth of May, my father was sitting in his chair to read the evening paper, but apparently had fallen asleep. It was a warm evening, as is usual in Bakersfield at that time of the year, and the doors and windows were all open except for the screens in them. Someone came in, struck him on the head with a short piece of pipe, crushing his skull, then dragged him into the bedroom and left him there on the bed. He was in the house alone. . . . The crime was discovered the next morning. . . .
>
> My parents' home was on Niles Street, next to the corner of Baker, the principal street of East Bakersfield. There was a gas station on the corner, thoroughly lighting the area, including the front portion of our home. The back yard was unlighted, and there was a grape arbor and some orange trees which contributed to its seclusion. . . . When authorities searched the premises after discovery of the murder, a piece of half-inch pipe about a foot long was found in the yard of our neighbor, where it could easily have been thrown from our back porch. It had blood and hair on it. . . . On examination of our own back yard, the place where the pipe had long been partially imbedded in the soil was found and also evidence of someone having relieved himself a short distance from the place where the pipe rested. My father's wallet had been taken, but nothing else in the house had been disturbed. . . .
>
> This was, of course, a tragic experience for the entire family. . . .
>
> My father's death must go down in history as one of the thousands of unsolved murder cases that plague our nation each year and cause such general apprehension for the security of our loved ones, ourselves and our homes.[93]

The deadened, matter-of-fact quality of Warren's prose in this passage suggests the years he had spent blanketing the pain of so arbitrary and shattering a loss. Methias Warren had allegedly grown eccentric in his last years, becoming a slave to his habits of frugality and quarreling regularly with neighbors and tenants of the several rental properties he owned. Several years earlier Chrystal Warren, his wife, had left him to live in an apartment in Oakland near her married daughter, and in her absence Methias spent much of his time in a few rooms of his Bakersfield house, reading, eating, and attending to his business affairs. A *Life* magazine reporter, interviewing Bakersfield residents six years after Matt Warren's murder, stated that Matt "was as hard-fisted in his financial affairs as he was personally abstemious, and he made many enemies." The reporter quoted "a local police officer" as saying the murder was an impossible one to solve because it was a "case of a thousand motives—so many people had reason to kill Matt Warren."[94]

But Matt Warren remained one of his son's heroes. The attorney general wept openly at a press conference in Bakersfield and dispatched some of his staff to aid in the unsuccessful search for Matt's killer. In 1963 Warren was to dredge up his father's death in connection with another arbitrary murder, this time of a president of the United States.

After 1938 Warren moved from the realm of local law enforcement to the realm of state politics. He did not anticipate so dramatic a change in focus. He did not aspire to be governor of California at all, and he had not envisaged that his position as attorney general would involve him so thoroughly in political issues. He saw the attorney generalship as an expansion, perhaps a culmination, of his career as a law enforcement official. He had not reckoned, in 1938, that a world war would break out in Europe, making preparedness and civil defense high priority issues, especially when Japan, with its large number of immigrants among the California population and its access to the Pacific coast, entered that war. Nor had he fully anticipated the effects of the Republican party's loss of the governorship the same year he was elected attorney general. As a consequence of that gubernatorial election Warren, a nominal Republican, became a member of the administration of Culbert Olson, a competitive, partisan Democrat. Olson was to remind Warren how little he liked being beholden to someone else in an office.

Little by little these unanticipated features of the attorney generalship were to result in Warren's becoming a full-time politician. But in 1938 his public life was still that of a law enforcement official. In eighteen years he had come remarkably far from the gregarious, conscientious, but seemingly undistinguished person who joined Ezra

Decoto's office; he had blossomed within the confines of the profession of law enforcement. As a law enforcement official he had discovered his gifts for efficiency and office management. Law enforcement crusades had channeled his capacity for moral outrage and allowed him to paint the world as he liked to paint it, in blacks and whites, stark shades of evil and good. Holding the office of district attorney had reinforced his desire to be his own boss, an independent, nonpartisan, incorruptible force. Involvement with law enforcement matters had brought him into contact with hitherto unfamiliar ideologies: His first response had been to label those ideologies "alien" and to juxtapose them against a nativistic paradisiacal vision of California.

The district attorney's office had also been a forum from which Warren refined his style of officeholding. His belief in nonpartisanship went hand in hand with his cultivation of the press and his attention to the vicissitudes of public sentiment. His commitment to activism and reform was balanced by a skepticism about the altruistic motives of legislators and other elected officials and by a sense of the evanescent nature of public opinion. His insistence on independence for himself was balanced by an insistence that his subordinates be, above all, trustworthy and loyal to him. His Progressivism was still intact, but he had become a Republican party worker. His memories of the Southern Pacific were matched with the more recent memories of Communist sympathizers in unions. Earl Warren had not fundamentally changed since Bakersfield and Berkeley, but he had assuredly grown. He was, in 1938, a special kind of California law enforcement official. Law enforcement had transformed his public life, and he had transformed law enforcement in California.

# 3

# Law Enforcement:
# The Attorney General

Warren's first acts as attorney general of California were consistent with the style of officeholding that he had developed as district attorney. He changed staff rules to make "the rules governing private practice . . . more stringent"; he appointed persons who were interested in full-time careers as state officials. He sought to "reorganize the office into a manageable unit," which meant the creation of a central filing system, the establishment of a litigation calendar, regular staff meetings that had as their primary purpose "supervision of the work of the deputies"; and the development of a "good system of fiscal accountability."[1]

In addition, Warren sought "to demonstrate immediately [the] new responsibilities of the attorney general." Activism as a means of perpetuating and expanding the power of an office had been a theme of Warren's district attorneyship; he intended to continue this activism as attorney general. Before long he and his deputies were shutting down illegal dog tracks, scrutinizing bookmaking on horse races, closing houses of prostitution, and padlocking speakeasies. By those activities Warren sought to demonstrate "that the commercialization of gambling, bootlegging and prostitution, all of which are so productive of official corruption, was not to be tolerated, and that under the new powers of the attorney general there was a force that could nullify any local arrangements which permitted these operations to exist." [2]

The movement of the attorney general's office to areas where local enforcement officials had previously been free to decide for themselves how and when to exercise prosecutorial powers raised some potentially prickly jurisdictional issues and posed a test of Warren's discretion. He responded by making use of his goodwill in state law enforcement circles. He attempted, as he put it, to have "frank talk[s] with the [local] authorities" before making a publicized foray into their areas. This technique, he felt, "enabled me to exercise the new powers of the attorney general in an unobtrusive manner, without alienating the law officers of the state." Warren's office also stressed the need for cooperation among the various levels of state law enforcement, dividing the state "into several regions," each of which contained "an organization of the chiefs of police, the sheriffs, the district attorneys, and later the fire chiefs." The groups in each region were asked to come to regularly scheduled meetings, which Warren attended, where "conditions in the area, . . . evidence of organized racketeering, and ideas for better crime prevention" were discussed. This system gave Warren "an intimate acquaintance with most of the responsible officers" in the state and provided, in his view, a "well organized" and "functioning" law enforcement community in California. [3]

Warren anticipated that his principal task as attorney general would be keeping organized crime from getting a foothold in California. His office was "well briefed on the rackets that had spawned . . . in eastern and midwestern cities"; kept watch over the California legislature "to make sure that no laws conducive to mob incursion would be enacted"; stressed "leading mobsters' methods" in "our police schools at San Jose State College"; and "preached the doctrine . . . that the racketeers . . . [had] no potency until they had established a foothold in the political structure." [4]

In 1939 Warren sought to test his power to suppress organized crime by closing down a gambling ship, the *Rex*, which was stationed off the California coast in Santa Monica harbor. The *Rex*, Warren wrote in his memoirs, was owned by Antonio Stralla, commonly known as "Tony Cornero," "a notorious rumrunner and underworld figure," who allegedly had connections with "the Al Capone crowd." [5] The location of the *Rex* posed a jurisdictional problem for Warren, since the ship was stationed more than three miles from the shoreline of Santa Monica and was therefore arguably not within California waters. Warren's office countered this claim by arguing that, where bays were part of the coastline, "California waters" extended three miles out from a mythical straight line drawn from a bay's north headland to its south headland,[6] and that the *Rex*'s use of state telephone lines and licensed state taxis made it a "public nuisance to the state."[7]

By July, 1939 Warren's office was prepared to raid the *Rex* and three other gambling ships. Having finally secured the cooperation of the Los Angeles district attorney and sheriff, Warren sent off a flotilla of twenty boats and about three hundred officers. The officers served an injunction on the ships that required them to close as public nuisances, and attempted then and there to take action to enforce the injunction. The *Rex* resisted, turning fire hoses on the flotilla and refusing the officers access. Warren ordered the flotilla to blockade the *Rex*, making it impossible for customers to leave the ship. Eventually the *Rex* capitulated and was towed to port, where its gambling paraphernalia and its proceeds were confiscated. Warren then compromised with Cornero and the other operators, allowing them to keep their ships in exchange for their closing their operations, forfeiting the confiscated money and paraphernalia, and paying the expenses of the raids.

Cornero resurfaced in 1946 with another gambling ship, the *Lux*, in Santa Monica harbor. Warren was governor at the time and was unable to secure cooperation from the attorney general, Fred N. Howser. Warren first considered activating the California National Guard against Cornero but then decided to seek cooperation from the United States Justice Department. Eventually federal agents seized Cornero's ship for a technical violation of federal navigation laws, and in 1948 Senator William Knowland of California successfully sponsored a bill prohibiting gambling ships in the coastal waters of the United States.

"I must confess," Warren wrote in his memoirs, "to an ingrained bias against commercialized gambling." Gambling for Warren was "corruptive, dishonest in operation, and often cruel in its consequences. I had seen at first hand all of these bad attributes as a youth in Bakersfield."[8] At one point during his tenure as Chief Justice, in response to a law clerk's statement that he thought gambling was not a serious matter, Warren responded: "Gambling is immoral, it preys upon the poor, it takes the paycheck out of the hands of the worker

before he can get it home so he can feed his family and clothe them
and educate them."⁹ Matt Warren's "lessons" appear vividly in that
statement. Warren the attorney general fought gambling and its associated vices not only because he wanted to demonstrate the new
jurisdictional reach of his office but also because he personally thought
gamblers were evil persons. Combatting gambling was the kind of
public task Warren had anticipated in the attorney generalship. He
had seen corruption, debauchery, and moral laxness before; now he
was prepared to roam the state as the most visible public opponent
of those vices.

Given this definition of his office, Warren felt that political activities would be distracting. He "resigned my Republican connections
and announced a nonpartisan approach to the Attorney General's
Office" when he declared his candidacy; he "maintained the same
neutrality throughout my term." He was, he recalled, "intensely interested in my job and wanted nothing more than to be re-elected as
attorney general."¹⁰ But politics was to filter into all phases of the
attorney generalship, and Warren was to emerge from his four-year
term as a candidate for governor of California. To understand how
a declared nonpartisan who "took no part in . . . any politics"¹¹
during his tenure as attorney general became a full-time elective politician, one must understand the character and influence of Culbert
Olson, California's Democratic governor during the years of Warren's attorney generalship.

ʞ

If Matt Warren and Hiram Johnson were the two most important
"positive role models" in Earl Warren's life, Olson was the first significant "negative model." Up to his successful campaign for attorney general, Warren conducted himself according to the principles
Matt Warren and Hiram Johnson had espoused; by 1938, the year
of Matt's murder, Warren had developed a distinctive style of officeholding that relied on "lessons" associated with his father and Johnson but was essentially his own. He "felt secure" as attorney general;
he saw his particular brand of law enforcement as a long-term ambition or as leading to comparable "opportunities in the legal field."¹²
But Culbert Olson reminded him of how much he chafed under the
authority of another person. Moreover, the techniques that Olson
employed to assert his authority over Warren were politically in-

spired. Olson saw Warren as a Republican and a potential political rival, and consequently as a man to be isolated. Warren was to respond by assuming the very stance that Olson had mistakenly attributed to him.

In Olson, Warren faced an adversary that in some respects resembled Earl Warren. Olson was, as was Warren, physically large and fair-haired, the son of a Scandinavian immigrant, a former employee of a powerful railroad, an admirer of Robert LaFollette, a bitter foe of "special interests," a former state chairman of a major political party, and a delegate of that party to one of its national conventions. He was also, as was Warren, a person whose capacity to bear grudges and settle scores was marked and whose rhetoric attempted, not always successfully, to conceal passion and intemperance behind a layer of moderation and blandness.

As Olson made his way up the ranks of California Democrats in the 1920s and thirties, he revealed himself as decisively located on the "Progressive" or "New Deal" points of the Democratic continuum in California. After voting for LaFollette in 1924, as did Warren, Olson made a relatively painless transition to the liberalism of the 1930s.[13] He became attracted to Upton Sinclair's "End Poverty in California" plan, which envisaged state-owned factories where previously unemployed persons, instead of going on the welfare rolls, would produce essential goods and be paid in scrip money, usable only to purchase the goods they produced.[14] Sinclair's plan was so attractive to the California electorate in the Depression-ridden 1930s that he changed his political affiliation from Socialist to Democrat and secured the 1934 nomination for governor. Olson was heavily involved in Sinclair's campaign; Warren, as noted, managed the campaign of Frank Merriam, the Republican who defeated Sinclair.

"End Poverty in California" vanished with Sinclair's decisive loss to Merriam, but Olson emerged relatively unscathed, and between 1932 and 1934 the number of California voters registered as Democrats surpassed those registered as Republicans. Olson was elected to the state legislature and emerged, between 1935 and 1938, as an effective opponent of "special interests," notably the oil industry. He also began to move toward the center of the California Democratic party, disassociating himself from Francis Townsend's controversial old-age pension plan in 1935 and equivocating on the even more controversial "Ham and Eggs" plan, which proposed to pay numerous classes of California residents half their salaries or benefits in scrip.[15]

In 1937 Olson announced that he would seek the 1938 Democratic nomination for governor, and after a campaign in which he declared that he had refused contributions from "underworld forces and [from] the agencies of special interests,"[16] won the nomination with forty-two percent of the primary vote. Then Olson, who under normal circumstances would have been a vulnerable candidate because of his self-description as a "representative of the party's left-wing"[17] and because of some factionalism among California Democrats, enjoyed some of the "luck" associated with Warren's career.

The Republicans, in 1938, engaged in a bitter primary struggle, with Merriam pitted against Lieutenant Governor George Hatfield, the latter backed by the notorious lobbyist Arthur ("Artie") Samish.[18] Samish, who represented liquor, horse racing, and transportation interests, and was once referred to as "for years the undisputed boss of the California legislature,"[19] was having his affairs investigated by a Sacramento County grand jury in 1938; Governor Merriam helped finance the Sacramento district attorney's investigation.[20] The resultant outcry by Hatfield, and subsequent washing of dirty linen by both Republicans, provided Olson with some helpful campaign propaganda.

Olson eventually defeated Merriam by a significant margin, winning nearly fifty-three percent of the vote and carrying all the populous California counties.[21] And Olson won without softpeddling either his partisan affiliations or his liberal views. He identified his campaign closely with the Roosevelt Administration, attacked "affiliated corporate interests,"[22] flirted with the "Ham and Eggs" movement, associated himself with organized labor, and called for public ownership of public utilities and state aid to the elderly.[23] He received almost no support from newspapers but had warm endorsements from labor unions and the tacit support of the Communist party, whose state chairman referred to Olson, before his primary victory, as "the leading progressive candidate for the Democratic nomination for Governor."[24] The Republicans called Olson's platform "irresponsible radicalism and declared Socialism" and put up posters warning voters to "Keep California Out of the 'Red',"[25] but were unable to command much voter attention. On one occasion a member of the Associated Farmers of California, an emerging anti-Communist lobby group, declared before a congressional committee that Olson fraternized with Communists, but that charge was overshadowed by the announcement of another committee witness that

Shirley Temple had allowed her name to be used as a Communist front.[26] Before a large gathering in Los Angeles Olson immediately said, "I am sorry Comrade Shirley Temple is not here. She should be here to aid us in plotting to overthrow the government of the United States of America."[27]

As Warren took over the office of attorney general, Olson began his governorship. Olson simultaneously proceeded on two fronts. First, he began to fashion "a state government in sympathy with the principles and policies of the New Deal,"[28] taking firm and outspoken liberal positions on social issues. Concurrently, he sought to establish the Democratic party as the predominant force in California politics. In pursuing both these goals Olson was to clash with his formidable "nonpartisan" attorney general.

One of Olson's first official acts was to pardon Tom Mooney, a militant labor advocate who had been convicted of murder in 1916 for planting a bomb during a San Francisco parade. Mooney's conviction was based on less-than-impeccable evidence. Warren was sharply critical of Olson's pardon, writing Olson that the action could "lend encouragement to those forces that are opposed to the enforcement of our laws and to the maintenance of security of life and property."[29] Olson also attempted to pardon Warren Billings, an accomplice of Mooney in the bombing, who had previously been convicted of a felony. Warren, as a member of the State Advisory Prison Board, exercised review powers, which he had in cases involving the pardons of convicted felons, to block Olson's attempt. Olson countered by appealing by letter to the California Supreme Court, which recommended that Billings's sentence be commuted.[30]

As noted, Olson and Warren also clashed over treatment of the *Point Lobos* defendants, who sought a gubernatorial pardon and were eventually paroled in 1941, in part as a result of pressure from Olson's office. In addition, Warren and Olson disagreed over what should be the scope of investigative powers for two legislative committees investigating Communist infiltration in state government. One of these committees, the Assembly Relief Investigating Committee, under the chairmanship of Samuel Yorty, began hearings in 1940 on the issue of Communist influence in the California State Relief Administration, an agency whose powers Olson had sought to revive. The Yorty Committee sought to subpoena lists of labor union members in order to secure testimony about the workers' Communist affiliations; Olson argued that this use of the subpoena power was overbroad. Warren, who later as Chief Justice was to declare constitutional limits to the legislative investigative powers of Congress, supported the Yorty Committee.

Those skirmishes between Olson and Warren were minor, however, when compared with the two men's estrangement over civil defense issues. Warren, we have seen, had become an advocate of preparedness and expanded civil defense as early as 1938. His perspective as attorney general was essentially that of a patriotic and xenophobic law enforcement official; he was convinced that both Communists and Fascists, as well as other "aliens," threatened the nativist paradise that was California. He determined, upon becoming attorney general, to make civil defense one of his principal concerns and set about this task in characteristic fashion, dividing the state into several law enforcement regions and drawing on his earlier efforts to professionalize and coordinate state law enforcement agencies. "By the summer of 1940, when France fell and the Battle of Britain began," he recalled, "we law officers of the state were . . . ready to expand our scope."[31]

Warren attended a national conference on law enforcement and civil defense in August, 1940, and on his return called a joint meeting of California law enforcement officials and representatives of the military. He lobbied for the passage of two acts in the legislature— the Mutual Assistance Act, authorizing local fire fighters to operate outside their jurisdictions, and the Uniform Sabotage Prevention Act, prescribing penalties for acts of sabotage in defense industries. Warren, in arguing for the latter legislation, referred to its origins in the 1940 national conference on law enforcement and maintained that it had the support of United States Attorney General Robert Jackson.[32] By 1941, the year of Pearl Harbor, Warren recalled, "we had a viable organization and program."[33] By "we," of course, he meant the attorney general's office and its cooperating law enforcement agencies, not Culbert Olson or his political supporters.

Olson had meanwhile made some efforts of his own with respect to preparedness and civil defense. He had been a pacifist in the 1930s,[34] but he declared in May, 1940 that he was "the kind of pacifist who believes that our Nation should be prepared at all times to resist encroachment of foreign dictatorships."[35] In 1939 Olson established a state council of defense, with himself as chairman and Richard Graves, a friend of Warren's who had graduated from the Bureau of Public Administration at Berkeley, as director. Early in

Graves' tenure Olson asked him "for a list of jobs and salaries" with the council of defense; Olson told Graves that "he was under terrible pressure" and "just had to find a way to take care of his people." Graves told Warren about Olson's action, and Warren advised him to resign, which Graves did. From that discussion on, Graves claimed in his oral history of his experiences, "[Warren] became in many ways an active candidate [for governor] early and [civil defense] became an issue."[36]

Beginning in 1940, Olson and Warren engaged in a series of efforts to outflank each other on civil defense issues. In a December, 1940 address to the legislature, Olson called for the establishment of a California State Guard, anticipating that the California National Guard would be incorporated into the federal armed forces as the United States became more active in World War II. The legality of a state guard was questionable, given language in the California Military Code, but Olson persisted, ignoring an impending lawsuit.[37] Moreover, he secured an opinion on the legality of the state guard not from Warren, the chief legal officer of the state, but from a Los Angeles law firm. Warren later said that "[Olson] would announce in the papers that he was [creating the state guard] because he had the opinion of Lawyer X and Lawyer Y of Los Angeles. . . . And those things were humiliating, and of course I resented them."[38] A state guard was eventually funded by the legislature in June, 1941, but Olson was to have some difficulty getting future allocations for the guard from the 1942 legislature, and Warren, after announcing his candidacy for governor, was to claim that "there has never been an honest effort by the Governor of this state to build up the State Guard."[39]

Olson's next effort to preempt the civil defense issue came a week after Pearl Harbor, when he declared California in a "state of emergency," and designated himself and other "duly constituted officers of the State" the fountainhead of authority on civil defense matters. The attorney general was not included among the "duly constituted officers." Warren, four days later, challenged the constitutionality of Olson's actions, pointing to a 1929 statute that prohibited governors from assuming emergency powers in "time of war." "I stated in the Governor's Office," Warren said, "that I would not abide by any such assumption of power by the governor. There was much hysteria. . . ."[40]

The legislature, after noting the irreconcilable positions Olson and Warren held and "the unanimity of police and fire services in opposition to the governor's position,"[41] did not give Olson his appropriation for the state council of defense and passed a rider to an appropriation for Olson's "state of emergency" fund stating that no funds could be used "in exercising any powers [under the 1929 statute] which might be interpreted as bestowing emergency powers."[42] Warren felt, however, that "I had won the battle, but came close to losing the war to the governor, because he refused to recognize me in the council, told me nothing about what he had done or intended to do, and would not have his subordinates consult with me."[43] The final indignity, for Warren, was when Olson refused to sign a bill appropriating $214,000 for the attorney general's office "to meet the additional costs . . . due to the war."[44] This action by Olson came at the end of February, 1942; Warren announced his candidacy for governor in April.

<br>

The Olson-Warren feud was to provide the context for Warren's opposition to the nomination of Max Radin, a law professor at Boalt Hall and a visible Olson supporter, to the California Supreme Court. Olson nominated Radin to the court on June 27, 1940. His nomination was rejected on July 22 by the State Commission on Judicial Qualifications, which confirmed supreme court appointments, and he withdrew his name from consideration on July 30. Warren was secretary (and unofficial administrator) of the commission and announced the rejection of Radin to the press. That announcement was the last public word Warren offered on the Radin affair. He did not discuss Radin in his memoirs or in any public communications he made as attorney general or governor. In December, 1971, in response to a question put to him during an oral interview, he said, "I don't think [the Radin nomination] is a matter of any importance. There have been judges that have been proposed and have been rejected. They try to put a partisan look on this, but it wasn't that, it was a normal procedure, and I don't think it's of any importance myself. I don't propose to write anything about it."[45]

Earl Warren, Jr., in his recollections of his father's career in California, called Warren's reaction to the Radin affair "a very old and sort of bitter thing" and suggested that the issue of Radin's nomination was "personal" to Warren.[46] In this brief comment Earl junior captured, perhaps unconsciously, the essence of the Radin affair for Warren. Warren was motivated to deny Max Radin the nomination out of deep personal animus. He believed that Radin was a dangerous radical, he identified Radin with the criticism the Alameda County District Attorney's Office had received in the *Point Lobos* affair, he associated Radin with a political enemy, Culbert Olson, and he meant to show his opponents that Earl Warren was a man to be reckoned with. The story of how Warren helped to scuttle Radin's nomination can be seen as an exercise in how accomplished politicians settle scores; it can also be seen as revealing the single-minded determination Warren concealed under his surface blandness.

Max Radin was in many respects an unusual nominee for a California judgeship. He had been born in Poland and had emigrated to the United States in 1884 at the age of four; his father was a rabbi. He came to California after thirty-five years in New York, where he had attended CCNY, NYU, and Columbia, receiving an LL.B. from NYU and a Ph.D. in classical languages from Columbia. He had been a Latin teacher in the New York public schools and a lecturer in Roman law at CCNY and Columbia before being appointed to the Berkeley law faculty in 1919. His scholarship ranged from Roman law and jurisprudence to contemporary issues in legal education. Radin was one of the few scholars attracted to the Realist movement of the 1920s and thirties who was a serious student of legal history, and one of the few legal historians of those decades to hold a Realist jurisprudential perspective. He said in 1936, for example, that "all judges . . . are influenced by unconscious and implicit premises created by their own temperament," and added that he had advanced that "theory of jurisprudence . . . in something like thirty or forty articles."[47]

All this suggests that Radin's professional home would be somewhere other than the California Supreme Court in the 1940s, but the composition of that court was changing because of the influence of Culbert Olson. Olson hoped to forge the same kind of coalition of intellectuals, workingmen, and traditional Democrats that the Roosevelt administrations had forged nationally, and, like Roosevelt, Olson regarded academic lawyers as prospective judicial appointments. Radin, who was himself outspoken on public issues and sym-

pathetic to the goals of the Olson Administration, became a member
of Olson's "Brain Trust," a circle of persons who advised Olson on
public policy matters. [48]

Although Radin, through his involvement in public affairs, had
identified himself with a number of controversial positions, such as
opposition to California's Criminal Syndicalism Act of 1919 and support for the defendants in *Point Lobos,* Olson tended to share Radin's
views and came to be persuaded that Radin would be an effective
judge. Radin actively lobbied for a nomination, [49] motivated to some
extent by financial considerations: The salary of a California Supreme Court justice in 1940 was eleven thousand dollars a year, while
Radin was receiving sixty-five hundred dollars from Berkeley. [50]

In the first two years of his governorship Olson had had the opportunity to make three appointments to the supreme court. His first
two nominees, named in July and August, 1939, came from the ranks
of his strong political supporters and were persons unlikely to arouse
much opposition. Jesse Carter, Olson's first appointment, was a California state senator; Phil Gibson, the second, had been a gregarious,
effective director of finance in Olson's administration. Radin was
considered for both nominations and mounted a strong campaign,
securing letters from persons such as George D. Kidwell, director of
industrial relations in the Olson Administration, and Roger Traynor, then a faculty member at the law school at Berkeley. [51]

Finally in June, 1940, with the death of Chief Justice William
Waste, Olson was given a third nomination. He replaced Waste with
Gibson as chief justice and, after some deliberation, nominated Radin to fill Gibson's vacancy. Gibson was himself a strong supporter
of Radin, and Radin secured additional endorsements, including those
of the prominent Realists Jerome Frank and Thurman Arnold. Olson had earlier referred to Radin as one of "our intellectual and most
competent liberals"; [52] on nominating Radin, Olson described him as
having "a legal mind of general brilliance and versatility." [53]

Meanwhile Earl Warren had kept close watch on developments
within the Olson Administration and within the Commission on Judicial Qualifications, of which he was the secretary and presiding
officer. The commission's history was linked to Warren's career in
law enforcement circles. It had come into existence as the fourth of
the constitutional amendments sponsored by Warren in 1934 and
was therefore in effect when he successfully ran for attorney general
in 1938. The commission, as its 1934 charter provided, was to con-

sist of three persons: the chief justice of the California Supreme Court, the senior presiding judge of the California District Courts of Appeals, and the attorney general of California, who was designated secretary.

The first active involvement of the commission in a supreme court nomination came with Jesse Carter's appointment. Warren, apparently playing politics,[54] argued that Carter was ineligible for the position because of a provision in the California constitution that no state senator could hold a nonelective office during his term. Warren's argument was sufficiently plausible to be passed on by the California Supreme Court,[55] which unanimously ruled that appellate judgeships were still "elective" offices after the 1934 amendments since judges were required by the initiative to "run against their records" in a popular election every twelve years. Warren made the formal case against Carter before the supreme court; Max Radin was one of the lawyers arguing for Carter.[56]

Having lost his first challenge to an Olson appointment, Warren routinely acquiesced in Gibson's nomination. When Radin's nomination came up, however, Warren began a concerted series of efforts to keep Radin off the supreme court. The immediate background to those efforts was a series of incidents that served to identify Radin, in Warren's mind, as an enemy.

Radin was sympathetic to the defendants in *Point Lobos* and was identified by the Tenney Committee on Un-American Activities as a possible contributor to the King-Ramsay-Conner Defense Committee. He eventually appeared before the committee and denied charges that he had helped finance Ben Sakovitz's escape. Radin apparently believed Conner to have been innocent of involvement in the murder, and he later said that "garbled versions" of remarks he had made about *Point Lobos* "seem to have reached Mr. Warren."[57] Jesse Carter, reminiscing about the Radin nomination in the 1950s, speculated that Warren "blackballed" Radin because of Radin's views on *Point Lobos*.[58] That feature of Radin's candidacy was very probably the most distasteful to Warren, but there were others. Radin had written a law review article defending the use of pardons as a follow-up to the Mooney controversy;[59] Radin had argued against Warren in the technical dispute over the Carter nomination; Radin had sided with Olson on the issue of extensive legislative subpoena powers.[60]

In addition, Radin had supplied Warren with a moral basis for opposing his nomination. On June 3, 1940, eighteen workers in the

Stockton office of the State Relief Administration were arrested and charged with contempt for failing to testify before the Yorty Committee. Radin's daughter was employed by that office, although she was not one of the workers arrested and charged. The case of these workers was set down for trial. That same day Radin wrote letters to two men who had been former students of his at Berkeley—Raymond Dunne, a close associate of the judge who was to preside over the contempt trial, and Irving Neumiller, the city attorney for Stockton. In the letter to Dunne, Radin said, "If [the SRA workers] should be found guilty, I wonder if it would not be proper for you to suggest to the judge that a nominal fine or a suspended sentence would fully meet the needs of justice."[61] In the letter to Neumiller he wondered whether Neumiller might "properly do something" about the case.[62] Radin was subsequently to call these letters "perfectly proper conduct by all legal standards,"[63] but Neumiller wrote him back quickly to say that he thought it "highly improper and unethical [of Radin] to discuss pending matters with any judge."[64]

The letters, raising questions of ethics, represented the kind of information Earl Warren liked to seize upon when confronting an opponent; Warren was to make good use of the letters in his effort to thwart Radin's nomination. First, however, he set the commission's review procedures in motion. The Commission on Judicial Qualifications technically had no formal procedures: It could consult with anyone it chose, or no one, in its consideration of a supreme court nomination. Its practice, however, since Warren had been its secretary, was to ask a committee of the California State Bar Association to prepare and submit a factual report on nominees to the supreme court. Immediately after Olson announced Radin's nomination, Warren wrote a letter to Gerald Hagar, the incumbent president of the California State Bar Association, asking him to have a report on Radin prepared. Warren, Hagar, and others had known, two weeks before Olson's formal announcement of Radin's nomination on June 27, that Radin was likely to be nominated.[65] Later Radin was to call Hagar a "bitter Republican partisan . . . and a confirmed witch-hunter."[66]

Hagar appointed a fact-finding committee made up of three members of the Board of Governors of the California State Bar, a procedure that had also been followed in the nominations of Carter and Gibson. Hagar made certain suggestions to the committee with respect to witnesses and areas of inquiry: Among his proposed witnesses was Samuel Yorty, and among his proposed inquiries was Radin's connections with the Communist party.[67] The same day that Hagar formally made these suggestions, July 2, Arthur Caylor reported in the *San Francisco News* that "the gentlemen of the Qualifications Commission" were "being very cagy about their operations—formal even," and that "Mr. Warren would have a nice chance to embarrass the governor."[68] Six days later, on July 8, the *San Francisco Examiner*, a strong backer of Radin, reported that "legal and political circles" were "confident he will get [the nomination]."[69] The day the *Examiner* story appeared, Warren wrote Hagar stating that the suppositions in the story were groundless.[70]

The committee interviewed witnesses, including Radin, between July 10 and July 12, collected unsolicited statements (primarily opposing Radin) on the nomination from outside groups, and secured the sources of a private investigator, who looked into Radin's past activities. Some of the witnesses were former Communists. One such witness, Don Martin, labeled Radin an "underground Communist" and charged, as another witness would later in testimony before the Tenney Committee, that Radin had contributed money to help Ben Sakovitz escape.[71] Radin himself was called as a witness on July 12, and he was questioned fully about his involvement with the *Point Lobos* defendants, his membership in organizations sympathetic to communism, and his letters about the Stockton contempt trial.

Radin made some public admissions to the committee that he had previously not made. He admitted that he believed Frank Conner to be innocent, although he denied ever commenting publicly on the *Point Lobos* affair. He conceded that he had become a member of the Lawyers Guild, an allegedly subversive organization, in 1939, "at a suggestion that a few people like we older men [Radin was 59 in 1939], . . . might calm some of those younger people down." He denied, to the best of his recollection, having "taken part in any organization that was ever associated with Communism" and said that he had "fought Communism before the millions of people who now talk about it knew what it was."

Radin also said that his name was "generally mentioned in radical circles with an extremely contemptuous expression: 'He is nothing but a mere liberal.' " He admitted that his brother, Paul Radin, was "an active socialist" but said that "I have always been a 'middle of the roader' " and accordingly "our points of view are pretty far apart." Finally, he categorically denied any ethical violation in the Stockton

SRA case, making a distinction between communications to a judge before and after a verdict had been handed down.[72]

The committee forwarded its report to the state board of governors on July 19. The report considered Radin's "training and experience," "general reputation," relationship with the Communist party, and involvement in the Stockton contempt trials. It also discussed an earlier controversy with which Radin had been identified, the 1935 "Modesto Dynamiters' Case," where Radin had been accused of sympathizing with perpetrators of labor unrest in Stanislaus County. (Radin's association with the defendants in the Modesto case was never documented.) The committee took no position on Radin's nomination, stating that "we do not have any evidence on which we can base a finding as to his judicial stability, or judicial balance, or judicial temperament." It recommended that the Commission on Judicial Qualifications make "the closest inspection and examination of all the attached,"[73] by which it principally meant the battery of evidence linking Radin to one or another controversial incidents. Radin himself later called the committee's findings "quite unexceptionable," and its report "perfectly fair and factual."[74]

The same day that the committee forwarded its report, the California State Board of Governors met in Los Angeles. Prior to the Radin nomination, the practice had been for the board to act merely as a courier between the fact-finding committee and the Commission on Judicial Qualifications. On this occasion, however, the board formulated two new "general principles" for the confirmation of judges, which it suggested should be applied to the Radin nomination and future nominations. The first principle was that a nominee for a California judgeship should not be confirmed if he had "given just ground to a substantial number of the public for believing that he is either a member of, or in sympathy with, subversive front party organizations." The second principle was that he should not be confirmed if he had "given just cause for a substantial number of the public to believe that he is lacking in financial or intellectual integrity." The second disqualification should pertain, the board stated, "regardless of the truth" of allegations about a nominee's integrity.[75]

The board of governors was composed of fifteen designated members. Ten voted for the resolution formulating these principles; three were absent; two opposed it. One of the two members opposing the resolution, Oscar Goldstein, wrote Phil Gibson that he "firmly believe[d] Professor Max Radin was most unfairly treated by the passage of [the] resolution," and that "the resolution was adopted for the obvious purpose, if possible, to prevent the confirmation."[76] The other, Grove Fink, a member of the fact-finding committee, wrote Warren that the resolution was "a complete repudiation of the plan pursuant to which the Bar Association undertook the investigation," and consisted of "a matter of prejudice only."[77] Radin subsequently charged that the resolution was designed to supply Warren with a politically acceptable basis for voting against his nomination.[78]

There is no evidence that Warren proposed any of the novel procedures that were used in the Radin nomination. He may merely have benefited from them, and he was, under existing practices, under no obligation to reveal the reasons for his eventual vote, which came on July 22, in a meeting in which the Commission on Judicial Qualifications rejected Radin's nomination by a two-to-one margin.[79] Radin may simply have had the misfortune of having Gerald Hagar chair the fact-finding committee and to have ten of the California State Board of Governors be convinced that he was at best a highly controversial candidate. Radin later referred to the "puerile and absurd" charges made about him to the committee, and called the majority of the board of governors "as fine a collection of putty-faced dodos as exists anywhere except in the glass cases of a Museum of Natural History."[80] Radin may have paid the price for being an outspoken liberal on social issues at a time of rising anticommunism in California. Earl Warren may have been just one of numerous obstacles to Radin's judgeship.

The difficulty with such an explanation is the amount of evidence suggesting that Warren acted out of passion and spite toward Radin. Warren had opposed an earlier Olson nominee, Carter, but his opposition had been much more technical and overt: There was a serious legal question about Carter's eligibility, and Warren, as attorney general, was the appropriate person to raise it. But Warren's actions in the Radin nomination and his calculated silence afterwards suggest deeper machinations. Certainly Radin himself believed such: He said, after his defeat, that Warren's point of view was "a compound of Ku Klux, antisemitism, witch hunting, Republican partisanship, and . . . general cussedness" and that Warren was "a thoroughly unreliable and slippery politician."[81]

A conversation Warren had in December, 1940 with Robert Gordon Sproul, then president of the University of California, serves to illuminate Warren's motives in the Radin affair. After learning of the

commission's action, Sproul, who had supported Radin in earlier
controversies, telephoned Warren for a copy of the State Bar Association Committee's report; Warren refused.[82] Subsequently Sproul
asked Radin for a formal denial of various charges that had been
made against him in the course of the commission's inquiry, and
Radin complied.[83] Then, in December, 1940, Warren met with
Sproul.

Sproul kept a file in his office marked "Special Problems," which
included a variety of sensitive matters involving Berkeley faculty and
which was crammed with items concerning Radin. When Sproul had
oral communications with others about "special problems" he would
sometimes reduce the substance of the conversations to memoranda
in his file. Warren, throughout his entire public life, did not commit
many of his thoughts to paper, especially when he felt strongly on
an issue. He preferred face-to-face conversations, such as he had on
this occasion with Sproul.

The subject of Max Radin came up, Sproul remembered, and
Warren "launched forth in a vigorous denunciation of Professor Radin, which very evidently contained a good deal of personal animus." Warren alluded to Radin's remarks about *Point Lobos*, "point[ed]
to the fact that Radin's brother and daughter are Communists," and
charged "that Radin constantly gives aid and comfort to Communists
and other radicals. . . ." Sproul then continued to summarize Warren's remarks:

> He also criticizes Radin for his pleas on behalf of the SRA work-
> ers at Stockton and meets Radin's defense that this was wholly eth-
> ical by saying that every bar association that knows about the act
> has condemned it. One thing that particularly infuriated Warren
> was the fact that Radin was given the Boalt Professorship when
> Professor McMurray retired. Warren thinks that this was "a profli-
> gation of a distinguished chair."
>
> I asked him toward the end of our conversation what he thought
> I should do about the matter, and he said that he did not expect me
> to do anything because it would be impossible to prove that Radin
> was a Communist, and because even if he were his dismissal after
> years of service and distinguished scholarship would merely aid the
> Communist party, but he hoped that I and others in the University
> would cease to favor Radin and glorify him at every opportunity.
> Under the heading of "glorification" he placed my statement that I
> did not believe Radin was a Communist.[84]

Warren and Sproul were not passing acquaintances. Warren had "marched directly behind" Sproul in the California band as an undergraduate at Berkeley,[85] was to urge Sproul to run as a Republican against Olson in 1942 before deciding to make the race himself, and was to side with Sproul against members of the board of regents during the loyalty oath controversy of the 1950s.[86] "We had a very happy association," Warren said of his relations with Sproul. "[H]e and I had splendid relations with both the school and the public."[87] The kind of intemperate comments Warren made to Sproul about Radin were comments he made sparingly and to those whom he liked and trusted. The Max Radin nomination was indeed a "bitter thing" for Earl Warren.

✍

By the denouement of the Radin affair, we have noted, Warren had become heavily involved in national security and civil defense issues. His conviction that preparedness and civil defense were vital concerns, combined with his nativistic provincialism and his strong commitment, especially after 1940, to the "war effort," were to push him into one of the thorniest controversies of his public career, the relocation and incarceration of Japanese Americans on the West Coast. For Warren, at this stage in his life, the presence of Japanese in California, like the presence of Communists and the potential presence of organized crime, was a threat to law-abiding native citizens. In making this judgment Warren was acting as a provincial law enforcement official: He was idealizing California as a law-abiding nativist paradise beset by "foreign" threats.

The Japanese threat, however, turned out to be qualitatively different from the threats of organized crime or communism. Japan was, after 1940, officially at war with Germany and Italy against the Allies. While the United States did not formally participate in World War II until December, 1941, American sympathies, and increasingly, American foreign policy, were supportive of the Allied war effort. From at least 1940, then, Japan was perceived as an enemy of America, and the Japanese in California were associated with that perception.

But there were Germans and Italians in California as well, and no relocation policy was ever implemented with respect to these groups, even after America's formal declaration of war. The Japanese threat was qualitatively different in another sense. Japanese were Orientals. Orientalist racism was one of California's traditions. Chinese immigration into California in the late nineteenth century had provoked nativist responses, in part reflecting an emotional reaction to physical and cultural differences between Chinese and whites and in part reflecting fears of job competition. Eventually, in 1902, Chinese immigrants were excluded completely from the United States, and xenophobic feelings came to be directed toward the Japanese, who did not start to come to California in significant numbers before 1891, the year of Warren's birth.[88]

In 1920, when Warren joined Decoto's staff, there were 71,952 Japanese in California, out of a total population of approximately 3,400,000. Of those, about half were engaged in agriculture, many with notable success. Others were domestic servants, small merchants, gardeners, florists, or commercial fishermen. Warren once advanced a general theory about the entry of the Japanese into California. "The whole minority development in our country," he said in 1972, "has stemmed from the search for cheap labor. . . . They brought the Japanese in . . . for farm labor." But the Japanese, Warren suggested, "were too smart, and they started owning the farms."[89]

Throughout the 1920s and thirties no influential segment of California political life was sympathetic to the Japanese. The Oriental Exclusion League, formed in 1919, included among its members representatives of groups as diverse as the California Federation of Labor and the American Legion. Hiram Johnson was openly antagonistic to the Japanese, but so were William Randolph Hearst and the Socialist writer Jack London. The *Los Angeles Times* declared in 1920 that "assimilation of the [Oriental and white] races is unthinkable. It is morally indefensible and biologically impossible. An American who would not die fighting rather than yield to this infamy does not deserve the name."[90]

The Japanese were excluded from trade unions, went to segregated schools, lived in segregated neighborhoods, and generally were barred from entry into non-Japanese communities. In 1923 the constitutionality of California's Alien Land Law, codified in 1913 and 1920 in statutes preventing unnaturalizable aliens from owning land, was challenged before the United States Supreme Court on the grounds of being racially discriminatory.[91] Warren's predecessor as attorney general, Ulysses S. Webb, successfully defended the statutes by arguing that they were premised on "the obvious fact that the American farm . . . cannot exist in competition with a farm developed by Orientals with their totally different standards and ideas of cultivation of the soil, of living and social conditions."[92] Despite these statutes the Japanese, by 1940, were said to own over five thousand farms in California, at an estimated value of sixty-five million dollars, and to have raised about half of the commercially marketable fruits and vegetables in the state.[93]

Warren's relationship to Oriental racism and to the Japanese in California was representative of that of his peers. His high school classes in Kern City had debated whether the Japanese should be segregated in the San Francisco schools and whether Japanese aliens should be allowed to become naturalized;[94] as noted, Orientals were not given names in the Bakersfield telephone directory. During his years as district attorney, Warren had joined the anti-Oriental Native Sons of the Golden West, whose membership included Hiram Johnson, Webb, and one of Warren's long-standing political supporters, newspaper publisher V. S. McClatchy. In his 1942 race for governor, Warren campaigned extensively in the company of his old companion from the army in World War I, actor Leo Carillo, who reportedly said that "[t]here's no such thing as a Japanese American," and that "if we ever permit those termites to stick their filthy fingers into the sacred soul of our state again, we don't deserve to live here ourselves."[95] As attorney general, Warren quickly came to believe, and declared publicly shortly after Pearl Harbor, that "the Japanese situation as it exists in this state today may well be the Achilles' heel of the entire civilian defense effort."[96]

In his memoirs Warren remembered the atmosphere that Pearl Harbor created in California:

> For a long time, we were blacked out at night because of the likelihood of air raids or submarine attacks. California was immediately declared to be a theater of operations after the Pearl Harbor attack, and we were defenseless. Incidents never reported in the news were known by enough people to keep the public at all times aroused as to the danger. . . .
>
> Many Japanese were commercial fisherman, and their boats, like those of other fishermen, went to sea at night. The flashing of lights on such boats, whether Japanese or not, was thought to be a signal to enemy submarines. Military intelligence assured us this was a danger. [Japanese submarines had surfaced off the California coast

at this time, one sinking a tanker off of San Luis Obispo, and the navy had informed Warren that only two destroyers were deployed to protect the entire Pacific coast.]

The atmosphere was so charged with anti-Japanese feeling that I do not recall a single public officer responsible for the security of the state who testified against a relocation proposal.[97]

Warren himself had no hesitation about recommending the relocation of Japanese from the California coast to detention centers in eastern California, Idaho, Arizona, Utah, Wyoming, Colorado, and Arkansas. He warned against "fifth column activities in this country," arguing that the Axis powers "would like nothing better than to create the same situation here that they developed in France, Denmark and Holland."[98] In January, 1942, he made his "Achilles' heel" comment, and on February 2, at a conference of California law enforcement officials, he secured the passage of a resolution calling for "[a]ll alien Japanese [to] be forthwith evacuated from all areas in the state of California to some place in the interior."[99] He also pointed out that since the internment and forcible evacuation of Japanese could only be effectuated by military authorities, he had explored the idea with the army and the navy.[100] Five days later, testifying before the Joint Immigration Committee of the California legislature, he suggested that "the political approach to [relocation] is just too cumbersome," and that "only one group . . . can protect this State from the Japanese situation [—] the armed forces of the government."[101]

In coordinating matters with the military authorities, Warren operated from a position of strength since his early concern about preparedness had resulted in his attorney general's office being well-informed on the legal consequences of various courses of action. In 1939 Herbert Wenig joined Warren's staff and was instructed to study exhaustively the relationship of military law to civil defense in California. After the United States had formally declared war on Germany, Wenig enlisted in the Judge Advocate General Corps of the army, and secured, through Warren's efforts, a job with General John L. DeWitt at the Presidio in San Francisco. Prior to his being assigned to DeWitt, Wenig had prepared for Warren memoranda on the legal consequences of various civil defense measures, including the internment and evacuation of enemies in wartime. Wenig continued to act as a liaison between DeWitt's office and the governor's and attorney general's offices while Warren was attorney general and after he became governor in early 1943.[102]

In sum, Warren was a vital moving force in the formulation of the Japanese relocation program. He did not issue the formal orders, or design the program's apparatus, or administer the program once it had been designed. But by more than fully cooperating with the military in the shaping of the program, he was the most visible and effective California public official advocating internment and evacuation of the American Japanese.

Moreover, he revealed, in the process of arguing for relocation, that he was a prisoner of the radical stereotypes that contributed to the relocation decision. On February 21, 1942, testifying before a House committee on "National Defense Migration," he said that "when we are dealing with the Caucasian race, we have methods that will test the loyalty of them; and we believe that we can, in dealing with the Germans and the Italians, arrive at some fairly sound conclusions." Earlier in that same month Warren had opposed evacuating Italians and Germans because "the first generation born Italians and Germans in this country are no different from anybody else."[103] The Japanese were another matter: "when we deal with the Japanese, we are in an entirely different field and we cannot form any opinion . . . [because of] [t]heir method of living,"[104] by which he meant their perceived unwillingness to shed their racial and ethnic identities.

In their efforts to treat Japanese as qualitatively different from other prospective "enemies," both Warren and the military authorities who implemented the relocation program indulged themselves in a remarkable argument about "fifth column" activity among the Japanese. On February 2, 1942, at the conference of law enforcement officials previously mentioned, Warren, in response to a question about whether any Japanese had ever supplied information about possible sabotage by other Japanese, said that "five or six [California sheriffs and district attorneys] . . . said some individuals had dropped in to give them some information."[105] Five days later, testifying before the California Joint Immigration Committee, he claimed that no Japanese had supplied information, and on February 21, in remarks before the House committee, he said the following:

> I had together about 10 days ago about 40 district attorneys and about 40 sheriffs in the State to discuss the alien problem. I asked

all of them collectively at that time if in their experience any Japanese, whether California-born or Japanese-born, had ever given them any information on subversive activities or any disloyalty to this country. The answer was unanimously that no such information had ever been given to them.[106]

Warren made these comments in the course of an argument that *because* no fifth-column activity and no sabotage had been reported, sabotage "has been timed for a different date."[107] This argument was repeated by Army Colonel Karl R. Bendesten, one of the military architects of relocation, in a speech in May, 1942. "Contrary to other national or racial groups," Bendesten announced, "the behavior of the Japanese has been such that in not one single instance has any Japanese reported disloyalty on the part of another specific individual of the same race." He called "this attitude" a "most ominous thing."[108]

Warren's original conception of Japanese relocation appears to have distinguished between aliens and citizens. The February 2, 1942 resolution of the law enforcement officials, as noted, referred to "alien Japanese"; at that same conference Warren suggested that the 1920 Alien Land Law might be used as a basis for removing some Japanese. But by February 7 Warren was suggesting that the military could remove "any or all Japanese" from combat zones.[109] On February 9, Tom C. Clark, who had been designated coordinator of the Alien Enemy Control Program by the United States Justice Department, stated, in a press release, that "the Army and Navy believe evacuation of enemy *aliens* from prohibited areas and their relocation in 'clear' areas within the state will be wholly adequate."[110] But by February 14 General DeWitt had recommended the evacuation of "all Japanese,"[111] and by February 17 the Justice Department had agreed.[112] President Roosevelt's Executive Order of February 20 gave the military discretionary authority to exclude "any or all persons" from designated areas, including the California coast.[113]

Between February and August, 1942, the evacuation of Japanese from California, and Oregon and Washington as well, was implemented. Japanese aliens and citizens, about two-thirds of whom had been born in America, were transported to "relocation centers" in seven states, including a center in eastern California.[114] The great bulk of the Japanese who had been evacuated—some West Coast Japanese avoided, successfully resisted, or were exempted from evacuation—remained incarcerated in relocation centers until January,

1945, when they were officially released. Warren, upon hearing of their prospective release, asked Californians to "maintain an attitude that will discourage friction and prevent civil disorder."[115] He had earlier said, in remarks at a 1943 governor's conference, "We don't propose to have the Japs back in California during this war if there is any lawful means of preventing it."[116]

The relocation centers could be said to have resembled concentration camps. They were enclosed with barbed wire and patrolled by armed guards. Privacy and independent family life for the incarcerated Japanese were reduced to a minimum, and the daily lives of the Japanese were controlled by their supervisors. The relocation centers were not, however, instruments of genocide or barbarism or even brutality; in this sense the term "concentration camp" incorrectly describes them.

The analogy to Nazi internment centers is appropriate in another sense, however, for the relocation camps were the instruments of a policy of racism. Japanese were interned; Germans and Italians were not. General DeWitt, in a statement justifying his decision to recommend a policy of mass evacuation, made it clear why Japanese were being singled out. "The Japanese race is an enemy race," DeWitt said, "and while many second and third generation Japanese born on United States soil, possessed of United States citizenship, have become 'Americanized,' the racial strains are undiluted."[117] Later DeWitt was to say that "you needn't worry about the Italians at all except in certain cases," or "the Germans except in individual cases," but that "we must worry about the Japanese all the time until [they are] wiped off the map. Sabotage and espionage will make problems as long as [they are] allowed in this area."[118] Warren, two months after DeWitt's remarks, was to state this attitude more succinctly: "If the Japs are released no one will be able to tell a saboteur from any other Jap."[119]

Warren's role in the Japanese relocation controversy was, however, not exclusively that of racist theoretician and advocate for mass evacuation. In an action that was characteristic of his capacity to embrace potentially contradictory positions without acknowledging the contradictions, Warren sharply dissented from a February, 1942 order of the California State Personnel Board disqualifying citizens who were "descendants" of alien enemies of the United States from holding state civil service jobs. The language of the order included Germans and Italians as well as those of Japanese ancestry.[120]

Warren's dissent charged that the order violated the state constitution by invidiously discriminating between "naturalized citizens and citizens by birth of the first generation," on the one hand, and "citizens whose forbears have lived in this country for a greater number of generations." Warren argued that "a substantial portion of the population of California consists of naturalized citizens and citizens born of parents who migrated to this country from foreign lands"— his own family was an example. Such persons, in Warren's judgment, "represent the highest standards of American citizenship," and it was "cruel" to "question [their] loyalty or place them in a category different from other citizens."[121] Ten days later Warren again protested a discriminatory measure aimed at the Japanese, the decision of the State Department of Agriculture to revoke summarily licenses held by "enemy aliens" to handle fruits and vegetables. He felt, probably correctly, that the summary nature of the revocation violated the California Constitution.[122]

Finally, Warren took steps to protect the safety of the Japanese returning to California after their release from the detention centers and refused to make face-saving gestures once public opinion began to question the incarceration policies as the course of war in the Pacific became more favorable. In December, 1944, after learning of the forthcoming release of the Japanese, Warren asked others to "join in protecting constitutional rights of the individuals involved," and to "comply with [the decision] . . . loyally and cheerfully."[123] The next day he was reported as saying of the evacuation, "[A]t the time of their exclusion not one of us raised a voice against it. We can't condemn it now."[124]

⚜

In assessing Warren's role in the Japanese relocation controversy, certain facts should be taken as established. First, Warren was one of the individuals most responsible for bringing the relocation program into being. He was in a position, as the official ostensibly in charge of California civilian defense, to influence all other policymakers, including the military and the federal authorities. He was, from January, 1942 on, a persistent advocate of some form of evacuation, and his skillful marshaling of arguments, some of them spurious and others based primarily on racial prejudice, significantly contributed to the decision to intern and evacuate the Japanese. Although Warren himself did not make that decision, each of the persons who contributed to its being made—Tom Clark, John J. McCloy, Allan W. Gullion, DeWitt, Bendesten, and even Franklin Roosevelt—relied substantially on evidence and arguments that came from Warren's office.

Second, among Warren's motives for excluding the Japanese from California was a provincial, xenophobic racism. Warren thought of Orientals in general, and Japanese in particular, as foreign, not easily assimilable, inscrutable, resourceful, and, especially after Pearl Harbor, treacherous. He did not believe disloyal Japanese could be distinguished from loyal Japanese because of the perceived cohesiveness and "foreignness" of the Japanese community. He had "methods" for testing the loyalty of Germans and Italians because he perceived those groups as having embraced Western values and traditions in a way that the Japanese in California did not.

In holding this perception about the Japanese in California, moreover, Warren lumped together very different classes of Japanese persons, ranging from American citizens of Japanese descent, some of whom had been born in the United States and spoke no Japanese, to aliens who had recently arrived from Japan and planned to return. Warren's perception of the Japanese in California was astigmatic in two respects: It failed to distinguish among classes of Japanese, and it arbitrarily singled out Japanese as the only "enemy" group to bear the burdens of internment and evacuation.

Thus Warren not only participated in but can be said to have engineered one of the most conspicuously racist and repressive governmental acts in American history. But he was by no means alone in his efforts or his attitudes. Congress immediately approved Roosevelt's executive order authorizing the evacuation and internment of Japanese and provided criminal penalties for violations of military directives under the order.[125] The United States Supreme Court twice sustained constitutional challenges to the relocation program.[126] Among those who defended the program and its constitutionality were Walter Lippmann, Harlan Fiske Stone, Felix Frankfurter, William O. Douglas, and Hugo Black.[127] Not a single California political leader opposed the decision to evacuate: Only Carey McWilliams, Olson's director of immigration and housing, proposed a policy that did not require mass relocation of the Japanese.[128]

Warren publicly confessed error for his part in the Japanese evacuations, stating in his memoirs that "I have since deeply regretted the removal order and my own testimony advocating it. . . . It was wrong to react so impulsively without positive evidence of disloyalty." While Warren's atonement reveals something about his capacity for growth and his ability to penetrate, occasionally, the curtain of his own defensiveness, its significance should not be overstated. Although one commentator has suggested that Warren had begun to regret his role in the relocation controversy by 1947,[129] Warren wrote an article in 1962, eight years into his tenure as Chief Justice, that defended the Supreme Court's 1943 and 1944 decisions supporting the internment of the Japanese. In that article he said that in wartime "actions may be permitted that restrict individual liberty in a grievous manner";[130] in arguing for evacuation of the Japanese in 1942 he had said that "in time of war every citizen must give up some of his normal rights."[131]

At the time of Warren's death in 1974 many of his close acquaintances and former employees believed that his position on the Japanese evacuation had remained unchanged. The author of a 1967 biography of Warren noted that "Warren has never publicly expressed regret or admitted error for his part in the Japanese evacuation," and quoted Warren's close friend Robert Kenny, who succeeded him as attorney general in 1942, as saying that "[w]e'd been brainwashed about the Japanese all our lives."[132] But by 1972, in an oral interview, Warren said that "I feel that everybody who had anything to do with the relocation of the Japanese, after it was all over, had something of a guilty consciousness about it, and wanted to show that it wasn't a racial thing as much as it was a defense matter."[133]

In his memoirs Warren attempted just such an explanation; he had, in fact, written a draft of the explanation before the 1972 interview. His account of the Japanese relocation controversy occupied five pages in his memoirs. Four and a half of those five pages dealt with "defense" matters: Japanese submarines off the Pacific coast, the absence of the United States fleet, propaganda balloons "floated over the Bering Strait by the Japanese . . . to fall on our forests and prairies";[134] Japanese military successes in the Pacific theater; the dual citizenship of many Japanese, which allowed them to travel freely between the United States and Japan; and, in general, the "fear, get-tough military psychology [and] propaganda" that marked the period, as well as his "responsibility for public security" as attorney general.[135]

The remaining portion of Warren's account dealt gingerly with the "considerable amount of racial prejudice" in California, which Warren believed "stemmed largely from some of our farming communities," and the "race antagonism occasioned by the war."[136] Warren himself "always believed that I had no prejudice against the Japanese as such except that directly spawned by Pearl Harbor and its aftermath." He had had "great respect for people of Japanese ancestry" while district attorney, he said, "because during my years in that office they created no law enforcement problems."[137]

Then Warren turned to a few other matters about the Japanese evacuation that had troubled him over the years:

> Whenever I thought of the innocent little children who were torn from home, school friends, and congenial surroundings, I was conscience-stricken. . . .
>
> At about the same time we were considering their removal for military reasons, I wrote a formal opinion to the State Personnel Board, telling its members that they could not constitutionally take away Japanese-Americans' Civil Service rights to their state jobs as the commission directed. *I know that it seemed ambivalent to protect their constitutional rights in this regard with one hand and deprive them of other rights by removing them from their homes.* However, I consoled myself with the thought that the latter was occasioned by my obligation to keep the security of the state.[138]

In this passage one sees the retired Chief Justice, who had been so committed to "constitutional rights" during his tenure, confronting the starkest contradiction in his public career. One also sees the family man, approaching eighty and taking great pleasure in his children and grandchildren, confronting the faces of Japanese children forcibly separated from their homes and friends. Here one finds Warren not trying too hard to conceal his feelings of guilt and shame, not papering over his distress save perhaps for his last appeal to "an obligation to keep the security of the state," which he undoubtedly felt was threatened. In this passage, a rarity in Warren's memoirs, one is allowed to penetrate the official, self-preserving prose and glimpse the embarrassment caused by a vivid memory. During the 1972 interview in which he discussed the "guilty consciousness" of those who had participated in the relocation, Warren mentioned the faces of the children separated from their parents; tears rolled down his face, and the interview was temporarily halted.

✍

The executive order beginning the relocation program was signed in late February, 1942. One morning in March of that year, as Warren recalled,

> I left my office one morning, went across the Bay to my home in Oakland, and told Nina [his wife] to prepare herself for a shock. She inquired about the nature of it, and I told her that, as a last resort, I had made up my mind to run for governor. It was, indeed, a shock, and she asked me if I thought I could win. I told her that was not the main consideration; that the governor and I were at loggerheads; that he would not permit me to do the things that the Constitution and the laws of the state required of me, and that I would not sit on the sidelines for a term as attorney general while we were in the midst of a war that threatened our very national existence.[139]

In 1972, in an interview not then intended for publication, Warren was to give a less varnished account. He began reminiscing about Olson and moved on to the decision to run for governor. Warren's remarks were as follows:

> [Olson and I] just didn't have any relationship. . . . I [only had one] conversation with Olson in four years. He just treated me as though I wasn't there . . . and that's why he wasn't governor a second time. . . .
>
> Governor Olson just treated everyone as an enemy if he didn't believe the same theory as [Olson] did. . . .
>
> There were a lot of things that Olson advocated that I could have supported . . . but I had no opportunity to be helpful to him. . . .
>
> Olson wanted to divide the state up into [civil defense] units and be in a position to send down his own people to replace the regularly constituted law enforcement officers of the state. . . . To me that's just an abolition of civil government, and it's getting to martial law. I just resented that deeply . . . I fought that bitterly. . . .
>
> I didn't want to run for governor! Why, I couldn't afford to run for governor. I had six children and they were all of school age . . . and I didn't have any means and I just didn't want to leave a legal position. . . . If I got into the governor's office and then was beaten, I'd be clear out of the law and maybe difficult to get back. . . .
>
> I never would have run for it if he hadn't bedeviled me in that way. But he just treated me as a common enemy. . . . I just made

up my mind and decided I wasn't going to sit out four years as attorney general and do nothing during the war. And that I'd run for governor and if I was beaten I'd do something in the war effort.[140]

There has been some commentary to the effect that Warren had intended to run for governor in 1942 virtually from the minute he became attorney general. Richard Graves's remarks, quoted earlier, referred to Warren as being "in many ways an active candidate" from 1939 on; a March, 1940 article in *Fortune* magazine identified Warren as having gubernatorial ambitions.[141] But in 1967 Leo Katcher, in the course of writing an early biography of Warren, quoted "an official who was close to [Warren]" as saying that he "wanted to be Governor and his ambition was to attain that office, but 1942 was ahead of his timetable."[142] Katcher also reported that Warren, as previously noted, had tried to persuade Robert Gordon Sproul to run against Olson. Warren also reportedly had a conversation with Mayor Fletcher Bowron of Los Angeles early in 1942 in which Bowron urged him to run, and Warren demurred, saying that he was "a prosecutor" who "didn't have that type of administrative experience."[143]

In speculating about Warren's plans after becoming attorney general in 1938 one needs to take into account his instinctive caution and his tendency to remain in situations where he "felt secure." Despite Warren's association with Republican politics and civic organizations, he still saw himself, while attorney general, primarily as a "prosecutor," i.e., a law enforcement official. The principal network of acquaintances and confederates that he had throughout the state was a law enforcement network; the principal issues, civil defense and anticommunism, on which he had clashed with Olson, were "security" issues. He had flourished in law enforcement, and he was a person who preferred security to risk-taking.

Warren ran for governor of California primarily because Culbert Olson "bedeviled" him. Warren's style of officeholding had rested heavily on an active, expansive interpretation of the powers of his office; Olson, by seeking to preempt Warren on civil defense, the major area in which the attorney general's office could demonstrate its activeness and visibility, was attempting, Warren felt, to make Warren "sit out four years as attorney and do nothing during the war." Warren's patriotism and his concern for national security reinforced his natural inclination toward institutional activism; Olson was seeking to repress Warren's desire "to be helpful." Olson, moreover, was thwarting Warren for reasons that Warren did not respect. He treated Warren "as a common enemy," in Warren's view, simply because of perceived partisan differences. Warren believed that he "could have supported . . . a lot of things that Olson advocated," notwithstanding their different political affiliations. Nonpartisanship had, after all, been one of Warren's principles of officeholding. But Olson acted as if Warren had declared his opposition to the Olson Administration by being a Republican.

Olson's view of Warren thus became a self-fulfilling prophecy. Olson acted as if Warren was a Republican partisan coveting his office and *because he acted in that fashion* Warren eventually did covet his office, a covetousness born of pique, humiliation, stubbornness, and competitiveness. Warren would respond this way in other situations later in his life. When he was confronted with a situation in which another strong person to whom he felt some responsibility treated him with a lack of respect, he became almost irrationally offended. Once so offended, he allowed a combination of feelings to fix and rationalize his position. He developed a deep suspicion and even animosity toward the other person; he made him into the "enemy"; he vigorously and regularly opposed that other person; he made his opposition to that other an all-out, us-against-them struggle—in Warren's eyes, a matter of principle.

Warren may well have been biding his time in the attorney general's office, seeking to broaden his political base with the hope of becoming governor at some later time. But not in 1942. He ran then because of a tendency to chafe under the restraints of authority, because of a tendency to personalize checks on the exercise of his powers, and because of a tendency to respond to perceived humiliation with decisive and channeled action. "When you put the pressure on him," an intimate once said of Warren, "he just became firmer and firmer." [144]

The wartime atmosphere allowed Warren to deemphasize his partisan affiliation in the gubernatorial campaign: Nonpartisanship was his principal theme. He filed for both the Democratic and Republican nominations; he ran on a platform that had stated that "partisan politics had no place and must be eliminated, so that we may give to President Roosevelt our unqualified support"; [145] he made his campaign slogan "Leadership, Not Politics." Carey McWilliams, no friend of Warren's, later claimed that Warren also refurbished his public image "under the talented direction of [public relations advisers] Clem Whitaker and Leone Baxter," changing from a "grim, humorless, prudish" district attorney to "a smiling, warm, friendly man with a charming wife and a bevy of attractive children."[146] Warren did, in fact, begin to expose his family to public view during the campaign. He also fired Whitaker and Baxter for not following his instructions concerning press releases. On one occasion Whitaker and Baxter erroneously informed the press that Warren had endorsed the candidacy of the Republican lieutenant governor, Frederick F. Houser; Warren, who made it a policy in all his campaigns not to endorse other candidates, was embarrassed by the incident. Warren's calculated nonendorsement of other candidates was to form the basis of his longtime feud with Richard M. Nixon.[147]

Olson's campaign was not well organized and was late in starting. By the time Olson picked up momentum with an effective performance in a radio debate with Warren in October, Warren had too formidable a lead. Some vivid rhetoric had passed between the candidates, such as Olson's comparing Warren's "Leadership, Not Politics" slogan with Hitler's "Fuhrerprinzip," and calling Warren "either . . . a political eunuch . . . or a political hypocrite,"[148] or Warren's referring to Olson's "arrogance, blundering and . . . selfish manipulation of State Government."[149] But on the whole the campaign reduced itself to an exercise in the skillful use of cross-filing. Warren, emphasizing his nonpartisanship and concentrating on defense issues, got 404,778 votes in the August Democratic primary, compared to only 514,144 for Olson. That showing, plus Warren's decisive victory in the Republican primary, his widespread newspaper support, and his cautious campaigning, decided the outcome well in advance. Warren eventually received fifty-seven percent of the votes for governor.

One piece of Warren "luck" emerged in the campaign, and proved a harbinger of things to come. As a law enforcement official, Warren's persona had been that of a conservative: He was the champion of law and order, the defender of native California, the relentless foe of radicals and foreign criminals and Eastern racketeers. Underneath, however, lurked Warren's own instinctive Progressivism, occasionally breaking through the surface in such instances as his speech on behalf of organized labor during the *Point Lobos* trial. But in 1942 Warren was thought to be the candidate of conservative, moneyed elements, and it was to these that his campaign turned for financial support. Then, as Warren later described it:

[T]he greatest thing, probably, that ever happened to me in politics
[was when] the money bags who had urged me for two or three
years to run for governor [declined to contribute to the 1942 cam-
paign] . . . Those were all the special interests in the state.[150]

The first public reaction of Governor Olson to my candidacy
was that the people of California would not want as their governor
a district attorney who was the creature of the moneyed interests of
the state. But [the action of the "moneybags"] had deprived him of
one of his appraisals of me, and, what was far more important, *it
had relieved me after my election of any pressure based upon campaign contributions from powerful . . . interests.*[151]

Here one finds the origins of Warren's political strategy as governor, and here one finds, at a deeper level, the beginnings of an
expanded definition of Warren's self-image as a public official. There
was genuine doubt and concern in Warren the law enforcement official for the prospects of Warren the governor and politician. We
have noted that when Warren became accustomed to a public role,
as he had become accustomed in the 1930s to the roles of crusading
district attorney and vigilant attorney general, he was reluctant to
abandon that role. He was not ambitious in the conventional political
sense: He did not generally use one office as a springboard for another, and when he did, as in the transition from district attorney to
attorney general, the offices were similar enough to make him comfortable.

In his concern and uncertainty about the governorship, Warren
had tentatively reached out for support from the very kinds of persons that his earlier experiences had told him to distrust: the "special
interests" and the "moneybags." He was, after all, a Republican and
seemingly a conservative; he was to run against a conspicuously liberal governor. But when Warren reached out, his prospective supporters rebuffed him. Warren never forgave that class of persons for
this slight, as he never forgave Olson for his. In fact, a similar set of
responses took over, fixing and justifying his rejection of those who
had rejected him.

Warren was never again to expose himself as any organized lobby's
"candidate"; he was resolutely to remain aloof from conventional in-
terest-group politics. And, most important, his vision of the political
world, put into renewed and sharper focus by this incident, was to
remain constant throughout the rest of his public career. He merged
his own sense of personal rejection, and the outrage that accompa-

nied it, with the rejection of the common citizen by special interests. Public life was indeed, he felt, the struggle of the individual against special interests; the struggle had the clearest kind of moral overtones. Warren's role, as he defined it for himself, was that of the Progressive crusader for nonpartisan, open, and honest government, representing the public against the Southern Pacific and its later counterparts. In this self-defined role one can find the origins of Warren the unpredictably liberal governor and Warren the militantly liberal Chief Justice.

# 4

# *Governing: Officeholding and Policymaking*

In the last months of 1942 Warren prepared himself for the third of the four major transitions he was to make in his public life. The first had been in 1908, when he left Bakersfield to attend college and law school at Berkeley; the second in 1920, when he entered Ezra Decoto's office; the fourth was to take place in 1953, when he was appointed to the Supreme Court. The move from district attorney to attorney general had been a succession, not a major transition. Warren had laid plans to become attorney general because he saw the office as an extension of his service as district attorney. As attorney general, Warren had become immersed in political issues, but he had not fully come to terms with that aspect of the job.

The move to governor of California, however, represented a significant change in Warren's life. Despite his active interest in politics, he had never before been a professional politician; despite his identification with the Republican party, he did not think of himself as a partisan. He was, moreover, at the age of fifty-two, a cautious man who liked to "feel secure" in his occupation and disliked abrupt change. He said in his memoirs that after becoming "engrossed in public service," he "soon realized I was a politician," but while he was "active in politics throughout my years as governor," he was active "in my own way." He had no permanent organization, "never tried to purge anyone from office," did not support or oppose can-

didates for the legislature, and took, especially in the early years of
his governorship, only a minimal interest in national politics.[1] In
short, Warren initially responded to the governorship by attempting
to continue the pattern of officeholding that he had established earlier. He sought to "maintain [my] independence in the performance
of [my] duties,"[2] deemphasize his partisan affiliations, expand the
jurisdictional reach of his office, and professionalize his staff. In establishing these goals he was emphasizing the continuity between his
past positions and his new one.

The office of governor was different from Warren's previous offices, however, and Warren recognized that fact. He was now identified as a politician, nonpartisan or not; he was thrust into national
Republican politics, despite his limited experience outside the state
of California; he was forced to negotiate with the California legislature, despite his insistence on not maintaining a "personal or trading
relationship"[3] with that body. He was holding down a new office,
with new demands and new possibilities, and he was a person who
disliked sudden change.

It was not surprising, then, that during this transition of his middle life Warren reached out for the support of two sets of persons
that he identified as sources of stability and security. One set was
the members of his family; the other, trusted staff members. Both
groups of persons were vital to Warren's accommodation to his new
public role. His family relationships provided him with a buffer
against change, an index that he had remained the same person despite his new public exposure. His staff relationships provided him
with the assurance that the same people on whose services he had
come to rely in earlier offices would be there to serve him in his new
office. Central to the emergence of Earl Warren in a new public role,
one that suddenly and dramatically expanded the vistas of his career,
was the presence of his family and his intimate staff.

K

As a young man Warren's gregarious qualities had not often been
extended to women. In high school he "was not interested in the
girls," who were "all very nice to me but in a more or less patronizing way"; in college he "spent [his] spare time with the men of the
University."[4] While a college classmate reportedly claimed that
Warren "would like to get married if he could find a girl to live on

air,"[5] as late as his thirtieth year he was deterred from considering marriage by his modest financial circumstances. "I would have felt humiliated," he said of his early years in Oakland, "if my wife had been compelled to work."[6] Nonetheless, in 1921 he met and fell in love with Nina Palmquist Meyers, a widow whose parents had also been Scandinavian emigrés. Nina Meyers at that time managed a women's speciality shop in Oakland, according to Warren's recollections;[7] she and Warren began dating "on Saturday evenings" and "part of . . . Sundays."[8] Two years later Earl Warren and Nina Meyers became engaged; after Warren was appointed district attorney in 1925 they "decided to be married at an early date."

Meeting Nina was "the greatest thing that ever happened to me," Warren said in his memoirs.[9] The Warrens' marriage was notable for its closeness, its mutual admiration, and the complementary roles played by the participants. Nina Warren "never made a speech for me in her life," Warren said in 1972. "She never asked anyone to vote for me in her life. She was just an observer, that's all. A quiet observer."[10] Earl Warren was the provider, the public figure, and the patriarch; Nina Warren was the household manager, the keeper of the hearth.

The Warrens' public and private lives were firmly separated. While Warren was Chief Justice, for example, the Warrens gave only one ceremonial dinner a year for the other justices, reserving their socializing for intimate family friends. They "accepted practically no invitations to private homes because that called for reciprocation"[11] and in general used their leisure hours for private family pursuits. This pattern of separation of Earl Warren's life into public and private spheres, already present during Warren's years as a law enforcement official, was formalized when Warren became governor of California.

When Warren was inaugurated in 1943, the Warrens had six children—James (Jim), Nina's son by her previous marriage, Virginia, Earl junior, Dorothy (Dottie), Nina Elizabeth (Honeybear), and Robert (Bobby). Their ages at the time Warren succeeded to the governorship ranged from seven to twenty-two; all save Jim were in public elementary or high schools. Since 1935 the Warrens had lived in Oakland, the Warren family remaining largely unknown to the public. Warren regularly took his children on extended outings on Sundays, visiting zoos or parks; one contemporary recalled him chasing after children "like . . . leaves in a windstorm" at the 1937 Berkeley alumni picnic.[12]

With the emergence of Warren as a political candidate, however, the potential of the large Warren family for winning votes was recognized, and Nina Warren and the children became participants on the campaign trail and the objects of feature stories in the media. John Weaver, recalling the impact of the Warren family in California politics, said that "for a generation of middle-aged Californians, the Warrens were the family next door."[13] Carey McWilliams, as noted, saw the emergence of the Warren family as a deliberate effort to humanize Warren's image.[14]

Beginning in 1944 popular feature magazines began to run illustrated articles on the Warren family. *Life*, using the occasion of Warren's dark horse candidacy for the 1944 Republican presidential nomination, maintained that "Warren's political personality" was "a composite not only of his own qualities, but of those of his family," and cited a "disgruntled California Democrat" as claiming that "the people of the state elected the Warrens *en masse*, as much for the warm and human picture they make in the Executive Mansion at Sacramento as for the statesmanlike qualities of the head of the house."[15] *Better Homes and Gardens* took a slightly different tack in 1947, citing Warren as "a rebuke to all American men who plead, in defense of their neglect of their father role, that they are 'too busy.' " Warren, according to the article, had "never been too busy to be a dad"; his intimates were "his wife and children" and "his hobby [was] his family"; he "jeopardize[d] the goodwill of important people, accepting as few dinner engagements as he [could] so that he [could] have his evenings at home." However Warren's career developed, the article asserted, "the structure of the Warren family will endure."[16]

Additional articles in the 1950s continued to emphasize the Warrens' rigid separation of public and private worlds. *Redbook* claimed in 1950 that "the Warrens, from Honey Bear in the pantry to the Governor in the Executive Office . . . manage to live normally and comfortably by meet[ing] their problems and dispos[ing] of them beforehand."[17] In 1951 the *Saturday Evening Post* declared that while "everyone knows the story of Earl Warren, politician, . . . few know how, while living in the political limelight, he and Mrs. Warren contrived to raise six normal, attractive youngsters." The article focused on the Warren's system of household management, "a combination of love and logic." It stressed the Warrens' conviction that "their children should be raised without physical punishment," their lack of tolerance for "put[ting] on airs around the household staff," their concern with "letting the children develop their own personalities and talents," and their insistence that "when you make a mistake, you take the rap for it." [18]

*American Magazine*, in 1953, named the Warrens "Family of the Month" for their ability to "take fame in stride" and "live so normally on the political merry go round" that they merited being called "still a typical American family." Friends of Nina Warren were quoted as calling her "more domestic than the Old Woman in the Shoe"; Earl Warren was said to be "never . . . too busy to be the closest pal and confidant of his kids"; the Warrens had "done everything as a group" and "had become a single strong unit, each always ready to help the other." [19]

The feature articles, in addition to stressing the cohesiveness, unpretentiousness, and efficiency of the Warren family, conveyed a picture of the Warrens' daily routine during their years in the California Governor's Mansion. As noted, all the Warren children save Jim were in elementary school, high school or a California college while their father was governor: During Warren's first term, five of the children lived in the Governor's Mansion. Domestic affairs were handled by Nina Warren. Nina "belonged to no clubs, committees or organizations whatsoever"; the writer for *Life* felt that "she would be astonished if anyone asked her about her ideology." [20] She had had five children in six years, she told two of the magazines, and didn't have time for much else. Having six children, she told one reporter, was "the wisest thing we ever did." [21]

Nina allocated chores and arranged schedules, packing lunches early in the morning or the night before. She took down business telephone conversations in shorthand; she kept the family accounts; she supervised the renovation of the Governor's Mansion, "touring San Francisco stores selecting wallpapers, draperies, and upholstery materials." [22] She took "possessive pride" [23] in the mansion's grounds; she selected individual Christmas trees for the children's rooms; she baked cakes at birthdays; she tolerated, reluctantly, teenage adventures in cars and on horseback and on football fields. Warren once characterized "the dreams of every good man and woman" as "the desire to have a home and a fireside, to have happy, healthy children, taught by a good mother."[24] "For Warren," *Life* said in 1944, Nina was "the perfect consort."[25]

The feature articles on the Warrens reveal the elaborate care taken by Earl and Nina Warren to preserve their established patterns of family life in the face of expanding public scrutiny. Warren did not resist increased visibility as it came between 1943 and 1953, but he and Nina did resist, successfully, any transformation of the Warren family's character that might have accompanied such visibility. In order to preserve the integrity of their family structure in the midst of public life, the Warrens adopted a strategy that rigidly separated family members from public scrutiny except on those occasions when members chose to participate. When the Warrens moved into the Governor's Mansion in Sacramento, Warren made "some definite rules, to which I adhered religiously, about how, where and in what manner the business of the state should be carried on." The consequence of these rules was that the "[Warren] family home life was completely separated from governmental affairs."[26]

The Warren family's telephone number was listed in the Sacramento directory, but when the number was called, a state police officer answered the phone. If the call was for Nina Warren or for one of the children, the caller was put through. If it was for Earl Warren, the caller was routed to the governor's office during working hours or to the capitol switchboard in the evening or on weekends. Persons pressing to see the governor after hours were told "that if they would disclose the nature of the business and it appeared to be of such importance, I would open the governor's office and come there; otherwise it must be handled in the regular manner the following day."[27]

The Warren children, under these rules, were given time to get acquainted with their father. Honeybear said that Warren "never got up from the dinner table to talk on the phone"; never "spanked or hit or yelled out [or] put down or belittled" his children; but at the same time conveyed to them a clear sense of "what was right" and "what was wrong."[28] Jim was struck that it was "an uncommon thing to have a father like that, of such prominence, who was also such a regular guy."[29] He spoke of discussions with his father in which Earl Warren was not overbearing or dictatorial, but "first thing you know I was arguing against myself."[30] Earl junior said that his father had "the ability to make you think that he has only your interest in mind at the particular time": He "just could make every moment absolutely full and absolutely efficient."[31] While his daughters, Warren recalled, "were photographed for the press and television more than I was," none of the Warren children made speeches for their father, and the boys did not even attend national political events. "Their mother and I understood their lack of interest in such matters," Warren said. "I am sure they went fishing and hunting on such occasions because I raised them to be outdoor boys."[32]

There is no undercurrent of awkwardness and no defensiveness in Warren's descriptions of his private life in his memoirs. The overwhelming impression of the Warren family conveyed by that account and others is that of a group of persons who succeeded in creating a harmonious, closely-knit, self-contained family unit. The Warren family assumed traditional roles for its parental figures: Earl Warren as the "provider" and "head of the household"; Nina Warren as the "homemaker" and orchestrator of internal family affairs. A stark separation of Nina Warren's and the children's activities from the activities of public life was not only consistent with these roles, but it also provided a ballast for Warren's career. When he needed to, he could retreat into the sanctuary of his family, and by taking responsibility for the "public" side of life, he could shield his wife and children from public exposure.

Public service of a more visible sort became a realizable possibility for Earl Warren with the adjustment of his family to life in the Governor's Mansion. That adjustment took place early in his first term, when the Warrens moved into the mansion, and remained in an essentially constant state for the duration of Warren's governorship. The Warren family's calculated response to public life can be seen as a symbol of the themes of continuity and change that were to characterize Warren's years as governor. His family's ability to preserve its established patterns in the midst of a new version of public life revealed to Earl Warren that he could be a politician and apparently not compromise his family's autonomy, just as he felt he could be an elected public official and not compromise his independence. He was free to expand his conception of public service, to define a more openly political role for himself, because his family— the basic repository of his values—was not going to change significantly in the process. The anxieties of a new career, calling for a different interpretation of one's working life, were lessened by the realization that Warren's dream of a "home and a fireside" was not going to be altered.

Among the values Earl Warren held dearest were family solidarity, closeness between husband and wife, and closeness between parents and children. He found such values an antidote to the tendency of public life to become corrosive; he believed, at the same time, that values learned at home, such as equality, unpretentiousness, and fairness, could be transplanted into public life. He became an unusual kind of politician, a politician who eschewed much of conventional politics, in part because he continued to see himself as an average family man for the duration of his public career.

﹄

In the course of Warren's fifty years of public service he had a remarkably small number of intimates among his staff, but at the same time he came to rely heavily on those persons. Four individuals fulfilled the role of staff intimate during Warren's governorship: William Sweigert, his executive secretary; Verne Scoggins, press secretary; Helen MacGregor, his private secretary; and Merrell F. ("Pop") Small, departmental secretary.[33]

William Sweigert was originally hired by Warren in 1939 to reorganize the attorney general's office; he became Warren's executive secretary upon Warren's inauguration as governor. "[Sweigert's] generous manner of dealing with everyone," Warren said, "soon dispelled any feeling of partisanship or sectarianism, and he supervised the staff work in a way that maintained harmony and progress."[34] Sweigert stayed with Warren until 1949, when Warren appointed him to a municipal judgeship. He was one of the chief witnesses to Warren's apparent evolution from what Sweigert called a "Hoover Republican and a conservative"[35] to a less predictable figure.

In the last years of Warren's attorney generalship, Warren and Sweigert began lunching regularly and discussing politics and social issues. Sweigert was "a lifelong Democrat and an ardent Catholic" who had been educated by Jesuits and whose family had been active in Progressive politics in San Francisco.[36] In the lunches with Sweigert, Warren would complain about the New Deal, partly because at that time he "was a great believer in states' rights" and partly because he felt that "there was an unholy alliance between [President Roosevelt and] the Democratic Party of the North and the reactionary, narrow minded Democrats of the Southern States."[37] The former complaint was a typical reaction of former Progressives to the

New Deal's apparent emphasis on the federal government as the principal locus of regulation. When coupled with Warren's Republicanism and his law enforcement persona, however, it appeared to Sweigert to be the complaint of a conservative.

In his conversations with Warren in the early 1940s, Sweigert repeatedly stressed the obligation on the part of government to help disadvantaged persons,[38] and in late 1942, after Warren had been elected and was preparing to assume the governorship, Sweigert wrote Warren a memorandum entitled "Social Consciousness in Government."[39] The memo vividly articulated positions Sweigert had taken in conversation with Warren, arguing that affirmative governmental action in the pursuit of humanitarian goals ought to be the first priority of the Warren Administration. Warren read the memorandum thoroughly, referred to it often during his governorship, and later made similar arguments in some of his public statements.[40]

Warren used Sweigert in a variety of ways as a staff member: In addition to helping with internal management and policy formulation, Sweigert wrote speeches, worked on campaigns, advised on appointments, and kept an eye on the California legislature. His multiple functions were characteristic of Warren intimates. As Warren put it, "I tried . . . to get these people in the various departments to sort of more or less dove-tail together as a committee, and I divided them up into committees. . . . I would have them talk things over before they presented them to me. . . . If they worked [matters] out satisfactorily, there wouldn't be any necessity of seeing me. . . . I really relied on the people around and left them to their own devices."[41] This reliance on key staff members was not the equivalent of delegation of responsibilities: Warren, as one staffer put it, "had the capability . . . to fill every . . . cranny of the office he occupied."[42]

But Warren's use of key staff freed him from intraoffice bickering and petty detail and separated Warren as the office leader from everyone else in the office. His theory of administration worked well as long as no contentious, less than dedicated, or excessively independent persons served on Warren's staff. When such a person appeared, which was rare, since Warren carefully screened his staff appointments, that person did not stay long. Warren "demanded perfection" from his staff, Sweigert said, and was "impatient with what we might call sloppiness or laziness or anything of that kind."

When Warren found a staff member on whom he could rely he

was reluctant to lose that person's services. Sweigert had been originally persuaded by Warren to join his gubernatorial staff by the promise of a state judgeship. Several appointments fell to Warren during 1943 and 1949, and on each occasion he declined to name Sweigert. Eventually Sweigert indicated that he would have to leave Warren's office to go into private practice, as the judgeship promise appeared not to have come to fruition. At this point Warren told Sweigert that the only opening was a municipal judgeship, and that he doubted Sweigert would be interested. Sweigert immediately said that he would take that judgeship if Warren would promote him to the superior court at the first opportunity.[43] In his memoirs Warren suggested that "characteristically, [Sweigert] asked to start modestly in the Municipal Court. . . . Shortly thereafter I appointed him to the Superior Court."[44]

Despite Sweigert's impatience with Warren on the judgeship issue, the two men were kindred spirits. Warren shared with Sweigert, the latter said, "a deeply held belief that public office is indeed a public trust"; Warren's "great contribution," for Sweigert, was that he "accepted the appellation of politician and strove to make it a title of honor."[45] He and Sweigert had "a kinship" in their contempt for such things as "special favors [and] . . . special interests."[46]

While Warren made it a practice not to cultivate special interests or to endorse partisan candidates, he did attempt to curry favor with the press, relying heavily on newspaper support during his governorship, as he had as a district attorney and attorney general. His relations with the press were channeled through Verne Scoggins, a former correspondent for the *Stockton Record*, who was particularly adept at making sure that no one reporter had exclusive access to Warren. Warren self-consciously pursued a policy of not playing favorites with the press by carefully managing his press contacts, and Scoggins was a redoubtable and dedicated architect of that policy. "[H]e had," Warren said of Scoggins, "the toughness to handle press in any emergency," and "[h]e knew the politics of the state. . . ."[47]

While Scoggins was around, from 1942 through the early 1950s, Warren held two press conferences a week, one for the afternoon papers and a second for the morning ones. He gave no individual interviews, and his office was instructed to avoid leaks, even to friendly sources. Warren occasionally conferred privately with a reporter at the request of a paper, and once in a while, when issues he perceived as truly significant for his relations with the public were raised, he would have a face-to-face talk with a newspaper publisher.

Warren had some strong supporters and close friends in the press. Joseph Knowland, publisher of the *Oakland Tribune*, had been an early backer of Warren, beginning with the 1925 campaign for district attorney of Alameda County. Otis and Norman Chandler of the *Los Angeles Times* also supported him until 1949, when Warren fell out with the *Times* over the loyalty oath controversy at the University of California. Warren was on particularly good terms with Kyle Palmer, the crafty and prescient political editor of the *Los Angeles Times*, who, according to one commentator, was "as intimate with Warren as any newspaperman ever got to be,"[48] and with Walter Jones, editor-in-chief of the *Sacramento Bee*, who speculated in an oral history memoir that Warren's "liberal attitudes," which "didn't show until after he was Governor," may have been precipitated by *Bee* editorials.[49] But neither Palmer nor Jones got many "scoops" from the governor's office. Scoggins later said, perhaps with tongue in cheek, that Warren "had a natural flair for expressing himself in terms newspapermen could understand and use";[50] Warren recalled that "[o]ur method of dealing with the press was simple and direct but well regulated."[51]

Alongside staff management and press contacts Warren placed internal efficiency high on his list of office priorities. The staff member responsible for that feature of his office was Helen MacGregor, who served as Warren's private secretary from 1935 through 1953, at which time Warren appointed her to the California State Youth Authority. MacGregor, who had a law degree and was a remarkably efficient and forceful administrator, preferred to remain in the background in Warren's administrations, although she obviously wielded considerable power. Earl Warren, Jr., said that MacGregor "was the A No. 1 Troubleshooter, and if anything really had to get done, she was sort of the second-in-command."[52] Warren suggested in his memoirs that MacGregor primarily "managed the many-faceted stenographic and clerical work,"[53] but she did much more than that. "She was always able," Earl junior felt, "to steer problems to the proper place for solution."[54]

Warren, as noted, was at once demanding and uncomfortable with discontent in an office. As MacGregor put it, "each of the offices [Warren] headed was . . . a taut ship"; Earl junior also remarked that his father "wouldn't tolerate" dissension among staff.[55] MacGregor, Warren said, was responsible for "a fine group of people who kept our office functioning harmoniously and effectively."[56] She also was involved in Warren's campaigns, was consulted on legislative strategy, took a special interest in penal reform, state aid to the elderly, health services, and welfare, and enjoyed Warren's "confidence and friendship."[57]

MacGregor also knew Warren very well and did not often speak candidly about him to outsiders. This characteristic was notable in most Warren intimates: He knew how to pick "loyal and discreet associates who could be trusted with intimate knowledge."[58] In a long oral history memoir on Warren, published in 1973, MacGregor revealed almost nothing of value about her years of service with Warren, retreating to platitudes such as he "strictly observed federal standards of evidence," or he had a "remarkable ability of picking splendid people."[59] In a tribute to Warren in the *California Law Review* in 1970, MacGregor's tone was similar. Warren "was determined to do all in his power to ensure that the individual would be safe in his person and his home"; he "was a leader who could be trusted"; he "cared about people, not only as fellow citizens but also with a warmth of knowledge for individuals."[60] While these statements were sincere, they were also strikingly unrevealing for a person who had worked so long and so intimately with Warren.

MacGregor knew Warren better than she let on. She listed "careful preparation for action," and a "sense of timing" among qualities that made Warren an effective governor; in commenting on his intellectual powers she referred to his "ability to absorb complex and unfamiliar material and remember it" and to his "phenomenal" memory.[61] She also supplied Leo Katcher with some information when Katcher was preparing a biography of Warren in the 1960s. Katcher quoted MacGregor as saying that Warren "wanted each of us [on his staff] to speak up so that he could always get all sides of a problem," and that Warren did not mind "if you disagreed with him and sometimes he was disappointed if you didn't."[62] MacGregor also alluded to Warren's strong interest in self-education and in broadening his outlook as he approached new situations, recalling that prior to campaigning for the vice-presidency in 1948 Warren "read everything on politics, economics, and foreign affairs he could lay his hands on," and "turned to the University of California for specific information and expert advice."[63]

For the most part, however, MacGregor saw her function as articulating official Warren Administration doctrine. On the Max Radin affair, for example, she said that "Governor Warren simply didn't believe that Max had the temperament for the bench. He had a tremendous sense of the importance of the judiciary and of the type of man who should serve on the Supreme Court. He didn't think that Max, by his statements and actions, had shown that temperament."[64]

MacGregor was responsible for internal administration; Merrell F. ("Pop") Small for external administration. Small was a former newspaper editor and publisher from Quincy, California who had been given his nickname by friends after the birth of a child. Warren, however, attributed Small's nickname in his memoirs to "certain kindly qualities that enabled him to probe into controversial matters without being controversial."[65] Small functioned as a liaison between Warren, staff, and executive departments: Warren called him an "ombudsman." If "there was a complaint against a department or something of that kind," Warren said, "[Small] would handle it. He'd just go over in a gentle manner and sit down and talk to them. Because he came from my office the door was always open to him."[66] Small would report to Warren periodically on the executive departments through written memoranda. One "veteran Sacramento official" once told John Weaver, a Warren biographer, that he thought "Warren not only read [the Small memos,] . . . he memorized them. He used to ask the damnedest questions, and he expected you to know everything that was going on in your department."[67]

The same source, in noting Small's physical resemblance to Warren, said that " 'Pop' not only looks a lot like Warren, but he also has Warren's uncanny ability as a judge of character."[68] Of all the Warren intimates Small has been the most forthcoming in his public comments on his former employer. Between 1969 and 1971 Small wrote a series of articles on Warren in the *Sacramento Bee*. The articles discussed some features of Warren's governorship, his nomination to the Supreme Court, and the delay in his confirmation as Chief Justice. Small's articles were remarkably detailed and candid: He revealed, for example, that Warren's "revolutionary" proposal for state health insurance, which he announced in 1945, was crystallized in a conversation with Sweigert, something neither Sweigert nor Warren publicly acknowledged.[69]

Small's oral history memoir of Warren was in the same vein. He talked freely about Warren's personality. ("He could get very angry with the people around him, but he never carried a chip on his shoulder"; he "always insisted that he was the number one person.") He described Warren's style of administration. ("His life consisted . . . of talking to people. . . . He was a personal contact administrator. . . . He could pick people's minds.") He characterized Warren as an "authentic liberal" who "wasn't originally [such]": Warren "grew," Small felt, "under the . . . incubation of his exposure to problems"; his liberalism was "a development; a flowering." Warren's speaking and writing style, for Small, was "a little bit pedestrian and inclined to indulge occasionally in generalities," but in conversation he could "put a problem in clear terms to other people and therefore get responsive action." Small was even a little indiscreet for an intimate, at least by Warren's standards: He revealed that Warren's legendary memory for names and faces was aided by a file of four-inch by six-inch cards in his office in Sacramento of "everybody he ever knew and a record of why he knew them."[70]

To the circle of Warren intimates in his years as governor should be added one person who was not officially associated with Warren's staff during those years, but was perhaps Warren's closest personal acquaintance outside his family for the great bulk of Warren's public career. Warren Olney III, a descendant of a prominent social and political family in California, had joined Warren's staff in the district attorney's office in the mid-1930s, subsequently becoming head of the Criminal Division in the attorney general's office. He led the 1939 raids on gambling ships. In 1946 he was appointed by Warren as chief counsel and executive officer of a newly created Commission on Organized Crime and was continued in that post in 1950. In 1953 Olney was recommended by Warren to be an assistant in the Criminal Division of the Justice Department, and in 1958 was appointed by then Chief Justice Warren as director of the Administrative Office of the United States Courts. In 1963, Warren, as chairman of the commission investigating the assassination of President John F. Kennedy, proposed Olney as counsel to the commission; other commissioners objected and Olney was not selected.

Warren, out of a combination of temperament and managerial strategy, was very rarely effusive in his praise of others who worked for him, even when he held them in high regard. Olney was an exception. Of all the persons, other than family members, mentioned in Warren's memoirs, Olney received the most consistent and outspoken praise. In the attorney general's office Olney's execution of the 1939 raids was described by Warren as having been "intelli-

gently planned and successfully carried out," and his investigations
of bookmaking were "thorough": Warren was "extremely fortunate"
to attract "able lawyers" like Olney to join his staff. When Warren
appointed Olney to the Commission on Organized Crime, "it was
not long before [Olney] was working around the clock," which was
"true to his nature." The Commission on Organized Crime "performed yeoman service" with Olney as administrator. When Olney
was being considered for the Justice Department, Warren testified to
his "selfless dedication and his absolute integrity, for which I could
vouch unequivocally in any situation." According to Warren, Olney
"served honorably in [the Justice Department] post longer than anyone who had held it to that time." [71]

In an oral history of his experiences in public life, Olney, following the pattern of persons with intimate access to Warren, revealed
very little about his relationship with his employer, associate, and
friend. Warren was principally described in platitudes: He set a tone
of "promptness in trial, thoroughness in preparation [and] restraint
and dignity in making arguments"; he "had a policy . . . of moving
the cases along as rapidly as it was possible"; he offered into evidence
"anything that the rules of evidence permitted." Only occasionally
did Olney depart from his "official" tone, as when he said of his
early years with Warren that "I had no consciousness *at all* that I
was working for anyone anybody could call an unusual man. I never
pictured him as Chief Justice of the United States or even as a governor." [72]

Earl Warren and Warren Olney were alter egos in Warren's law
enforcement years. They fought vice, crime, and corruption together; they each worked "around the clock"; they believed in integrity, loyalty, and dedication; they shared both a nominal Republican
political affiliation and a deeper interest in nonpartisanship. Olney,
some fourteen years Warren's junior, was there when Warren needed
him, willing to interrupt his own career plans to advise Warren on
an informal basis, to do some hard-nosed investigating or apply some
administrative pressure if necessary, to define Warren's interests and
his as essentially synonymous, to keep confidences to himself. He
possessed many of the characteristics of the archetypal Warren staff
intimate.

The final example of a Warren staff intimate comes from Warren's career as Chief Justice of the United States. When Warren was
appointed to the Supreme Court, he said, it was "the most awesome

and the loneliest day of my public career."[73] As governor he had come to rely on his staff: In addition to the aforementioned intimates he could command batteries of departmental personnel. One of Warren's law clerks recalled his talking about the transition from governor to Chief Justice, and mentioning the "shocking contrast with his former way of life" as far as staff support went. "Warren as Governor I gather," Benno Schmidt said, "had been able to delegate . . . [He] had delegated an awful lot in California. . . . [T]here was just nobody to delegate to when he got to the Supreme Court."[74]

When Warren arrived at the Court his "entire personal staff" consisted of three law clerks, all appointed by his predecessor, Fred Vinson, "two elderly messengers, one of whom died during my first term, the other of whom was retired . . . the following year," Margaret Bryan, "from my Governor's office, to help with California mail,"[75] and one other person. That other person was Margaret McHugh, who had been Vinson's executive secretary at the Treasury Department and had come with him to the Court. To Warren's "great good fortune," McHugh "remained to hold [the position of executive secretary] with me throughout my time in office." McHugh, Warren felt, "knew every facet of staff relationships"; she "was never absent on a business day throughout my sixteen years on the Court";[76] in addition to Court business she assisted Warren and the Warren family in a variety of ways, including hiring limousines for social functions and securing tickets for sports events.

McHugh played a role at the Court roughly comparable to that of Helen MacGregor in the governor's office: the troubleshooter. Her most significant function was serving as a barrier between Warren and outside pressures. As Chief Justice, Warren gave no interviews; even after his retirement he consented only to pre-arranged discussions with known and trusted persons on specified topics. His chambers were not completely accessible to other justices or law clerks: McHugh screened all visitors unless otherwise instructed.

On matters of internal administration, to which Warren paid close attention as Chief Justice, McHugh served as the liaison between Warren and other Court personnel. As MacGregor had done, McHugh "managed the . . . stenographic and clerical work,"[77] contributing to the supervision of persons assigned tasks by the Chief Justice in his capacity as Court administrator. While the number of persons directly or indirectly responsible to Warren as Chief Justice was far fewer than the number of persons responsible to him as governor, McHugh's liaison opportunities were considerable. A blunt, imposing, and sometimes imperious figure, she was generally regarded in internal Court circles as having created, out of her legacy of trust with Warren, a position of formidable personal power. McHugh has never commented publicly on her experiences with Warren.

Together Sweigert, Scoggins, MacGregor, Small, Olney, and McHugh form a composite of the Warren staff intimate. The intimate staffers were notable for their loyalty to Warren, their discretion, their subordination of their personal ambitions to those of Warren, and their conviction that Earl Warren was an inspirational or at least a significant person with whom to be associated. Warren, in turn, thought each of his intimate staffers reliable, industrious, trustworthy, and dedicated to the "right" values. They protected him and allowed him to be himself; he listened to them and relied upon them.

Yet Warren remained, to his staff, a loner and at times an autocrat; a wall of reserve existed even between intimate staffers and their boss. They remained to him staff, even though in some cases they also served as political and even social companions. Warren's staff relations portray a classic picture of an administrator with instinctive qualities of leadership: He could not do without his staff, but they, even more, could not do without him. In their mutual relationships Warren's staffers took Warren as they found him; it was they who accustomed themselves to him, not the reverse. By the time Warren became Chief Justice, leadership and staff management were automatic reflexes, the products of a long executive experience.

The appointments of his critical staff members were made by Warren in his first term; all his California intimates, save Sweigert, remained virtually to the end of his governorship. With the Japanese relocated and then resettled, Olson defeated and his career as a figure in California politics cut off altogether, and the internal structure of the Warren Administration extant, Warren could turn to matters of policymaking. The transition period was over; the governor was emerging out of the law enforcement official.

↙

Warren began his tenure as governor being characterized as "a personification of Smart Reaction"; [78] five years later, in a national magazine, he was identifying himself with "true Liberalism."[79] His early years as governor were marked by such policies as reducing taxes and establishing a "rainy day fund" by which the state avoided spending surplus in its treasury; in his later years he was identified with prison reform, public health improvements, increased support for higher education, and compulsory health insurance. One close friend and observer of Warren, Richard Graves, said that "he came into office with the conventional idea that governments could be run like private businesses," where "the factors that control costs can themselves be controlled." But Warren "learned quickly," Graves felt, that "the same rules didn't apply" to "state needs." During the years of Warren's governorship, 1943 to 1953, California underwent a massive influx of population and faced attendant "health, education and welfare needs," as well as "new transportation problems." Warren, Graves noted, responded to these needs through programs that increased the size and the spending habits of state government.[80]

As governor, Warren has been regularly characterized as having moved from a "conservative" to a "liberal" policymaking stance.[81] The accuracy of this characterization is limited. His early policies of frugal governmental spending, reductions in taxes, and the creation of revenue surpluses were "conservative": They were implicitly designed to restrict the role of government as a means of distributing largesse or providing social services. By contrast, his later support for governmental involvement on behalf of the elderly, the mentally ill, or those severely injured or beset with serious illnesses can be said to have constituted an endorsement of policies that were "liberal" in that they envisaged the state as an affirmative, humanitarian force providing support for disadvantaged persons.

But to characterize Warren's political stance during his governorship as moving from conservative to liberal is to ignore his own descriptions of the perspective from which he addressed social issues and his distinctive methods of policymaking. Warren is best characterized, during the period of his governorship, as remaining within the basic philosophical framework of California Progressivism, retaining many of the original social assumptions on which Hiram Johnson and his followers based their approach to policymaking while abandoning others.

Progressivism, as practiced by its original adherents in California, emphasized, as we have seen, honesty and openness in government, opposition to "special interests" in the name of the "public interest," and close attention to the "white light" of public opinion. It also emphasized, notably in racial and ethnic issues but also to some extent in issues involving organized labor, the idea that California was a nativist paradise, an island to be preserved against alien persons and ideologies. One strand of California Progressivism was thus potentially supportive of affirmative governmental action on behalf of disadvantaged persons, at least insofar as the needs of those persons were regarded as personifying the needs of the public at large, and some Progressives did advocate social justice programs.[82] The great bulk of California Progressives, however, believed that affirmative governmental action ought to be selective and cautious. California government should not reward "undesirable" persons, as defined by a nativist conception of California society, or resemble, in its ideological approach, the "undesirable" ideologies of alien cultures.

As the Hiram Johnson version of California Progressivism was replaced by Culbert Olson's version of liberalism, an issue dividing "old" and "new" Progressives surfaced. That issue involved the scope of affirmative governmental action on behalf of disadvantaged persons: Should it be selective or universal? Early California Progressives envisaged affirmative government as a restorative moral force, the purpose of which was to purge California political culture of corrupt special interests. Government was more of an admonitory than a paternalistic agent: The idea that citizens were entitled to certain economic and social benefits, notwithstanding their status or condition, was foreign to early California Progressive thought. Indeed, early Progressives felt that certain persons, by virtue of their ethnic or ideological affiliation, were presumptively not entitled to governmental largesse; they were not included in the Progressive ideal for California. Thus the "old" Progressives could not be called civil libertarians as that term has come to be understood, for they embraced neither a universalistic philosophy of social justice nor universalistic protection for minority rights.

In 1938, during his campaign for attorney general, Warren had written a letter to Robert Kenny, a visible Democrat who had been an Olson supporter and chairman of the Southern California Committee of Progressives for Roosevelt.[83] Warren wanted Kenny's endorsement and had approached him at a luncheon about giving Warren his support. When Kenny asked Warren some sharp questions about the *Point Lobos* case, Warren responded by saying that he would write Kenny a letter detailing his position on civil liberties. In the letter, which Warren later made public, Warren said that he was "unalterably opposed" to majoritarian repression of minorities and that "majorities . . . should be willing to fight for the same rights for minorities no matter how violently they disagree with their views." He also said that "the American concept of civil rights should include not only an observance of our Constitutional Bill of Rights, but also the absence of arbitrary action by government in every field." [84]

In taking these positions Warren appeared to be moving from a selective to a universalistic definition of minority rights. To be sure, the letter was prompted by the *Point Lobos* case, which had raised doubts about his concern for civil liberties, and by Warren's campaign for the attorney generalship, in which he was attempting not to offend influential Democrats. But the letter associated Warren with a view that identified majoritarian government as a protector of minority rights. This was a view that early Progressives had not endorsed when the minorities in question were "undesirable"; this was a view that Warren was to depart from in the Japanese relocation. It was a view that ultimately served to distinguish Progressivism from liberalism. [85]

In 1948, again in the context of a political campaign, Warren wrote an article in the *New York Times Magazine* in response to the question, "What is liberalism?" The article was a significant index of Warren's thinking in two respects. First, he identified liberalism with "civil rights, representative government, and equality of opportunity"; second, he associated "true liberalism" with a "progressive . . . conception" of politics, which rejected both "indiffer[ence] to the welfare of others" and "alien tyranny." Liberalism, for Warren, was Progressivism with a special emphasis on "civil rights" and "equality of opportunity." [86] He saw nothing potentially contradictory in that statement.

Repeatedly in his memoirs Warren identified his governorship with a "progressive conception" of politics. "I did not care," he wrote, "to be categorized as either a liberal or a conservative . . . I believed in the progressivism of Hiram Johnson." In describing his first administration, notable for its lack of reform measures, Warren said that "it was something of a letdown to learn at first hand how much the war restricted progressive programs" and in summing up his years as governor, Warren said that he "attacked all the issues that confronted us . . . realizing that a start should be made and any advance in the right direction represented progress upon which to build." Because of that approach, he suggested, "my administration finished with the approbation of some for being 'liberal' "; persons "of the Right Wing called it a surrender to the New Deal." "I would like to believe," he said, that "it was a progressive administration."[87]

But if "true liberalism" and Progressivism were fused in Warren's mind, the two ideologies did not base identical views on the relationship of affirmative government to minority rights. Progressivism can be said to have been a protest on behalf of a hitherto silent, disadvantaged majority (the "public") against the control of government by vested minorities ("special interests"). Most Progressives did not number among their paramount concerns the protection of minority rights, and most Progressives did not envisage creating a governmental apparatus to facilitate that protection. The governmental apparatus they sought to create was principally designed to restore open, moral, majoritarian government.

In some instances, however, the political culture in which special interests had predominated had functioned to disadvantage other interest groups in addition to the public at large. Occasionally, where special disadvantages had occurred, Progressives, at least in California, were prepared to favor affirmative governmental action on behalf of the disadvantaged. During Hiram Johnson's first administration, for example, workmen's compensation legislation was enacted, women laborers were limited to an eight-hour working day, and state pension plans were established for public school teachers.[88] These measures met with opposition, both among Progressives and elsewhere, and were limited in their impact. But they reflected the fact that in certain situations Progressivism envisaged not merely governmental intervention to regulate or to expose special interests, but affirmative governmental support for other interests that had been disadvantaged by a corrupt political culture.

Warren's 1948 identification of liberalism with "civil rights, representative government, and equality of opportunity" was an extension of this tendency in Progressivism. Warren's statement assumed, as had William Sweigert's 1942 memorandum, a "social consciousness in government." The protection of "civil rights" was as much a province of government as was the restoration of openness and honesty in politics. And "civil rights" came to mean, in the later years of Warren's governorship, entitlements to a "decent" way of life as much as protection from overt discriminations.

Yet Warren was not prepared to abandon altogether a selective interpretation of the prospective beneficiaries of governmental support. He objected to "liberals who would . . . permit no change unless it was on their often unrealistic terms"; he did not approach issues "on the basis of an ideology," but "pragmatically as they arose"; he was not distressed because his administration "did not neatly fit, in all its actions, into some ready-made leftist ideology."[89] From 1943 to 1953 he occasionally supported paternalistic or redistributive governmental programs, the beneficiaries of which were, in many instances, disadvantaged persons. At the same time, however, he remained fearful of "alien" ideologies; his civil libertarianism was selective.

A brief review of the policymaking highlights of Warren's administration reveals that, for the most part, his values were those of early California Progressives, but that to a limited extent he had moved beyond Progressivism to embrace a fuller conception of the affirmative role of government. Warren's typical administrative response to a perceived social problem, for example, was to solicit the views of "disinterested experts" as opposed to established politicians. He appointed a bipartisan commission to consider reducing taxes and submitted its proposal to the legislature. He attempted to improve public health by appointing as director of the Public Health Service "a person who was trained and experienced in public health administration,"[90] as distinguished from a political appointee. He secured a director of the Department of Mental Hygiene by appointing "a committee of men and women knowledgeable in the field and commission[ing] them to find for me the best available man in the county."[91] He established "a committee of experts on prisons, finance, and legislative action" to "study penal affairs in California" with the objective of reforming the state's prisons.[92] In these instances he was following the Progressives' technique of interjecting nonpartisan experts into government. "I believed," Warren said, "that with a proper institutional structure, with enlightened administration by dedicated people, and with . . . a nonpolitical atmosphere, we could make real progress."[93]

Prison reform was made possible by another technique characteristic of Warren, also borrowed from Progressives such as Franklin Hichborn: the skillful use of public sentiment as a reformist force. When he became governor, Warren "was convinced that our prisons educated the inmates in crime and hardened them toward society" and assigned one of his staff members "the task of preparing a bill to completely reorganize procedures for handling prison matters."[94] Warren's office consulted with outside experts, including the Director of Federal Prisons and prison chaplains, and developed a reorganization plan, which included the report of the aforementioned "committee of experts." Warren then filed the plan away, waiting for an opportunity to attract public attention.

In 1944 Warren received a telephone call from the San Francisco chief of police, Charles W. Dullea, that a notorious "yacht bandit," Lloyd Sampsell, who had combined bank robbing with frequenting "one of the exclusive yacht clubs"[95] in San Francisco, had been spending weekends in a San Francisco hotel while purportedly incarcerated at San Quentin. Warren "urged [Dullea] to arrest [Sampsell] the next time it happened and to give the event the utmost publicity." He then "immediately called the legislature into special session for the purpose of advising it and the public of the sordid conditions in our prisons." The report of the "committee of experts" was released at the same time. By taking advantage of the timing of events, Warren was able to enlist "the newspapers of the state, the law enforcement officers, and the women's clubs" as "unofficial lobbyists" in the cause of prison reform, and an "integrated prison system with an expert penologist in charge of its administration" came into being.[96]

Other features of Warren's administration reflected Progressive blueprints for enlightened policymaking. He "repeatedly told the voters that I aspired to be the governor of California, not its political boss" and "never built up a political organization of my own." He devised a "town hall method . . . of keeping in touch with the people," which consisted of two- or three-day conferences in Sacramento on such subjects as public health, conservation, old age pensions, and highways. "Concerned citizens . . . without regard to political affiliation"[97] would be encouraged to attend these conferences, which amounted to informal public hearings. By this device Warren gave the impression that his administration was studying public issues in the open and paying attention to the views of "concerned citizens."

Finally, Warren's battles with the oil and water power lobbies were classic struggles against "special interests." In 1947, in supporting a three cent increase in the state tax on gasoline to create revenues to expand the state highway system, Warren referred to "the slick lobbyists of the oil companies" that were "overwhelming the capital with false propaganda" and stated that he "wish[ed] the public could actually be here to witness the power that these oil lobbyists are exercising over our legislative processes."[98] In 1949 he called for the development of hydroelectric power through "public agencies," so that development projects would foster other "public purposes," such as "flood control, navigation, salinity control," and other "municipal concerns."[99] A spokesman for private utilities responded by demanding "freedom from governmental interference, discriminating governmental taxation, and unfair governmental competition."[100] By 1952, an oil lobbyist, in opposing Warren's candidacy for the Republican presidential nomination, accused him of "abandon[ing] Republicanism and embra[cing] the objectives of the New Deal."[101]

These impressions of Warren suggest that he had given indications, by the later years of his term, of moving closer to liberalism. In his 1951 inaugural address, after being elected to a third term, Warren listed a variety of areas in which he thought state government should intervene. These included unemployment insurance, support for senile and infirm persons, child care, workmen's compensation, and air pollution. Warren also repeated his proposal, first made in 1945, for a state Commission on Political and Economic Equality that would investigate "thoughtless or deliberate . . . differences of treatment" based on race.[102]

The event most commonly identified with a change in Warren's ideological point of view has been his 1945 proposal for a limited form of compulsory state health insurance. One commentator has written that Warren "astounded his political friends—and enemies" by his medical insurance proposal.[103] Warren's support for health insurance, however, should not have been astounding to those who knew him well, nor was his proposal revolutionary for California in the 1940s. The fight over health insurance, however, constituted a significant "lesson" in Warren's public life and may have been a catalyst for change.

✒

In the fall of 1944 Warren developed a kidney infection and was hospitalized; the incident apparently stimulated him to reflect upon the potentially catastrophic personal and financial consequences of illnesses. One of his vivid childhood memories had been his father's tale of the death of Earl's uncle, Ole, from tuberculosis one Christmas Eve in Chicago. "[H]aving lost the one most near and dear to him and being broke in a strange city," Earl Warren recalled, his father "then swore that as long as he lived he would never be broke again."[104] There were other incidents "in my own family," Warren later remembered, "to indicate where, but for the grace of God, all of us might have been." He summarized:

> My dear mother, over a period of years while I was district attorney, had one serious operation after another. I had watched the daughter of my widowed sister fight death for six years from a hospital bed. I had seen my sister's son invalided home from the Marine Corps in the Pacific . . . destined to spend some years in hospitals.
>
> The only reason my immediate family was not reduced to the extremities that afflicted so many other people was that my Father, by living a Spartan life himself, had been able to take care of medical expenditures. . . . The cost of hospital and medical care had . . . advanced beyond the ability of the average family to pay. . . . I came to the conclusion that the only way to remedy this situation was to spread the cost through insurance.[105]

Warren had had some earlier exposure to the idea of medical insurance. In 1935, 1937, and 1939 the California Medical Association introduced bills in the legislature to require compulsory health insurance; Warren was aware of these efforts because he had been consulted by a group of Oakland doctors about the legality of a compulsory insurance fund. Warren said in his memoirs that the bills, "ostensibly for the benefit of the working people," were actually designed to "enabl[e] doctors to get their money from the government instead of their impecunious patients."[106] In 1943 the California Federation of Labor, whose secretary, Cornelius Haggerty, was a close friend of Warren, lobbied for a bill that would extend unemployment insurance, which had been extant in California since 1936, to include health insurance. And the 1944 Republican party platform, drafted at a convention where Warren had been the keynote speaker, had included a plank that anticipated states making medical and hospital services available for "needy" persons.

After recovering from his illness Warren had a conversation, one day in late 1944, with William Sweigert. Sweigert and Warren had been discussing conditions in the state's mental hospitals and prisons;

Sweigert suggested that improving the physical being of incarcerated persons was part of a general philosophy, also illustrated in California by workmen's compensation legislation, by which government took some responsibility for the health of its citizens. Warren spoke of his warm support for workmen's compensation, and Sweigert asked him if he wanted to "go a little further." Warren asked Sweigert "what he had in mind"; Sweigert proposed some form of compulsory state health insurance, and one of Warren's most controversial measures as governor took shape.[107]

Warren's program was a relatively modest one. It did not advocate compulsory health insurance for all California citizens; it was directed at employed persons and their families. Not all of its costs were to be borne by the state. The fund out of which health insurance claims would be paid was to be created by a three percent payroll tax, contributed to equally by employers and the employees. The political thrust of the plan did not represent a radical departure from positions Warren had previously expressed. One of his arguments for the plan, for example, was that the federal government was about to enter the health insurance field and that this measure would preserve "states' rights," a theme expressed in the 1944 Republican platform and one that he had repeatedly sounded in criticizing the New Deal.[108] Warren did not anticipate that the plan would be very controversial: He had consulted with the California Medical Association before commencing it and had been led to believe, apparently, that the CMA would support the plan. Finally, the device of using payroll taxes to create a statewide insurance program was not new, having been the basis of California's unemployment insurance fund, which had developed a surplus by 1945.[109]

Warren, we have seen, supported medical insurance principally because he and other members of his family had had disabling illnesses, and he had become fearful of their high cost. Medical insurance, in this sense, represented an elaboration for him of Matt Warren's principle of saving for the future. There was, however, another dimension to Warren's health insurance program. The idea of compulsory medical insurance was based on an assumption that the state could intervene in the economic relationship between a profession and its clientele, and by its intervention affect the ability of a profession to control its price structure. This assumption had been held by some Progressive social theorists, who had come to believe that all businesses, left unregulated in a mass society, would come to serve their own interests to the detriment of those in society least able to protect themselves.[110] Applied to the area of health services, this line of reasoning suggested that persons who could not afford medical care were "innocent" victims of unregulated professional conduct because they were being denied the benefits of treatment for their illnesses. Medical care, under this analysis, became a kind of birthright, and thus a concern of the state: If one were sick or injured, one had a presumptive claim to medical services regardless of one's purchasing power. The state intervened, through compulsory insurance, to make certain that such claims were being entertained.

The above dimension of Warren's medical insurance plan enabled its opponents to characterize it as "socialized medicine."[111] In fact, the plan as proposed did not constitute state interference with many of the private choices of the medical profession, such as the establishment of rates, the selection of patients, the process of specialization, or the location of a practice. What the plan did was to create a newly solvent class of prospective patients, persons who could not afford medical care at "free market" rates but were now being subsidized in part by their employers. That class of persons was by no means the underclass of the world of medical services. The only persons who qualified for compulsory health insurance under Warren's plan were those who were eligible for employment and who could thereby contribute to their own insurance.

Nevertheless the plan assumed that solvent persons in American society had a responsibility to ensure that insolvent persons were not deprived of medical care. In this sense the plan was "socialistic." It rejected an individualistic nineteenth-century theory of suffering, one in which the combination of poverty and lengthy illness in a person was deemed to be bad luck, a flaw of character, or an object of private charity. The theory of suffering implicit in Warren's plan defined the combination of poverty and lengthy illness as a "social problem" and proposed that it be alleviated by the state. In alleviating the problem the state's role was that of beneficent paternalist with respect to those who could not afford to pay for their medical expenses. It was also that of coercive distributor of wealth with respect to three classes of persons: employers who were compelled to pay for the health costs of their employees, employees whose contributions to the health insurance fund were used to pay for the health costs of their co-workers, and employees who were "forced to save" toward their own possible health costs.

While "socialistic" seems an inflated term to describe a plan with such minimal state intervention, when juxtaposed against an individualistic theory of suffering the term appears more apt. Warren's plan was "socialized medicine" because it assumed that the health care of some members of society was a responsibility of society generally and could not be based solely on one's ability to pay for medical services. His plan was consistent with Sweigert's notion of "social consciousness in government" in that it was premised on the belief that individualistic solutions to human problems, in a mass society, could engender suffering and that the state had some responsibility to provide relief.

But Warren, in advocating health insurance in 1945, did not explicitly link health insurance to "new" Progressive or "liberal" social theory. Indeed he continued, into his second term as governor, to criticize collectivist solutions to social problems and to advocate a minimum of governmental interference in the affairs of individual citizens.[112] Moreover, some observers believed that his support for health insurance was lukewarm and politically motivated. Carey McWilliams claimed that Warren had fought for health insurance "firmly enough to give an appearance of advocacy, but not with sufficient energy to force a show-down."[113]

Warren remembered his struggle with the California legislature, which defeated his health insurance proposals three times between 1945 and 1949, differently:

> I believed then and now that our state would have reaped great benefits from it. No state had such a program, and yet I believed something of this kind was inevitable.
>
> By 1945 . . . [d]octors were doing a thriving business and were being well paid for their services. As a result, they had no . . . interest in the patient's problem of paying. . . . [T]hey stormed the legislature with their invective, and my bill[s] [were] not even accorded a decent burial.
>
> Since my experience with these matters, there has been a growing awareness of the need for a national health program. . . . The mounting cost of doctors' fees and hospital care today has reached a point which is prohibitive for the average family, and most people are now convinced, I believe, that we must have a normal health program for everyone, based on the insurance principle.[114]

Health insurance may have once appeared as a potentially "safe issue" that attracted a cautious politician, but it had also appeared, to Warren, as an issue that transcended politics. He saw two principles at stake: the patent unfairness of catastrophic illness being coupled with disadvantaged financial circumstances, a situation that on its face called for redress; and Matt Warren's belief in saving for a rainy day, which Warren saw embodied in insurance plans. The indifference of the California legislature to health insurance plans convinced Warren that these principles were being subordinated to the selfish short-run goals of lobbyists. Once again special interests and their representatives stood as barriers to the full realization of the needs of ordinary citizens. Warren had warned in 1947 about "domination of the state by any group whose interests are less extensive than the people as a whole,"[115] and he had seen, he believed, such domination vividly confirmed in the health insurance fights.

As in the case of his unsuccessful appeal to the "moneybags" for campaign support in 1942, an embarrassing short-run defeat for Warren had been converted, in his mind, to an enduring moral lesson. In one sense the repeated defeats of his health insurance plan were minor setbacks for an overwhelmingly successful governor, who a year after his plan was first sabotaged won the gubernatorial nominations of both the Republican and Democratic parties. In another sense, however, the scuttling of his health insurance plan was a confirmation for Warren of the nature of the political process, in which advocates of programs based on humanity and common sense were pitted against selfish, vindictive special interests. Earl Warren had a well-developed capacity to hate: His hatreds endured and simmered over time. He hated most fully, and effectively, when he convinced himself that issues of principle were at stake and that his enemies were accordingly unprincipled and evil. The health insurance fights had put social welfare programs in a different light; they were crusades on behalf of the people against the sinister forces that corrupted government. A philosophy of affirmative government seemed all the more attractive because the "special interests" opposed it.

If Warren's philosophical stance shifted during his years as governor, the shift was not only gradual, it was uneven. While there were episodes, such as the health insurance fights, in which he seemed to have boldly cast off an earlier "conservative" persona, there were other episodes in which his attitudes were more cautious and more ambivalent. A feature of Warren's personality was that while his instinctive reactions to social issues were often charged, his public style was cautious. When his emotions were engaged he was moved to take action, but his political instincts regularly told him not to act impulsively, to bide his time, to achieve his goals covertly or indirectly.

There were two noteworthy side effects of Warren's ability to channel his emotions. One was that when he did finally take a visible, strong stand on an issue, his action emboldened him to act again, even if he was met with opposition. His actions in the health insurance controversy were one example of this tendency; his estrangement with Olson over civil defense was another. Another consequence of Warren's mixture of intensity and caution, however, was that when he was genuinely divided on an issue, or when his thinking lacked internal consistency, he proceeded with superabundant caution, and his actions were sufficiently opaque to approach being disingenuous. We have seen evidence of this tendency in some of his ambivalent statements about the "rights" of the Japanese during the relocation controversy; the loyalty oath controversy in California in 1949 and 1950 furnishes another example.

In 1950 George Stewart, a professor of English at the University of California at Berkeley, published, in collaboration with a number of anonymous faculty members, a brief account of the loyalty oath controversy at Berkeley. In the spring of 1950, when the book was in preparation, Stewart wrote, "[T]o the shame of our state and of our University it must be said that we felt it necessary to organize for the writing of this book as the French organized their Resistance during the years of the Nazis, with radiating lines of responsibility and with no one knowing all the others who were involved." The loyalty oath controversy had taught Stewart and his colleagues "something about how suspicion arises, and mistrust, and fear. Before then these were only things we had read about in books as having happened years ago or in other countries." [116]

Warren gave an account of the loyalty oath controversy in his memoirs, which he began to write after completing sixteen years on the Supreme Court. During those years his support for civil liberties had come to approach a state of militancy, his opposition to loyalty oaths and to broad-ranging legislative investigations into political beliefs had become a matter of public record, and he had been repeatedly denounced as being soft on communism. Warren remembered the loyalty oath controversy from the perspective of an old warrior for academic freedom. He spoke of "the witch-hunting disease" that affected "all governmental institutions" in the 1950s, the "hysteria about Communism" that was rampant at the time, the atmosphere of "fear, distrust of neighbors, bitterness, and persecution" in which the loyalty oath controversy arose, and the "McCarthyism" of certain members of the Board of Regents of the University of California and of the press. He pictured himself as an opponent of loyalty oaths and as a defender of the Bill of Rights. "Everyone," he recalled, "gave lip service to the Bill of Rights as a political philosophy but too many people were unwilling to apply it except where an infringement of it adversely affected them. . . . The academic and scientific communities . . . had an equal concern about their right to pursue knowledge and teach the truth as they find it."[117]

But Warren's memory was flawed. He had contributed to the "anti-Communist feeling" that he later described. While he had repeatedly voted against the imposition of an oath on University of California employees, he himself proposed, and signed, a superseding loyalty oath, this one for all state employees, including members of the University of California faculty. He was as militant an anti-Communist as those he associated with McCarthyism, and he found his anticommunism warring with his affection for the University of California. His performance in the loyalty oath controversy was a confused and ambivalent one.

As governor, Warren was the *ex officio* chairman of the university's board of regents. He did not regularly attend meetings of the board, however, and between March 25, 1949, when a loyalty oath for faculty members at the University of California was first proposed, and January, 1950, he attended no board meetings.[118] He seems not to have been particularly aware of the growing controversy surrounding the oath until November, 1949, when University of California President Robert Gordon Sproul telephoned him to inform him that the regents and the Berkeley faculty were seriously at odds on the oath issue. Sproul, as we shall see, had proposed the oath himself in the March, 1949 meeting, but by the fall of 1949 he had reversed his position. Sproul had also nominated Warren for the presidency of the United States at the 1948 Republican convention. Warren said in his memoirs that Sproul soon "realized that on the oath issue he had been misled into an untenable position."[119]

The loyalty oath was originally conceived as a ploy to neutralize the possibility of more stringent legislative restrictions on the University of California faculty. The Tenney Committee, dormant during the war years, had had a rebirth with the advent of the Cold War, and in January, 1949, a bill was introduced in the Tenney Committee to give the legislature, rather than the board of regents, power to assess the loyalty of University of California employees. The comptroller of the university, James Corley, became aware of this bill and other Tenney Committee bills attempting to widen the scope of legislative power to investigate or to suppress subversive activities. Corley proposed to Sproul an oath for university employees that would dissassociate them from support for "any party or organization that believes in, advocates, or teaches the overthrow of the United States Government by force or violence.[120]

Meanwhile two incidents occurred that hastened consideration of the oath by the board of regents. On January 26, 1949, Harold Laski, a member of the British Labour party who admitted to being a Socialist but denied being a Communist, was invited to speak at UCLA. And on February 1 two University of Washington professors were dismissed from employment because of membership in the Communist party; two weeks later one of the professors was invited to UCLA to debate the question whether one who had membership in the Communist party could be "an objective teacher and impartial researcher."[121] Both of the invitations were approved by the UCLA provost, Clarence Dykstra, but not without misgivings, and the question of whether the University of California should allow its facilities to be used by persons allegedly sympathetic to communism was informally discussed at the board of regents' March 25 meeting.

According to Sproul's account of the meeting, the regents as a collective body decided, in an informal discussion during the morning session, to consider passing some form of loyalty oath in response to the proposed Tenney Committee bill and the two incidents at UCLA. The regents' attorney, Jno Calkins, was instructed to draft an oath, with Corley's assistance, during the lunch hour. Sproul, using Calkins' draft as his text, proposed the oath at the afternoon session; the draft, after some stylistic changes, was unanimously passed. One member of the board of regents then asked Sproul "if you will have an opportunity to present that new oath to men who are already members of the faculty," and Sproul said that the oath would "be in the new contracts[s]" for the academic year 1949–50.[122]

Employees of the University of California were already required to swear an oath of allegiance to the Constitution of the United States and the Constitution of the state of California. The new oath added Corley's phrase, previously quoted, which disassociated signers from support for parties or organizations dedicated to undermining the United States government. University employees did not learn of the board's action until May 9, when a statement in the university's *Faculty Bulletin* informed them that a new oath of allegiance would be included in 1949–50 contracts and that checks could not be released until the oath had been signed and notarized.[123] The text of the oath was not made available, and at the June 7 meeting of the university faculty senate's North Section, which included Berkeley, most faculty members did not know its contents.

At the June 7 meeting a decision on the oath was postponed for consideration at a special meeting a week hence.[124] Between the two meetings an assistant of Sproul's was quoted in the press as saying that "in the face of the cold-war hysteria we are now experiencing something had to be done," and Sproul was quoted as denying unequivocally that the board's passage of the oath had any "connection with Senator Tenney."[125] At the June 14 meeting the senate voted to delete or to revise the new oath and to delegate that task to its advisory committee, possibly subject to senate approval.[126]

A power struggle now began between the Berkeley faculty, the principal force behind the university senate's actions, and the board of regents, with President Sproul caught in the middle. On June 24 the regents met again and attempted a compromise. They declared that "membership in the Communist Party is incompatible with objective teaching," and that "no member of the Communist Party shall be employed by the University." They also amended the new oath, however, to read, "I am not a member of the Communist Party, or under any oath, or a party to any agreement, or under any commitment that is in conflict with my obligations under this oath."[127] This amendment had the support of the senate's advisory committee. A group of Berkeley faculty members, to be called the "nonsigners," then repudiated the amended oath and denied the advisory committee's power to represent the faculty. Sproul's office, after apparently first stating that salary checks would be sent out to employees regardless of whether they had signed the oaths, then issued contracts for the 1949–50 academic year only to signers.[128] The faculty senate of the university did not reconvene until September, 1949, at which point a majority of its members repudiated the June 24 compromise and demanded that the amended oath be deleted.

At a meeting of the regents on September 23, Sproul declared that "present enforcement of the requirement of a special loyalty oath [in the form of the June 24 compromise] is neither practical nor wise," and urged that 1949–50 salaries be paid to nonsigners.[129] Several members of the board, notable among them John Francis Neylan, a San Francisco attorney whom Warren had reappointed to the board of regents in 1944, now began to see the loyalty oath issue as raising two distinct problems: the tolerance of the University of California for the employment of Communists and the tolerance of the university faculty for the board of regents' authority. Neylan had originally been opposed to any loyalty oath,[130] but from September, 1949 on, he became a persistent and resourceful advocate for the proposition that, as he later put it, "a condition precedent to employment . . . in the University shall be [an oath or] affirmation that the appointee is not a member of the Communist Party, or under any . . . commitment . . . that is in conflict with the policy of the Regents excluding Communists from membership of the faculty of the University."[131]

Meanwhile the university senate had moved from a previous position of ambiguity with respect to both the new oath and the regents' authority, to one where a majority of its members refused to approve the regents' policy excluding Communists from membership in the university community, and asserted that "in the making of decisions affecting the conditions crucial to the work of teaching and research," such as "qualification for membership" on the university faculty, the university faculty was to be the ultimate authority.[132] It was at this point, which was reached by November, 1949, that Warren was first informed of the situation. When the regents and the faculty still seemed to be at loggerheads in early January, 1950, Frank Kidner, a professor of economics at Berkeley, an opponent of the oath, and a former adviser to Warren in his 1948 presidential campaign, went to Sacramento, informed Warren about the current state of the controversy, and implored him to attend the next meeting of the board of regents, scheduled for January 12.[133]

Warren attended that meeting and all subsequent meetings of the board in which the oath was discussed. He entered the controversy with strong personal ties to Sproul and to Kidner and with a residue of goodwill toward the University of California. He also entered the controversy with his forthcoming candidacy for a third term as governor of California in mind, with his hatred and fear of communism undiminished since the 1930s, and with an awareness of the momentum of anticommunism in both his party and the nation generally.

On February 24, 1950, a majority of the board of regents, increasingly provoked by Sproul's apparent disinclination to press for an oath and the faculty's growing militancy, passed a resolution drafted by Neylan that reaffirmed an oath (or "affirmation") requirement, asserted the regents' authority to set conditions of employment for the university, and stated that failure to sign an oath or the equivalent before April 30, 1950 would constitute "sever[ance of the nonsigner's] connection with the University as of June 30, 1950." [134] The vote on the resolution was twelve to six, with Warren dissenting. The *San Francisco Chronicle* reported the next day, however, that Warren had "headed" the regents majority, which had "laid down a flat policy to reluctant professors: Sign the anti-Communist oath or no job." [135]

Warren waited a week, then publicly declared his opposition to the board's resolution. He called the oath one "any Communist would take—and laugh about it." [136] From that point on, Warren was consistently opposed to the oath requirement and to the regents' resolution of February 24. His position approximated that advanced in March, 1950 by a special committee of the university senate, chaired by Professors Malcolm Davisson of Berkeley and J. A. C. Grant of UCLA, which opposed any specific oath disclaiming sympathy for communism; agreed to abrogate tenure for proved members of the Communist party, but insisted on preserving the tenure principle in all other cases; contended that no member of the academic staff of the university should be dismissed without a hearing; and endorsed an already existing university regulation that distinguished between the "dispassionate duty" of seeking knowledge and using an academic position "to convert, or make converts." [137] In addition, Warren supported a 1940 policy declaration by the regents that the university would not employ Communists.

The fullest statement Warren made of his position during the controversy came at a March 31, 1950 meeting of the board of regents. At that meeting the board voted on whether it should rescind its February 24 resolution, which had created a furor in the university community and had been widely publicized in the press. Warren asked for an opportunity to state his reasons for supporting a motion to rescind. The following is an excerpt from his statement:

> I am an alumnus [of the university] myself, and have three youngsters in the University today. God willing I will have two more in two or three years on one of the campuses of the University. I would cut my right arm off before I would willingly submit my youngsters to the wiles or infamy of a Communist faculty.
>
> [But] I don't believe that the faculty of the University of California is Communist; I don't believe that it is soft on Communism, and neither am I. I believe that in their hearts the members of the faculty of our University are just as sincere on the things they represent against Communism as any member of this Board. . . . I have absolute confidence in the faculty of the University of California.
>
> I don't mean they don't have a Communist . . . some place in there and may have a Communist in other places, and you will find them infiltrated in business to a point where you know nothing about it. You may have some working for you, you men in business would be surprised to find out about.
>
> I don't believe there is any difference of opinion between the individual members of the Regents and between the Regents and the faculty in their desire to get rid of Communists. If I thought that my course would be entirely different. . . . The only thing the people of this State are interested in is our seeking to keep Communists out of the University and, believe me, I am interested in that too.[138]

The motion to rescind garnered ten votes, but ten votes opposed it, and it consequently failed. Between the March 31 meeting and the end of April both sides desperately tried to avoid what the regents' February 24 resolution had seemingly made inevitable—the dismissal, without any hearing, of tenured members of the university faculty solely because they had failed to sign the new oath.

Finally, on April 21, the board passed a resolution that changed the situation in three respects. First, the specific oath requirement first proposed in March, 1949 was scrapped and replaced by a "prescribed form" in which the signer stated that he or she "was not a member of the Communist Party or any other organization which advocates the overthrow of the Government by force or violence." Second, nonsigners were given the right to petition the president of the university for a review of their cases, including "an investigation or a full hearing on the reasons" for their failure to sign. Third, the new procedures were to go into effect on July 1, 1950, and persons who had never signed the new oath could secure their employment for the 1949–50 academic year by signing the "prescribed form" by May 15, 1950.[139] Twenty-one out of twenty-two regents present voted for this resolution. The lone dissenter, Mario Giannini, a Warren appointee, resigned from the board, stating that as a result of the board's action, "the flag would fly in the Kremlin."[140] Warren had vigorously supported the resolution.

The April 21 "compromise," however, did not settle the controversy. A group of regents came to realize that a person could conceivably be appointed to the university simply by not signing the form, petitioning for review, and convincing the president and the reviewing Committee on Privilege and Tenure that he or she was not a member of the Communist party. This, in the view of those regents, seemed to invite persons not to sign the form and to undermine the regents' authority to set conditions for employment. Meanwhile a group of faculty nonsigners remained resolute in their conviction that any statement conditioning academic employment on a statement of political affiliation was a violation of academic freedom. Eventually, on August 25, the board voted, twelve to ten, to dismiss thirty-one persons who had not signed the form and who had been favorably reported on by the reviewing Committee on Privilege and Tenure. Warren dissented.[141]

There the situation stood in September, 1950. Then, as the dismissed nonsigners were preparing to mount a legal challenge to the board's action, perhaps the most remarkable development of the whole controversy occurred. Earl Warren, speaking before a special session of the California legislature that he had called on September 21, proposed that the legislature require all public employees of the state of California to sign, as a condition of their employment, a special oath stating that they did not support or belong to "any party or organization, political or otherwise, that now advocates the overthrow of the Government of the United States or the State of California," and that they would not advocate or belong to such a party or organization during their employment.[142]

With Warren's proposal the loyalty oath controversy came full circle. A special oath for university employees had been proposed by persons within the university community as a means of forestalling more drastic legislative action. The regents of the university, for a variety of reasons, had reacted enthusiastically to that proposal. Members of the university faculty had sharply questioned the desirability of the oath, and some had refused to sign it. Earl Warren had repeatedly and strongly supported those who questioned the oath and had opposed the need for a special oath itself. In part because of Warren's efforts, the regents had modified the form of the originally proposed oath, although they had not abandoned the idea of requiring some statement of loyalty. Now Warren was encouraging the California legislature to produce the very kind of special loyalty oath that the original proposal to the regents had been designed to forestall. The legislature responded by passing an act imposing an oath five days after Warren's proposal; Warren signed the act, popularly known as the Levering Act, on October 3, 1950.

Edward Tolman, one of the original nonsigners of the regents' oath, commented on these developments a year after Warren signed the Levering Act. "As to content," Tolman said, the Levering Act "seems even worse than the Regents' declaration." While a group among the regents, led by Warren, was now arguing for withdrawal of the April, 1950 prescribed form, Tolman noted, this group "backed their arguments . . . largely in terms of saying that the Levering Act oath was stronger and had more teeth in it!"[143] Warren himself did not give any public reasons for his September 21, 1950 proposal to the legislature and at that time sidestepped the question whether the state oath would apply to university employees.[144] Ten days after Warren signed the Levering Act, Thomas Kuchel, state controller and a close friend of Warren, was reported as stating that university employees "must sign the new State loyalty oath or go without pay."[145]

The preamble to the Levering Act stated that the oath was a response to the "present emergency in world affairs": The Korean War had broken out in June, 1950. Some commentators have speculated that Warren's proposal to the legislature was a hasty response to the Korean crisis, which had seemingly given a concreteness to fears about Communist aggression.[146] Others have tended to minimize Warren's call for a state loyalty oath, suggesting that his stand on the issue of an oath for university employees was more significant.[147] Warren himself, in his account of the loyalty oath controversy in his memoirs, declined to mention the Levering Act or his support for it. His tone was that of someone who had repeatedly and consistently opposed loyalty oaths as a matter of principle. At one point in the account he said that "the protesters believed they were mostly protecting the freedom that was guaranteed them under the Bill of Rights."[148]

But Warren's actions in the loyalty oath controversy cannot be attributed entirely to short-run political considerations. To be sure, his opposition to the regents' oath had been made public before the Korean War broke out. Moreover, as a candidate for a third term as governor, and possibly for president in 1952, he was aware of prominent California Republicans, especially Lieutenant Governor Goodwin Knight and 1950 senate candidate Richard Nixon, who were virulent anti-Communists and strong supporters of the regents' oath. As late as August 25, 1950, however, Warren had voted not to dismiss the nonsigners of the Berkeley faculty. This was one of a series of actions, beginning with his public opposition to the regents' oath in February of that year, that consistently protested against the use of the regents' oath as a condition of employment at the university. In addition, Warren filled three vacancies in the board of regents in 1951 with persons who supported his position that even the "compromise" prescriptive form adopted in April, 1950 imposed too heavy a burden on university employees.[149] Finally, Warren voted on October 19, 1951 to rescind the regents' oath altogether.[150]

It is hard to avoid the conclusion, given Warren's consistent opposition to the regents' oath and his call for a state oath, that he was making a distinction between an oath applied against employees of the University of California and an oath applied against all state employees, even if those included members of a university community. This distinction was, in fact, the "official" rationale that Warren's office gave for his simultaneous support of the Levering oath and opposition to the regents' oath. Helen MacGregor, in a conversation with Leo Katcher, a Warren biographer, in the 1960s, said that "as far as the Governor was concerned, the issue was the right of anyone to impose a special oath on a single group. He was not against a loyalty oath *per se*. . . ."[151] Warren had also taken this position publicly.[152]

But if the rationale of singling out a class of persons unfairly was the heart of Warren's opposition to the regents' oath, it is strange that he did not advance this rationale in his discussions before the board of regents or in his summary of the loyalty oath episode of his memoirs. The latter omission is perhaps less surprising, since Warren's account of the loyalty oath controversy contained several er-

rors. He claimed, for example, that the idea for an oath had originated with three members of the board of regents,[153] whereas it had
originated with Sproul and Corley. He also said that he had opposed
the regents' oath at "the next meeting"[154] after it had been proposed.
In fact, six meetings of the board had taken place after the oath
resolution, none of which Warren attended; he attended a seventh
meeting before making his opposition known; and he made his opposition known only after a majority of the board had voted to impose the "sign or get out" policy.

At the time the regents' oath was being discussed by the board
and the university faculty, Warren's announced objections to it were
threefold. First, he thought the oath was probably unconstitutional,
since one section of the California Constitution required that the university be insulated from "political and sectarian influence"[155] and
another declared that "no other oath [except allegiance to the California Constitution] shall be required as a qualification for any office
or public trust."[156] Second, he thought the oath was both unnecessary, because the 1940 regents' solution had already committed the
university to a policy of not hiring Communists, and ineffective, since
persons falsely swearing allegiance to the oath could not be prosecuted for perjury, the oath being unauthorized by statute. Communists could therefore take the oath and laugh. Third, he believed that
the University of California faculty, on the whole, was not "soft on
Communism." He later added a fourth objection, that the oath would
principally affect nonsigners "who conscientiously believed it would
be an abridgement of . . . their academic freedom."[157] Only in this
last argument, made after his service on the Supreme Court, did
Warren couch his objections in terms of the principle of academic
freedom.

The pattern of Warren's behavior during the loyalty oath controversy is clarified by an analysis of his statement at the board of regents meeting of March 31, 1950, quoted earlier. We have seen that
two themes run through the excerpt from that statement: Warren's
antipathy to communism and his emotional attachment to the University of California. "The only thing" he and "the people of this
State are interested in" is "seeking to keep Communists out of the
University"; Communists "infiltrate in business to a point where you
know nothing about it"; a Communist faculty would possess "wiles"
and be infamous. At the same time Warren was an alumnus of the
university; he "had three youngsters in the University today," and

"God willing . . . will have two more in two or three years"; he had "absolute confidence in the faculty of the University of California." For Warren these two themes were self-reinforcing and not inconsistent with one another because the great majority of university employees were not only not Communists but opposed to communism; "the members of the faculty of our University are just as sincere on the things they represent against Communism as any member of this Board." Warren could be both the militant anti-Communist and the loyal, deferential alumnus.

But the crux of the loyalty oath issue was that militant anticommunism and autonomy for the University of California were incompatible positions. Repeatedly Sproul, the regents, Warren, and others involved in the controversy tried to strike a consensus, consisting of a reaffirmation of the university's opposition to the employment of Communists. But that opposition was already university policy and needed no loyalty oath, "prescriptive form," or other affirmation from the faculty to remain university policy. The regents' oath and the Levering oath asked for much more than a reaffirmation of an established practice. They asked for a political statement of nonbelief as a condition of university or state employment.

An individual who refused to sign either oath might genuinely believe that the principles of the Communist party were incompatible with academic freedom or even with the basic values of American society. But he or she could still oppose the oath on the ground that a political belief should not be made a condition of employment in a university, where the course of one's career is easily susceptible to being affected by one's beliefs, or in state government, the ostensible function of which is to serve a public constituency composed of diverse beliefs. Warren himself was to advance this position, during his tenure as Chief Justice, in a case involving the interaction of anticommunism and academic freedom.[158]

Warren simply was not prepared, in 1950, to resolve the contradiction between two instinctual reactions he had toward the loyalty oath controversy. He believed in the University of California because of the importance that institution had played in his life, and he could not square that belief with the idea that Communists had infiltrated the university faculty. If they had, however, he was not prepared to endorse the principle of academic freedom to the point where it could permit employment at a university of persons who endorsed the stated beliefs of the Communist party. He hated communism as much in 1950 as he had during the Radin affair.[159] Communists, however, were different from non-Communists: Warren thought it was unwise and unnecessary for the regents to force university faculty members who were not Communists to make an affirmative statement of that belief or lose their jobs. His principal objection to the oath was its ineffectiveness as an administrative tactic.

Warren first devoted his efforts, in the discussions of the regents' oath, to building a consensus that simply announced unanimous opposition to the presence of Communists on the university faculty. When those efforts failed, as evidenced in the regents' August, 1950 decision to dismiss non-Communist nonsigners of the oath, Warren decided to create the "statute authorizing . . . a new oath," and he charged that the regents' oath had been unconstitutional from the outset. With the passage of the statute, the Levering Act, he could declare his concern for "security for the state and the nation," voice his conviction that "Russia has been in conflict with us ever since V-J Day in 1945," reaffirm his resistance to the Communist menace, and blunt opposition from militant elements in his party.

He could also render the regents' oath moot. The Levering oath, he knew, would be applied to university employees, was more stringent than the regents' oath, and was to be made a condition of employment. The Levering oath, however, was an "emergency" measure: In wartime, he said, "loyalty comes first."[160] He could see the Levering oath as merely codifying the principle that in wartime emergencies the loyalty of persons in offices of public responsibility was a matter of paramount public concern. In addition, the Levering oath avoided conflicts between the regents and the university faculty: It was a state oath that applied to all state employees.

The Levering oath thus can be seen as an effort on Warren's part, created by the pressures of politics, to paper over the conflict between anticommunism and academic freedom by declaring that in national emergencies, when academics who worked for a state were compelled to declare their opposition to communism, no such conflict existed. That solution to the dilemma was so cosmetic and so clearly responsive to contemporary political events that Warren shortly abandoned it. Seven years after signing the Levering Act he declared that academic freedom was a First Amendment right that presumptively could not be infringed upon by state legislatures.[161] By 1967 he had joined Court decisions invalidating, on First Amendment grounds, state loyalty oaths and state efforts to suppress "subversive" activity in colleges and universities.[162] By the time he came to write his memoirs, he saw himself as a champion of academic freedom, and he chose not to mention his support for the Levering oath.

&

The oath controversy indicated that Warren, by the beginning of his third term as governor, still resisted any characterization as an ideologue. He was as virulent an anti-Communist in 1950 as he had been during the *Point Lobos* case fourteen years earlier, but he had begun to sense a distinction between persons who were genuinely committed to the principle of academic freedom and "Communist sympathizers." He had, if anything, become more visibly identified with the Republican party during his gubernatorial years, but by the election of 1950 he was openly at odds with conservative Republicans in California and was still conducting a campaign free of partisan endorsements or rhetoric. He had moved from a critic of the New Deal to a modest supporter of affirmative government, but he continued to articulate the fears of nativist Californians about "aliens" of all types.

Warren's gubernatorial administrations in California had been Progressive administrations at a time when Progressivism had ceased to be an influential political ideology. He had retained features of California Progressivism that appeared to identify him, in the postwar context of internationalism and a heightened concern for civil liberties, as a conservative provincial xenophobe. He had modified other features of Progressivism to endorse, on occasion, the principle of affirmative, humanitarian governmental action. He was difficult to characterize in part because there was no influential ideological base for his positions; he was also difficult to characterize because he found that for him a pragmatic rather than an ideological approach to issues worked as well when he was governor as it had done when he was district attorney and attorney general. He had learned that for him carefully timed activism in office was far more effective than rhetoric; he had learned that the fact that one cared deeply about issues did not mean, in political life, that one inevitably articulated one's views.

The imperfect fit of Warren's Progressivism into the political context of the 1950s and his administrative technique of divorcing activism from partisan rhetoric provide two of the bases for the widespread misunderstanding of his public career at the time he was nominated to the Supreme Court. A number of persons, including the president who nominated him, believed that Warren would be a "middle-of-the-road" or "moderate" Republican Chief Justice and were shocked when Warren turned out to be something altogether different. But that characterization of Warren was a superficial, and highly misleading, appraisal.

Warren had, of course, repeatedly engaged in rhetoric that did not commit him to partisan or ideological positions; such rhetoric seemed to suggest that he was a "moderate." He had also, by 1952, twice campaigned for the presidency as a Republican. The succeeding chapter will demonstrate, however, that Warren had almost no constituency and no organization as a national Republican; he was influential in national Republican politics principally because he controlled the votes of a populous state and was identified with a rapidly growing region.

Moreover, Warren had not been "moderate," in terms of his approach to office holding, as a California public official. He had been an activist, continually seeking to use the powers of his office to further goals that he personally believed in, regardless of their partisan consequences. The fact that Warren had been cautious in the language he used to justify those goals, or in the timing of their implementation, did not mean that his commitment to them lacked intensity or that he endorsed institutional moderation.

Warren's pattern of governing in California had been one that resisted characterization in the conventional political terms of the 1940s and fifties. It was a pattern of continuity and change within a distinctive, highly individualized theory of officeholding. Just as he had been as a law enforcement official, as governor he was essentially concerned with solving problems, "making progress," deemphasizing partisan affiliations, keeping his office active and efficient, battling "special interests," retaining his independence, and resisting the alien enemies who would destroy California's island paradise. He was an independent Progressive at a time when American political culture was dominated by conservatives and liberals, Republicans and Democrats.

Had Warren not sought to extend his theory of officeholding to the national political arena, it is unlikely that his pattern of leadership in California would have constituted the basis of so many misunderstandings about him. But Warren did become, first reluctantly and then enthusiastically, a candidate for the only national office that seemed consistent with his theory of officeholding—the presidency. And in so doing he not only paved the way for his nomination to the Supreme Court, he stimulated others to characterize in him in the conventional political terms of the time, terms that when applied to him were not germane.

# 5

# Governing:
# National Politics and
# the Supreme Court Nomination

Of all the oddities in Warren's public life, the circumstances of his being nominated in 1953 as Chief Justice of the United States were perhaps the oddest. Dwight Eisenhower, who nominated him, and Warren had become acquainted during the 1952 Republican national campaign. Prior to that campaign, neither man had ever demonstrated much interest in national office or in Republican party politics: The two men were notable, in fact, for their lack of partisan affiliation. Warren, while nominally a Republican and Thomas Dewey's reluctant vice-presidential running mate in 1948, had been conspicuously nonpartisan in California; Eisenhower, after being appointed commander of NATO in 1950 by President Harry Truman, had not declared his party affiliation, and as late as 1951 was reported to be uncertain whether he was a Republican or a Democrat.

Eisenhower nonetheless appointed Warren Chief Justice of the United States because he thought of Warren as a "middle-of-the-road" national Republican. Had Warren not appeared as such to Eisenhower and to Herbert Brownell, Eisenhower's attorney general and chief consultant on Court nominations, Warren would not have received the nomination. And had Eisenhower not perceived himself and Warren as congenial apostles of moderate Republicanism, he would not have been as disappointed in Warren's eventual performance on the Court. Warren later quoted Eisenhower as calling his appointment of Warren to the Court the "biggest damn fool thing I ever did"; [1]

Eisenhower's mistake stemmed from a fundamental misperception about Warren. Warren's involvement with national politics was one of the ironies of his public career. He was not particularly attracted to national politics and was not particularly effective as a national Republican. Yet when he was appointed to the Supreme Court many observers found him a surprising choice because of his image as a national politician. In fact, Warren was far better prepared than most national politicians to serve on the Court, but he was appointed because of his affiliation with national politics, not because of that preparation.

✍

"While I was attorney general," Warren wrote in his memoirs, "it was easy for me to remain free from any connection with national politics, and I did so." [2] Prior to his election as attorney general in 1938 he had occasionally ventured into the national political arena. He had been a visible supporter of Herbert Hoover in 1932 and a Republican national committeeman in 1936, and in the latter year he had headed a slate of delegates to the Republican convention who were opposed to William Randolph Hearst's efforts to pledge the California delegation to Alf Landon, the eventual Republican candidate.

The Warren delegation, while reportedly interested in fostering the candidacy of Hoover, was actually sponsored by California newspaper publishers, such as Joseph Knowland of the *Oakland Tribune*, and Harry Chandler of the *Los Angeles Times*, who resented the Hearst papers' efforts to dominate state politics. In a California rump convention Warren's delegation defeated the Hearst-Landon delegation in a startling upset, but then Warren, who had been a candidate only because he was chairman of the state Republican party, released the delegates, all of whom eventually voted for Landon, the 1936 Republican presidential nominee. Warren made speeches for Landon during the presidential campaign, calling the New Deal "a totalitarian state wherein men are but the pawns of a dictator" [3] and a "regime of impractical experimentation and . . . unbridled waste of public funds." [4]

"I was a Republican," Warren said in his memoirs, "simply because California [had been] an overwhelmingly Republican state." In the 1930s, however, "the Republican Party was at the nadir of its existence, and party jobs such as [state chairman] were mostly caretaking in character." Warren described his interest in national affairs at that time as "slight" and his political views as "not well formulated."[5] When Warren announced his candidacy for the attorney generalship, he "resigned . . . as Republican national committeeman, stating that the office of attorney general was not a legitimate area for political partisanship."[6]

Warren may have been only too happy to deemphasize his Republican affiliations for more practical reasons as well. The changed population demographics in California had been accompanied by a dramatic increase in the number of registered Democrats. As attorney general and governor, Warren had demonstrated a tendency not only to campaign as a nonpartisan but to deemphasize partisanship in his office: Sweigert, his chief policymaker, was a Democrat. Nonetheless Warren, by virtue of his striking talent for winning elections and the growing political significance of his state, had become a visible presence in national Republican politics. He had defeated Olson by nearly three hundred fifty thousand votes in 1942, and in 1946 he outpolled the Democratic gubernatorial candidate, Robert Kenny, not only in the Republican but also in the Democratic primary, thereby reducing his opponents in the general election to a Prohibitionist and a Communist. His 1946 victory was regarded as stunning by national commentators. *Newsweek* commented that "Warren . . . scored a triumph that electrified GOP leaders from coast to coast"; Raymond Moley stated that Warren's "sensational victory . . . puts him in the front rank of 1948 presidential eligibles."[7]

The source of Warren's national political stature was both simple and deceptive. He was a Republican in a period when long years of Democratic incumbency were beginning to grate upon the public and the New Deal coalition was beginning to dissolve. He was a Republican from the Far West in a period when the population and consequently the political strength of that region was dramatically increasing. And he was, to all appearances, remarkably successful at winning votes.

In one sense, Warren was a prominent figure in national Republican circles because his obvious popularity in a politically strategic region could not be ignored. In 1944 he was chosen to be keynote speaker at the Republican convention and was Dewey's first choice for vice-president. Warren declined, stating that accepting the position would be betraying "certain commitments and . . . obligations to my state which are yet to be fulfilled."[8] In 1948 he reluctantly agreed to run as Dewey's vice-president, feeling that "there were no other public positions that I looked forward to," and that the vice-presidency "might be a satisfactory way of topping off a public career."[9] Dewey had chosen Warren primarily because he thought Warren's presence would strengthen the ticket in the Far West.

In another sense Warren was an outsider as a national Republican. He was, first of all, an old-fashioned kind of Republican. Despite rhetorical attacks on the New Deal[10] and some partisan comments that he made, without much enthusiasm, in the 1948 campaign,[11] Warren was not a standard Republican of the Dewey or Robert Taft variety. He had first become a Republican because most registered voters in early twentieth-century California were of that affiliation and because he identified the party with the reforms of Hiram Johnson. He voted for Woodrow Wilson in 1912 and Robert LaFollette in 1924; he thought of himself as an advocate of good government, a proponent of state and local autonomy, and a believer in progress through beneficent government intervention. Those were "Progressive" Republican doctrines, more easily identified with Theodore Roosevelt than with any segment of the national Republican party in the 1940s and fifties.

Second, Warren had been a distinctly provincial candidate in California. From 1938, when he resigned as Republican national committeeman to run for attorney general, through his third term as governor, Warren operated on the principle that "broad questions of national party politics" bore no application "to the problems of state and local government in California."[12] He cross-filed in all his campaigns; he did not endorse other Republican candidates or campaign with them; he did not make use of the Republican party organizational apparatus, such as it was; and he had no continuing political organization of his own. Warren developed, we have seen, a style that was distinctive to the political culture of California—a style that substituted for party organization nonpartisan opposition to special interests in the legislature and cultivation of the press and the general public.

Thus, as a national Republican politician in the years after World

War II, Warren was a curious phenomenon. The core of the Republican party platform from 1946 to 1952, it might be argued, was partisan opposition to the New Deal, the Fair Deal, and the philosophy of affirmative participation by the federal government in social and economic affairs. Republicans were expected to be critics of affirmative government, defenders of free enterprise, and party loyalists. Warren was not strongly identified with any of those positions. His criticisms of the New Deal had been from the point of view of an advocate of paternalistic action by states and localities. He supported "decentralization of authority," he said in 1948, "because the strength of the republic depends largely on the ambit of the state and local governments."[13] While he mouthed free enterprise doctrines, he had sought, as governor, to tax oil companies, regulate the medical profession, and increase public control of utilities. And he had been the opposite of a party loyalist in his campaigning, his appointments, his staffing, and the conduct of his offices. He was admirably suited, it seemed, to serve as a provincial, eccentric figure in national Republican politics, the opposite, in many respects, of his fellow Californian Richard Nixon, an outspoken partisan and a staunch critic of a philosophy of affirmative government.

A microcosm of Warren's role as a national Republic politician can be seen in his unsuccessful 1948 campaign for the vice-presidency. Warren had been a reluctant candidate for the vice-presidential nomination: In his memoirs he recalled that

> I never gave a thought to [the vice-presidency] because I did not cherish the idea of merely presiding over the Senate . . . nor was I anxious to leave California, where life and the people had been so good to me.
>
> The night before the morning session that was to nominate a vice-president, I went to bed about midnight, and was awakened from a sound sleep about 2:30 a.m. by a ring of the telephone. Tom Dewey was on the line. He asked me if I could come to his motel headquarters. After dressing I went directly there. He was waiting for me, and told me that earlier he had gathered together a number of Republican leaders from distant parts of the nation. They were agreed that I should be his running mate, if I would accept, and he said that would please him also. He told me he could understand why I would not relish merely presiding over the Senate, and that if I accepted and we were elected, he could make the job meaning-

ful by having the vice-presidency play an important role in his administration. We talked the matter over for a half hour or so, and also some of the family problems involved in such a move. Finally I told him I would accept.[14]

Warren then returned to his hotel and informed Nina that he had accepted Dewey's offer. Nina, while "characteristically" making "no complaint," expressed doubt "whether I could be happy with Tom Dewey directing my affairs." Nina Warren had been "as interested in my independence as I had been through the years," and considered the vice-presidency "a potential inroad on it."[15] Her reservations turned out to be prophetic.

The vice-presidential campaign, for a person as accustomed to making his own decisions and as stubborn in his beliefs as Warren, was a depressing experience. He later told Irving Stone, a campaign biographer, that he had accepted the vice-presidency "to speak . . . for all the things that I fought for all my life" and to counter the "influence [that] would be exerted from Wall Street" in a Dewey Administration. If Warren believed that he "could do a better job inside the administration than outside,"[16] the campaign disabused him of this belief. Dewey and his strategists, who were overwhelming favorites against the incumbent Truman Administration, designed a packaged campaign, filled with white papers on policy issues, devoid of controversy, and entirely prearranged. Warren's role in the campaign was to appear in places that Dewey did not feel obligated to visit and to make formal addresses reinforcing Dewey's views.

At the beginning of the campaign, shortly before Labor Day, Warren was assigned one of Dewey's speech writers "for whom he had no place in his campaign" and "two Hollywood writers . . . whose forte was writing gags." Warren, who valued his independence as a candidate and whose speaking style was rarely entertaining, had no use for any of the writers. The Warrens campaigned by means of "one whistle-stop campaign train," making eight to ten stops a day and then embarking at "some different auditorium" for Warren's nightly formal speech.[17] Warren was "entirely separated from the main campaign." He recalled having two conversations with Dewey during the duration of his whistle-stopping, once to complain to Dewey about the treatment the Warren entourage received in Buffalo, New York, the second to press Dewey to take a position on the use of hydroelectric power in the arid lands of the West.[18] Dewey was sympathetic about the Buffalo reception but implacable on water policies. The Dewey-Warren ticket lost overwhelmingly in California's Central Valley.

During the campaign Warren began to call old California friends and associates to express his discomfort and dissatisfaction with the role he had been assigned. Walter Jones, editor of the *Sacramento Bee*, remembered Warren calling him and saying, "I'm so low in this campaign. I can't say what I want to say."[19] In desperation, Warren summoned William Sweigert, saying that Dewey had "sent me a lot of those gag writers," that "I just can't live with them," and that "I've got to have somebody along here that can interpret me and help me."[20] Sweigert joined the Warren campaign train, and recalled that "my life was a life of misery . . . trying to construct the basis of a speech everyday to give every night." Sweigert would write a speech, Warren would "go over it and pull it to pieces," and "the process would wind up with [Warren] giving [Sweigert] an idea here and there, and my drafting it, and sometimes with him writing on a yellow sheet a paragraph or two." Sweigert recalled that "there was always the feeling that you were not to say anything unless it meshed in with what Dewey was saying." He felt that Warren was in "a very unhappy situation" in the campaign.[21]

Dewey and Warren lost, to the astonishment of most knowledgeable political observers, but not to the surprise of Nina Warren, who had predicted defeat,[22] or Earl Warren, who had told his staff two weeks before the election that Truman had "gotten through to the people" and who was quoted in the *Los Angeles Times* two days after the election as saying, "It feels as if a hundred-pound sack had been taken off my back."[23] Warren had "submerged himself to Dewey in the campaign," according to the *New York Post*, conveying an impression that "Earl Warren doesn't know the answers, but that Tom Dewey will find them."[24] While Warren's acquiescence in his subordinate role was consistent with his pledge when nominated to "give to [Dewey] every bit of loyalty and help in my make-up,"[25] it was also a testimony to his marginal status as a national Republican leader. The independence and nonpartisanship that had helped him in California and suited his temperament was a source of distress as a vice-presidential candidate.

Warren accepted the vice-presidential nomination in 1948 because he could not do otherwise and retain any standing with Republican regulars. "If I hadn't taken it this time," he told his aide, Merrell Small, "they'd never consider me for anything again." [26] His presidential candidacy was never a strong one either. He recognized that "while my nonpartisan state administration was a source of great strength in California, it limited my horizon because the other states, with strong party organizations, were not favorable to such policies." [27]

In addition, Warren's candidacy had not excited the national media. The *New York Times* called him "an attractive vote getter," but a "plodding, routine . . . District Attorney," and "an equally unspectacular Attorney General." A writer for the *American Mercury* said that Warren was "decent" and "hard-working," but that "it is doubtful that he has it in him to be a great leader of men." *Time* magazine said that Warren's critics felt that he was "a bull-headed, plodding mediocrity who never says or does anything out of the ordinary," and that his admirers could only characterize him as "solid, patient, dependable." [28] And John Gunther, in his 1947 edition of *Inside U.S.A.* gave a description of Warren that was to be quoted with amusement in the years of his Chief Justiceship. Gunther called Warren someone who

> will never set the world on fire or even make it smoke; he has all
> the limitations of all Americans of his type with little intellectual
> background, little genuine depth, or coherent political philosophy;
> a man who has probably never bothered with an abstract thought
> twice in his life; a man . . . with little inner force. [29]

As late as 1949 Warren had not even decided whether or not to remain in state politics after his second gubernatorial term expired. He "had no idea how a third term would be looked upon by the voters"; [30] he had a number of attractive offers in the private sector. But once again his response to attempts to pressure him was to resist. "Assorted groups and sundry forces," Kyle Palmer wrote in the *Los Angeles Times* at the opening of the 1949 legislative session, "are now surveying and exploring all possibilities for a concerted and organized attack [on Warren] . . . Sacramento hotels are filled with . . . individuals who are working both openly and secretly to accomplish the Governor's administrative eclipse and his political demise." [31]

Warren's own lieutenant governor, Goodwin Knight, was encouraged by Republican regulars and the Hearst papers to oppose

Warren for a third term, and floated a trial balloon in September, 1949.[32] Warren responded publicly by saying that "there is too much to be done as governor for me to be concerned about candidates,"[33] but in his memoirs he recalled that "I might not have run for a third term had it not been for the intransigence of the lieutenant governor." Knight's "insistence that I was not a sound administrator and his criticism of my programs," Warren wryly said, "led me to believe that he would not have completed my postwar plans."[34]

Thus Warren announced his candidacy on February 2, 1950 and went on to defeat James Roosevelt, the Democratic candidate, by over a million votes, the largest majority in California history. Once again talk of national office surfaced in the wake of Warren's triumph, the *New York Times* calling the vote "beyond the greatest expectations of most Republican leaders," and claiming that it had stimulated "talk of Mr. Warren as a serious contender for the Republican nomination in 1952."[35] After the 1950 election Warren decided to give "real thought to the national situation."[36] He began to pay more attention to national issues, favoring federal aid to education, pressing for a national health insurance program, and disassociating himself from Senator Joseph McCarthy. Warren did not deviate too far from his anti-Communist line: He claimed that the federal government had an internal security problem, and he called for a commission to investigate allegations of Communist subversion in Washington.[37]

Warren also became more outspoken on international issues, endorsing the concept of a United Nations and reviving his enthusiasm for civil defense and preparedness as the Korean War broke out. He visited Japan in 1951 and was briefed on strategic problems in the Far East by General Matthew Ridgeway; he praised Truman's secretary of state, Dean Acheson, for his work in securing a peace treaty with Japan that same year. By the time he announced his candidacy for president on November 14, 1951, he was the leading declared Republican alternative to Taft. A Gallup poll indicated that voters would prefer him in a presidential election over Truman by 55 to 33 percent.

But two events combined to dampen Warren's prospects severely. Shortly after the announcement of his candidacy he had "a serious operation for abdominal cancer." The illness was not made public, the press being informed that Warren's appendix had been removed. Warren received a written medical report "assuring me that there was no malignancy left in my system," but his original plan for an aggressive campaign to establish himself as preferable to Taft had to be modified while he recuperated. Taft supporters took the opportunity to circulate rumors that Warren was dying of cancer. In his memoirs Warren related a story where former President Hoover, a strong Taft supporter, said "at a dinner with thirty or forty national [Republican] figures present, 'You don't have to be concerned about Warren. I know the doctors who operated on him. They opened him up, took a look, and sewed him up again.' "[38]

The second and more devastating event was Dwight Eisenhower's announcement, on January 7, 1952, that he was a Republican and a candidate for the presidency. From the time of Eisenhower's announcement Warren's chances to win the nomination, except after a deadlock, were doomed. Eisenhower's positions, especially on domestic issues, were relatively unknown, but he was perceived to be eminently electable. Warren's strengths—electability, moderation, the ability to make nonpartisan appeals—were also Eisenhower's, and while Warren had an attractive family and had been called by *Time* magazine "the best campaigner yet" on television,[39] he could not compete, at the height of the cold war, with the nation's most revered military hero.

It may seem surprising, in this context, that Eisenhower's subsequent appointment of Warren to the Supreme Court has often been regarded as the payment of a political debt. Eisenhower and Warren had very little contact with one another before the convention, Warren did not release his delegation's votes for Eisenhower on the first ballot, and making political deals was foreign to Warren's nature, as he repeatedly stated. Eisenhower, for his part, said in his memoirs that "the truth was that I owed Governor Warren nothing."[40]

Eisenhower did, however, accumulate a small debt to Warren at the 1952 convention. Warren's strategy for 1952, after Eisenhower announced his candidacy, was to hope for a deadlock between the Taft and Eisenhower forces on the first ballot. In the event of such a deadlock, Warren's chances were promising, largely because of his perceived appeal to Democratic or uncommitted voters. The possibility of a deadlock was a serious one. Warren recalled that "the Taft forces were in complete control of the convention," dominating "seating arrangements as well as speakers," and that Taft had offered him "anything I desired" in a Taft Administration if he would release California's delegates.[41]

Taft's strength, however, rested in part on the votes of the Texas,

Georgia, and Louisiana delegations whose credentials were under challenge by other delegations from those states. Sixty-eight of the pro-Taft delegates were subsequently approved by the convention's credentials committee. Then the Eisenhower forces, claiming that those delegates were still tarnished, asked for a convention vote on a "fair play" resolution that would bar those delegates from voting on the remaining contested delegate seats. This resolution, if passed, would have had the effect of depriving the Taft forces of sixty-eight votes on the remaining contested seats, virtually assuring that those seats would go to Eisenhower delegates. However, if the Taft forces defeated the resolution, Taft delegates would probably be in control of the seating of the contested delegates, and Taft would then, apparently, have enough delegate strength to block Eisenhower's nomination on the first ballot. The vote on the "fair play" resolution looked to be extremely close, and the consequences were certainly very significant.

The votes of the California delegation, pledged to support Warren on the first ballot, turned out to be crucial. Seventy votes were in the delegation, and if all voted for the resolution (an "Eisenhower" vote), Eisenhower was virtually assured of the nomination. If the delegation split votes, however, with some delegates voting against the resolution, Eisenhower's chances for a first ballot victory were reduced. Indeed, if all the California delegates voted against the resolution, Warren was most likely to benefit, since this greatly increased the likelihood of more pro-Taft delegates being seated, the consequence of which would be a closer vote on the first ballot.

It was at this stage that freshman Senator Richard Nixon, ostensibly pledged as a delegate to support Warren but previously committed to Eisenhower,[42] spoke at a caucus of the California delegation. Senator William Knowland, the delegation's chairman, who was close to the Taft forces, had urged that the California votes on the "fair play" resolution be evenly split between approval and disapproval, a tactic that was intended to ensure the defeat of the resolution. Nixon then portrayed the "fair play" resolution as a moral issue, arguing that the contested delegates would taint any nominated Republican. Warren, who was present at the caucus, and who had already endorsed the "fair play" resolution in principle,[43] did not support Knowland's proposal and requested that the resolution be supported or defeated by a majority vote within the delegation.

The California delegates overwhelmingly voted to support the resolution, giving the "fair play" advocates seventy more votes; the "fair play" resolution ultimately won by ninety votes. Without the contested delegates' votes, Taft received 500 delegates on the first ballot, to Eisenhower's 595, Warren's 81, Harold Stassen's 20, and General Douglas MacArthur's 10. Some of Eisenhower's delegates were picked up after the "fair play" resolution; had it been defeated Taft and Eisenhower might have been close enough so that the California delegation would have been able to throw the nomination to either candidate or to force a second ballot. Thus Warren's acquiescence to the California delegates' support of the "fair play" resolution was unquestionably against his own interests and in support of Eisenhower.[44]

These events, at the 1952 Republican convention, are also of interest because they brought together two of California's then most well-known Republicans, Earl Warren and Richard Nixon, whose careers would, over a number of years and a series of incidents, intertwine frequently, almost always at cross-purposes. While some of the history of the relationship between Earl Warren and Richard Nixon lies outside the chronology of this chapter, the relationship serves to locate Warren on the spectrum of national Republican politicians in the 1940s and fifties.

Nixon, from his first campaign for Congress in 1946, had been a different breed of Republican from Warren. He was intensely partisan in his political dealings, and in his campaigns, he aggressively went after his opposition. His interests were also different. From the beginning he showed a flair for and fascination with international affairs and the ideological issues of the cold war. Warren, in contrast, was a nonpartisan, platitudinous, relatively provincial, and nonideological campaigner, who as late as 1946 continued to insist that his principal concern was with California affairs and politics, and that the California experience could not easily be extrapolated to a national context. It was characteristic of Warren that he was not interested in running for the Senate, preferring to be a state executive. It was characteristic of Nixon that four years after his election to Congress in 1946 he ran a partisan, aggressive, and ideological campaign for Helen Gahagan Douglas' senate seat.

Nixon's interest in partisanship and aggressive debate, plus his ideological cold warriorism, which posed a sharp contrast to Warren's muted stance, made him an attractive candidate to Republican partisans who were increasingly suspicious of Warren's Progressivism and annoyed at his independent style of campaigning. The first evidence that Warren's and Nixon's different styles would make them political adversaries came in 1946, when Warren, running for a second term as governor, continued his practice of not endorsing any Republican candidates for office. Nixon, engaged in a close race with Jerry Voorhis, resented Warren's lack of support, and in his Senate campaign in 1950 attempted to force Warren to endorse him.

In 1950, while Nixon was running against Douglas, Warren was running against James Roosevelt, as noted, and the Nixon strategists attempted to secure Warren's endorsement by pressuring Helen Douglas to endorse Roosevelt. A member of the California Young Republicans, Joseph Holt, was stationed at Douglas' rallies and meetings, and repeatedly asked Douglas whom she was supporting for governor. Finally four days before the election, Douglas stated that she "hope[d] and praye[d] that [Roosevelt would] be the next Governor." The press immediately asked Warren for a comment, and he said that "as always I have kept my campaign independent from other campaigns," and that Douglas' statement "does not change my position." He went on to add, however, that "in view of her statement . . . I might ask [Douglas] how she expects I will vote when I mark my ballot for United States Senator." Murray Chotiner, Nixon's principal campaign aide, immediately announced that "Earl Warren intends to mark his ballot for Dick Nixon on election day."[45]

Nixon, who apparently resented having to court Warren's support, as well as Warren's reluctance to give it, responded by undermining Warren's strength in the 1952 Republican convention, thereby enhancing his own candidacy for the vice-presidency and establishing an adversarial posture toward Warren that was to characterize the relationship between Warren and Nixon for the next two decades. In addition to endorsing the "fair play" resolution, Nixon, by most accounts, lobbied for Eisenhower within the California delegation while still pledged to support Warren. Prior to the convention Nixon had commented, upon hearing Warren's announcement of his candidacy, that "Senator Taft and General Eisenhower" were "the front-runners" for the 1952 nomination, and that Warren "lacks strength among the people who nominate outside of California," partly "because of his reputation for liberalism," and partly because "the Republicans . . . want someone who will hit and hit hard on major issues."[46]

While Warren, in his memoirs, stopped short of stating that Nixon had personally attempted to undermine his strength in the delegation, he maintained that "the Nixon delegates . . . held caucuses and urged other delegates to vote for General Eisenhower on the first ballot," and that when Warren made a ceremonial visit to Eisenhower's suite, "the doorkeeper who admitted me . . . was Murray Chotiner, one of the managers of my train." Warren also declined Eisenhower's subsequent staff invitation "to sit down with a group of prominent Republicans and select a vice-presidential candidate," because he believed that Nixon had already been selected.[47]

The adversarial character of Warren's and Nixon's relationship continued well past the appointment of Warren as Chief Justice. While Warren made a practice of never commenting publicly on political matters, he allegedly said, in an off-the-record conversation, that Nixon "ha[d] to be stopped" in his effort to become governor of California in 1962. To that end, Earl Warren, Jr., at that time practicing law in Sacramento, publicly switched his political affiliation from Republican to Democrat and joined the campaign staff of Edmund G. ("Pat") Brown, Nixon's opponent. Earl junior reportedly said, at a press conference, that "Nixon, through backdoor politics and for political gain for himself, pulled the rug out from under us in 1952. He wronged my father and the whole state."[48]

Nixon's turn came in 1968, when, after a startling comeback from his defeat in California, he won the Republican presidential nomination and the election, using as one of his campaign issues the alleged sins of the Warren Court, which he accused of protecting "criminal forces." Not only was Nixon successful in this effort, he was able, while president, to make four appointments to the Court, none of whose jurisprudential views, at least as far as they were discernible before appointment, seemed to square with those of Warren. One of Warren's law clerks recalled hearing Warren speak of Nixon, several years before Watergate, "in terms that would ordinarily be reserved for someone who has proved to have engaged in serious violations of criminal law and ethical conduct." Warren characterized Nixon as "untrustworthy, a scoundrel, a liar, completely unprincipled, and an exceedingly dangerous person."[49] The last twists and turns of Warren's and Nixon's relationship, which included Warren's retirement from the Supreme Court, the Watergate scandal, and Nixon's actions during Warren's final illness, are reserved for a later chapter.

ⱪ

In 1952 Nixon seemed to have temporarily stolen a march on Warren, helping to deny him the opportunity for a deadlock at the Republican convention and emerging as Eisenhower's vice-president. But "the Warren luck" had not yet run out. In late September, 1952, the existence of a secret fund for Nixon's private use as a senator, contributed to by a group of wealthy Californians, was revealed. Nixon's place on the Republican ticket and his electability were suddenly thrown into jeopardy, and Warren's open support for the Republican slate came to be regarded as imperative by the Eisenhower campaign. Warren was summoned to meet with Eisenhower's advisers, and on September 26, after Nixon's emotional "Checkers Speech" on national television, Warren announced that since Nixon had made a full account of the fund, and Eisenhower had vouched for its credibility, he regarded the matter as closed.[50] Republican headquarters then announced that Warren would actively campaign for the national ticket, and he began a cross-country speaking tour.

The "Nixon fund" incident provided Warren with the kind of leverage he had not fully received at the convention. Warren was regarded as being interested in some kind of federal appointment: He was not expected to seek a fourth term as governor and had made no other plans. Immediately before election day Warren appeared with Eisenhower on a televised broadcast where he reassured Western voters about Eisenhower's positions on hydroelectric power, soil conservation, and other land use issues, and reported that Eisenhower had called development of the West "the unfinished business of America."[51] After Eisenhower's election victory, according to one of Warren's biographers, he offered Warren a Cabinet position as secretary of labor;[52] Warren allegedly turned it down but indicated that he would be interested in going on the Supreme Court.

Warren's memoirs, however, gave a different account:

> As soon as the election was over and General Eisenhower was chosen, I returned to my knitting and stuck pretty close to my governor's job. . . . I was not a part of the new Administration's central organization and only occasionally would receive a call from Eastern headquarters concerning someone under consideration for appoint-

ment. The only Cabinet appointment I was asked about was that of a new Secretary of the Interior . . . The other inquiries made of me had to do with sub-Cabinet positions for individuals living in the Far West. . . .

I heard nothing more from the Administration until one morning in early December, [1952].[53]

Warren then recounted a conversation with Eisenhower in which the latter said that he "intend[ed] to offer [Warren] the first vacancy on the Supreme Court," and that the offer was a "personal commitment." Warren told Nina, who asked the salary and wondered how the Warrens could live in Washington on twenty-five thousand dollars a year. "Little more was [thereafter] said about the President's call," Warren noted, "and not much thought was given to it, because I had often heard of newly elected officials who promised positions in the indefinite future, only to forget when the jobs actually became open for appointment."[54]

Nine months later Earl Warren was appointed Chief Justice of the United States. The story of his appointment can be seen as an example of the fortuitous process by which individuals come to sit on the Supreme Court. Before the details are examined, however, perhaps it is best to raise the question contemporary commentators raised repeatedly after the character of Warren's performance as a judge had become apparent: Why did Eisenhower appoint Warren?

Eisenhower, we have seen, certainly did not owe his nomination and very likely did not owe his election to Warren. Nor did Warren, as his long-term Republican critic in California, Lloyd Wright, claimed, run for president in order to increase his chances of getting a Cabinet position.[55] Eisenhower won the Republican nomination, and defeated Adlai Stevenson in the 1952 election, because he was an ideal political candidate at the time. He was nonideological, affable, and not identified with political patronage, unpopular political judgments, or other features of a career in politics; he was a military hero come to save America from the politicians who had allowed communism to infiltrate American government and make inroads in Korea. Warren put up his name against Eisenhower's because he sincerely believed he could win the nomination and because he had confidence, after three successful terms in California, in his ability to govern.

But Eisenhower would not have appointed Warren had the latter not been a prominent Republican who was known to be receptive to an appointment on the Court. Warren was fortunate, in this instance, to have a nominating president who regarded Supreme Court seats as the equivalent of Cabinet positions and thus considered party affiliation as an important criterion for a nominee. Warren was also fortunate to have Herbert Brownell as Eisenhower's attorney general. Traditionally, the attorney general has been the member of the Cabinet most influential in screening potential Supreme Court nominees, especially in twentieth-century Republican administrations. In Brownell, Eisenhower had an attorney general who was also a partisan Republican and an experienced campaigner. Brownell knew Warren, knew of his support for the ticket in 1952, especially his assistance after the Nixon fund incident, and did not regard the fact that Warren was currently not practicing law and had not done so for ten years as disqualifying.

Brownell also knew that Warren was interested in being on the Court. In late May, 1953, Warren had met with Brownell in Washington, and had agreed to be considered for the office of solicitor general of the United States, sometimes a resting place for eventual nominees to the Court. In June, Warren, who had been appointed one of the delegates to Queen Elizabeth's coronation, left with his wife and family for an extended vacation in Europe, to begin after the coronation. The vacation itself was a signal that Warren would most likely not seek reelection as governor: Hitherto he had been reluctant to leave California for any extended period, fearing the political costs of leaving the state in the hands of Lieutenant Governor Knight, who was reportedly planning to oppose Warren in a primary should Warren run for a fourth term. As the Warrens traveled throughout Europe, Warren "meditated long and seriously" about the solicitor generalship. He finally concluded that he would accept the position if offered it and wired Brownell on August 3, 1953 that he would accept.[56]

The solicitor generalship offer had been tendered both because Eisenhower was having difficulty filling the post and because Brownell believed that the solicitor generalship would help alleviate the "considerable lapse of time since Governor Warren had actively argued cases and practiced law." Warren and Brownell both understood that the solicitor generalship was to be an interim job until Warren went on the Court. When Warren returned to California in late August, 1953, he began "preparing to close out his Governor-

ship and move to Washington."[57] On September 3, 1953, he announced that he would not be a candidate for a fourth term as governor.

Why was Warren interested in a position on the Court? That
question might seem not worth pursuing, since Americans regularly
rank the office of Supreme Court justice as the one for which they
have the most admiration and respect. In Warren's case, however,
going on the Court meant leaving California, where he had lived his
entire life; assuming an entirely new kind of office, an action he was
temperamentally disinclined to take; reducing his income, since the
governor of California was then paid as much as Supreme Court
justices and had rent-free use of the Governor's Mansion and other
fringe benefits; and assuming a position where he would no longer
be the dominant executive in an office.

A passage from Warren's memoirs sheds some light on his thinking about the possibility of going on the Court. As the European
trip indicated, Warren had "made up my mind to leave political life"
after his third term as governor expired in 1954 and was planning to
announce his decision on his return from Europe. He "had made no
plans," however, for the future, although he "was obliged to make a
decision because . . . it would be necessary for me to be gainfully
employed in some new capacity," since the Warrens relied on his
salary as the basis of their income. Warren then described his thinking:

> I had no desire to enter the business world for the purpose of mak-
> ing a fortune, although through the years I had been offered posi-
> tions that paid much more money than I had ever received as a
> public servant. I had no real desire to practice law privately, be-
> cause I had been in public service nearly all my adult life, usually
> as the head of an office which afforded me an opportunity to decide
> public matters on their merits rather than private matters for the
> fee involved. I was not sure I could be happy in a transition from
> one to the other. . . . There was no other political office that inter-
> ested me.[58]

In this excerpt one can see the features of an office that counted
for Warren: being "the head of an office"; having an opportunity to
deal with "public matters" in "public service"; feeling able to make
decisions "on their merits" rather than "for the fee involved." The
office of Supreme Court justice embodied most of these features.

While Warren would not be "the head of an office"—neither he nor Eisenhower was considering the chief justiceship at this time—he would be occupying a position that provided life tenure and was notable for its independence. He would be continuing in public service, considering "public matters," and deciding issues "on their merits."

Finally, appointment to the Supreme Court symbolized, for a person who had seen himself, for much of his public career, as a lawyer and especially as a litigator, a professional recognition that was gratifying to Warren. He had himself argued twenty-five cases before the Court as attorney general of California, and part of the reason he had agreed to accept the solicitor generalship was that he conceived being "the principal lawyer for the United States Government in its most important litigation" as "a challenge of enormous proportions."[59] While others[60] thought that Warren did not have the professional training to serve on the Supreme Court, he disagreed with that assessment.

Warren was not only to serve on the Court, he was to be its Chief Justice, a role that provided more continuity with his previous positions than he or anyone else had suspected. Five days after Warren announced his prospective retirement as governor, Chief Justice Fred Vinson died of a heart attack. Well before Vinson's death, Warren had been given assurances by both Eisenhower and Brownell that he would be appointed to fill the first vacancy on the Court. While the assurances were, of course, not binding, both parties considered them sufficiently serious for the Warrens to leave California and join the Department of Justice; by the time Warren made his announcement about the governorship, the Warrens were preparing to leave for Washington. Herbert Brownell, in an oral history interview given in 1974, said that Eisenhower considered Warren "part of the Eisenhower team," and confirmed Eisenhower's part in the "December," 1952 conversation with Warren, which Brownell placed on November 25. Brownell also confirmed that General Lucius Clay, who was actively involved with the Eisenhower campaign, may well have mentioned the possibility of a Supreme Court nomination to Warren before the 1952 election.[61]

Eisenhower's inclination to nominate Warren for the "first vacancy" on the Court seems, then, to have stemmed from a combination of factors: gratitude for Warren's help in the last stages of the campaign, especially in California; confidence in Warren's political judgment, at least on California matters;[62] continued respect for Warren and his views after his election; and the aforementioned conception of a Supreme Court appointment as the equivalent of a Cabinet post, not necessarily requiring skills different from Cabinet positions. There was no political pay-off or deal between Eisenhower and Warren, and no necessity for making one, but the context of the relationship between Eisenhower and Warren was Republican party politics.

When Vinson died, however, Eisenhower was not in a hurry to nominate Warren Chief Justice. He had not anticipated, nor had Warren, that the first vacancy would be the chief justiceship: Associate Justice Felix Frankfurter's seat was the one regarded as most likely to be vacant.[63] Brownell, who found the vacancy "totally unexpected," now began a three-week review of the available candidates for the chief justiceship, including Warren. The first stages of Brownell's discussions with Eisenhower in this period centered on "whether any of the incumbent Associate Justices should be appointed." Robert Jackson and Harold Burton were both given serious consideration and rejected. Eisenhower eventually decided that "the appointment should go to someone not then on the Supreme Court." Brownell stated that "there was no one on the Court that the President wanted to appoint."[64]

Brownell then listed for Eisenhower "outstanding individuals who would be qualified for the Chief Justiceship." In recalling his discussions with Eisenhower, Brownell said that "it was clear to me . . . that President Eisenhower did not feel that his prior discussions with Governor Warren had related to a vacancy in the post of Chief Justice, but that he had had in mind an Associate Justiceship in his original call to Governor Warren about a 'first vacancy' "[65] In his memoirs Eisenhower said that he had told Warren, prior to Vinson's death, "that I was considering the possibility of appointing him to the Supreme Court and that I was definitely inclined to do so if, in the future, a vacancy should occur." In the next sentence he said that "neither [Warren] nor I was thinking of the special post of Chief Justice, nor was I definitely committed to any appointment."[66]

Brownell began to brief Eisenhower on prospective candidates, including Arthur Vanderbilt, Chief Justice of the Supreme Court of New Jersey, and John J. Parker, chief judge of the U.S. Court of Appeals for the Fourth Circuit, who had been unsuccessfully nominated for the Court by Herbert Hoover in 1929. Eisenhower expressed interest in selecting "someone who was held in high esteem by the American people for past public service," and apparently, unbeknownst to Brownell, asked John Foster Dulles, his secretary of state, whether he would consider taking the position. Eisenhower also discussed with Brownell "the desirability of having the Chief Justiceship held by a person of broad administrative experience." [67]

Eventually Eisenhower told Brownell that "while he had not made up his mind definitely, he considered Governor Warren as the leading prospect." Eisenhower suggested that Brownell meet with Warren "to determine whether he would prefer service first as Solicitor General as previously agreed upon," or whether he would want to go directly to the Court. [68] Eisenhower had also been informed by sitting justices that the Court was eager to have its full membership intact by the opening of the term, scheduled for October 5, 1953. This required that the position go to a recent appointment, since Congress was not in session, and Brownell's office prepared a memorandum arguing that a justice could hear and vote on cases prior to being confirmed. As it was, Warren was not to be confirmed by the Senate until March 1, 1954.

On September 25, 1953, Brownell tracked down Warren, who had gone deer hunting on Santa Rosa Island, off the coast of Santa Barbara, as "kind of a hideout" from the press. There was no telephone on the island; its "only communication with the mainland was through a ship to shore radio used mostly by coastal vessels and fishermen." Warren received a message—"everybody out there in the Pacific could hear it"—that "Herb Brownell wanted to talk to me." [69] Warren radioed to the mainland, a small plane was sent over, and Warren called Brownell from Santa Barbara. They agreed to meet on Sunday the twenty-seventh at McClellan Air Force Base near Sacramento, where Warren's state plane was kept. The meeting was not publicized.

Brownell and Warren met at eight in the morning, in a lounge at the airport, for approximately one hour and a half. Eisenhower, according to Brownell, was uncertain as to whether or not Warren wanted to go on the Court immediately and "whether Warren would consider that the talk he'd had the previous November with Eisenhower . . . meant the first vacancy, regardless of whether it was associate justice or chief justice." Eisenhower was also interested "as to whether Warren was generally sympathetic with the ideology . . . of the Eisenhower administration."

Warren "took the position," according to Brownell, "that he was ready to accept the Chief Justiceship, . . . that he did feel that he had a commitment for the next vacancy, and that this *was* the next vacancy; and that he had followed Eisenhower's presidency and was in agreement with his policies and program." Warren also told Brownell that he could arrange to be in Washington for the opening of the Court's term. "When I left," Brownell remembered, "I left telling [Warren] I would report as completely as I could to the President everything that we had talked about, and [Eisenhower] would let him know. There was no commitment at that time." [70]

Brownell then reported to Eisenhower, who decided to appoint Warren, but hedged about making a formal announcement. Brownell suggested an informal news conference at Brownell's house where Warren's name would be "floated" to certain reporters. This would result in stories that Eisenhower was "thinking" of appointing Warren, and the Eisenhower Administration could gauge the reaction. Unfortunately, for this strategy Brownell "was a little too definite" to the reporters, who immediately wrote stories that Warren was to be nominated for the chief justiceship. At this point Eisenhower instructed Brownell to call Warren and confirm the appointment, and held a news conference in which he said that Warren's appointment "may not come as a shock to some of you." Warren learned of his appointment on Wednesday, September 30. [71]

Several rumors have sprung up about the process of Warren's appointment, none of which can be confirmed by reliable evidence, and some of which are clearly false. Merrell Small, in a document in the Earl Warren Oral History Project, claimed that Warren knew as early as September 15, 1952 that he was to get the first vacancy on the Court. Small cited a telephone conversation on that date between Warren and Eisenhower, the contents of which Warren related to Small in March, 1953. [72] Warren, Brownell, and Eisenhower all insisted that the first mention of a Supreme Court vacancy came in later November (or early December) 1952, and it seems highly unlikely that Small's recollection is correct, since Eisenhower had not been elected at the time of the call. What seems more likely is that Eisenhower, in the course of soliciting Warren's campaign support for the Republican ticket, may have made some allusions to a federal appointment, and that Warren, in later discussing the September conversation with Small, may have confused that conversation with the November 25 conversation.

Small also claimed that when Warren and Brownell met at McClellan Air Force Base, Warren "insisted . . . upon being appointed Chief Justice."[73] Small had written to that effect in a newspaper article in June, 1970,[74] and John Weaver, relying on Small, had written in his 1967 biography of Warren that "the Governor said . . . it was to be Chief Justice or nothing."[75] Warren said in his memoirs that this story was "positively not the fact."[76] Brownell also said that the story was "not accurate," that Eisenhower merely "felt that Warren might have some reluctance to move into the Chief Justiceship without having any previous judicial experience."[77] In the process of writing his memoirs Warren became annoyed with Small for having circulated the story, and complained, through Warren Olney, about Small's account being included as part of the collection of documents in the Earl Warren Oral History Project.[78] Small eventually conceded that he could not remember "whether Warren ever actually said to me that he insisted with Brownell upon being Chief Justice."[79] On balance, Brownell's account of the conversation between himself and Warren seems authentic.

Other, far wilder rumors have been raised: that William Knowland and Brownell discussed the possibility of appointing Warren on the "campaign train" in 1952 (Brownell said that he was not even on the train);[80] that Knowland had effectively pressured the White House for Warren's appointment in exchange for Warren's supporting the Eisenhower Administration's policy of a Korean truce (Brownell called that rumor "perfectly cockeyed");[81] and that Nixon was eager to have Warren appointed to the Court in order to remove him from California politics. Brownell doubted that Nixon or "any of the California people" had "serious discussions with [Eisenhower] about [the Warren nomination] . . . before I went to California."[82]

Warren's nomination to the Court, in sum, was a fortuitous and relatively prosaic process. For reasons principally having to do with Warren's support for the Republican ticket at the late stages of the campaign, but also having to do with Eisenhower's favorable impressions of Warren's judgment, Eisenhower resolved to give Warren a federal appointment, and the Supreme Court came to mind, probably because Warren was known to be interested in it. The offer of the solicitor generalship to Warren was conceived of as a "grooming" gesture: Both parties regarded it as such. Warren's inclination to accept the offer illustrates how seriously he took Eisenhower's pledge to name him to the "first vacancy" and how eager he was to sit on the Court. He had not previously expressed interest in leaving California for any other purpose.

When Vinson suddenly died, Eisenhower regarded his options, despite the pledge, as entirely open, but he apparently made no clear distinction between the chief justiceship and other justiceships, except, as we have seen, to wonder a little more about Warren's lack of recent experience as a lawyer. Eisenhower's central concerns seem to have been whether Warren would find the chief justiceship overwhelming and whether he continued to support Eisenhower's policies as he had during the campaign. This last concern might seem remarkable for a president appointing a Chief Justice, since partisan politics is ostensibly no business of the Supreme Court. It has been, however, a repeated concern of presidents when making nominations: At the press conference announcing Warren's appointment, Eisenhower spoke of Warren's "middle-of-the-road philosophy." [83]

Warren came on the Court because he had been a national Republican politician. That fact lent a double irony to his appointment. The first irony was that Warren's visibility as a Republican had come from his identification with the liberal wing of the Republican party. Yet Brownell, in reviewing Warren's candidacy, had seen him as "a very tough prosecutor" without "any record . . . in anti-trust, and really not much in civil rights." Eisenhower's surmise that Warren was "a middle of the roader in the Republican party in the same sense that Eisenhower was," Brownell felt, was accurate. [84] Even though Warren had written Brownell in November, 1945, arguing that the Republicans "ought to have a definite program on social security, a program for improving the health of our people . . . and an anti-monopoly program," [85] neither Brownell nor Eisenhower sensed his instinctive Progressivism. Warren was appointed as something he was not: an Eisenhower Republican.

The second irony was that Warren was named to the chief justiceship. Warren's experience and temperament were far more suited to the chief justiceship than to an associate justiceship. Warren had learned as an executive in California to lead, to manage, to set a tone, and to get results. He was asked to summon up those skills as Chief Justice; as an associate justice his position would have been far less managerial and would have afforded him far less opportunity to display his strengths as a "personal contact administrator." Yet Eisenhower and Brownell were given pause at appointing so "inexperienced" a person Chief Justice, and Warren, for his part, would have taken any seat on the Court. As we shall see in subsequent chapters, the fact that Warren was the Chief Justice rather than an associate justice of the Warren Court was to add considerably to his impact on his fellow justices.

✍

With Warren's appointment to the Court his formal governing ended and his informal governing began; he left professional politics and took up his own style of judicial politics. With Warren the Chief Justice in mind, a last look at Warren the governor seems appropriate.

Warren had entered and left the governorship a patriot, an advocate of civil defense and preparedness, a vigorous opponent of communism, and a proponent of loyalty oaths and other internal security measures. These positions were related to his provincial, nativistic conception of California society, a conception that defined California as a refuge from alien persons and ideas. But alongside that conception, late in his gubernatorial years, Warren had developed a growing interest in international affairs, especially insofar as they could be regulated by law. This interest was ultimately to make him a strong advocate of the "world peace through law" movement. Law, he came to feel, could transcend provincial fears and animosities.

Warren had campaigned against Culbert Olson as an opponent of the New Deal, an advocate of states' rights, an apostle of law enforcement, and an architect of the Japanese relocation program. By 1953 he had proposed state-supported health insurance, endorsed social security, and attempted to stake out a political position to the left of center on the continuum of postwar Republicanism. He had also shown no signs of modifying his views on Japanese relocation and had proposed a loyalty oath for state employees. He had presided over the partial reform of California's prisons and mental hospitals and had opposed oil and trucking interests in his campaign for public highways; he had made mild overtures toward civil rights and he had defined "liberalism" as incorporating a concern for civil liberties. He had been a Progressive in his stance; his stance was neither fully that of a postwar liberal nor that of a postwar conservative.

Of the major themes that were to characterize Warren's years on the Court—opposition to racial segregation, support for the reapportionment of state legislatures, opposition to wide-ranging legislative investigations in the name of national security, the abolition of prayers in the classroom, strong support for organized labor and for the economic regulatory powers of the federal government, and increased legal protection for criminals and persons suspected of committing crimes—only a handful surfaced in Warren's governorship, and he was not prominently identified with any.

Warren proposed a Fair Employment Practices Commission to the California legislature in 1948 and again in 1951 but was not vocal on race relations issues or civil rights for racial or ethnic minorities. As late as 1948 he had defended the then-existing apportionment of the California state senate, which gave each county one representative regardless of its population. He had defended state loyalty oaths. He had opposed, for the most part, intervention by the federal government in economic affairs, although he had supported state intervention in such areas as health and education. He had lobbied for prison reform but had continually supported law enforcement. He had had no opportunities to consider church-state issues. He had equivocated on issues involving organized labor.

These positions, taken as a whole, may suggest that Warren's eventual posture on the Court represented a dramatic shift in his attitudes. But to judge Warren by his rhetorical stance on issues was treacherous. He had presided over California during dramatic changes in its composition, and he had come to see that in a crowded, heterogeneous, complex environment affirmative governmental action often seemed necessary. Despite his Republican affiliation, his sympathy for persons oppressed by special interests ran deep: He was a Progressive first and a Republican second. His latent concern for disadvantaged persons and his established animosity to special privilege were to become, during Warren's tenure on the Court, components of a conception of American life as a struggle by disadvantaged individuals against entrenched forces—an impersonal government, powerful interest groups, economic conglomerates, guardians of racial prerogatives.

Moreover, Warren's interest in structuring his offices to take decisive action to "make progress" and to establish their power had been a greater concern for him as a California public official than the rhetorical basis of that activism. Only in his political rhetoric was Earl Warren a "middle-of-the-roader." Beneath that rhetoric he was passionately interested in using the powers of his office to solve problems that he thought were deserving of solution. When he came to the Court he confronted a different set of problems, but his penchant for activism and his moral passion continued.

# TWO

## *THE CHIEF JUSTICE: PRESENCE*

# 6

# The Crucible of
# Brown v. Board of Education

Of all the journeys made by Earl Warren during his public career, that from the governorship of California to the Supreme Court of the United States was the most wrenching. There was, first of all, Warren's realization that he was about to occupy a position that must have seemed, for most of his working life, unfathomably remote and munificent. A Bakersfield iceman's delivery boy, Ezra Decoto's assistant, Culbert Olson's frustrated attorney general, and Thomas Dewey's spear carrier was becoming Chief Justice of the United States.

There was, in addition, the stark contrast between the office Warren was relinquishing and the office he was about to occupy. The Supreme Court sat in Washington, a city with which he was not familiar; its members engaged in tasks that he had apparently not performed before; the Court personified a professional world in which he had not immersed himself for ten years and with whose uppermost echelons he was barely familiar. On becoming attorney general and governor he had brought along his own staff to smooth the transition: In October, 1953 he was alone, not even, after his induction, accompanied by Nina, who returned to California to supervise the move east.

Warren remembered the experience of becoming Chief Justice in his memoirs:

I approached the high office with a reverential regard and with a profound recognition of my unpreparedness to assume its obligations in such an abrupt manner.

It was completely different from the other offices I had held. . . . I was not acquainted with Washington or even with members of the Supreme Court. . . . Since becoming governor, I had not been engaged in handling legal matters. . . . As attorney general, my work was largely administrative. . . . Most of my practice at all times had been in the state courts. My experience with federal courts had been very limited. With the independent regulating agencies . . . it was almost nil. All of this lack of experience weighed heavily on my mind. . . .

[In my office were] Mrs. Margaret McHugh, the executive secretary, . . . three law clerks, . . . [and] two elderly messengers. . . . That was my entire personal staff. I told [these persons] they were welcome to remain with me, and they all consented to do so, aiding in the transition for me. But it was a painful transition.[1]

The characteristic caution with which Warren approached new experiences surfaced as he assumed the office of Chief Justice. He asked Hugo Black, the senior associate justice, to "manage a few of the conferences until I could familiarize myself with proceedings."[2] He observed as the rest of the Court disposed of certiorari petitions that had been filed over the summer, taking no part in the deliberations. He assigned himself a case "of no notoriety"[3] for his first opinion, interpreting the Federal Longshoremen's and Harbor Workers' Compensation Act for the benefit of workers whose job-incurred injuries had not been adequately reported.[4] He was "perfectly miserable"[5] for the first two months, while Nina was still in California, and he responded to his sense of dislocation by remaining relatively passive as Chief Justice.

Within a few months, however, Warren's presence as Chief Justice emerged rapidly and decisively. This development was striking, because Warren was not able, as Chief Justice, to make the typical changes in an office he had made as a California public official. While he retained a keen interest in such administrative matters as the Court's docket, the schedule of arguments and conferences, and the internal workings of the Court's permanent staff, he could not replace older, incompatible justices with new ones more aware of his executive style, as he had done with personnel in California. He could not cultivate the press; judges did not hold press conferences.

He had no powers of the purse, being beholden to Congress for appropriations, no ability to expand his personal staff, few perquisites of office.

Warren's position as Chief Justice gave him no formal powers that his associates lacked, only informal opportunities to exercise leadership. He had no experience in being a judge; he had given little attention to the principal work of the Court, deciding complicated issues of constitutional law. He had no reputation as a legal scholar; he was expected to be a conciliatory, "middle-of-the-road" chief, overshadowed by such influential associates as Hugo Black, William O. Douglas, Felix Frankfurter, and Robert Jackson.

Despite these obstacles, and the relatively low expectations of performance that accompanied Warren's appointment, he soon became a formidable presence on the Court. The most important feature of Earl Warren's chief justiceship, in fact, was his presence. By the time of his retirement Warren was ranked with John Marshall, Roger Taney, and Charles Evans Hughes as the most influential Chief Justices in the history of the Supreme Court. This ranking was all the more surprising because, unlike those other occupants of the chief justiceship, Warren was not regarded as a judge possessing considerable intellectual talents or conspicuous analytical abilities. He was regarded as one of the great Chief Justices in American history because of the intangible but undeniable impact of his presence on the Court.

The episode that enabled Warren to establish his presence on the Supreme Court was the decision in the five segregation cases, which were handed down under the name of the Kansas case, *Brown v. Board of Education*,[6] on May 17, 1954. The segregation cases, which had been set down for reargument in June, 1953, were reargued in December of that year, after Warren had become Chief Justice, and voted upon in March, 1954. Warren's opinion for the Court was approved by all the justices on May 15. The story of Warren's role in the *Brown* case is now a familiar one;[7] my interest here is in examining *Brown* as a means by which Warren established his personal imprint on his new office and as a formative experience in his career as a judge.

Warren had been neither an outspoken supporter nor a vocal opponent of segregation during his California career. As noted, he had been a member of the Native Sons of the Golden West, who were explicitly anti-Oriental, and he had figured prominently in the relocation and internment of Japanese Americans in 1942, basing his arguments for relocation partially on racist assumptions. A black state legislator remembered Warren during his years in the Alameda County District Attorney's Office as being conscious of color differences and felt that he "never had a broad association with Negroes."[8] Earl Warren, Jr., believed that in the Bakersfield of his father's youth whites regarded blacks as "definitely something below second class citizens";[9] Warren remembered, perhaps conveniently, that "I sat along side of black children in the public schools . . . and I never thought a thing about it."[10]

In 1938, the year that Warren made a strong public statement on civil rights in his letter to Robert Kenny during his campaign for attorney general, he also attempted to persuade Walter Gordon, a black, to join the attorney general's staff. By 1945 he had appointed Gordon head of the California Adult Authority and, as noted, had unsuccessfully proposed a Fair Employment Practices Commission to the state legislature. One black supporter of Warren, state legislator William Byron Rumford, said that as governor, Warren had "encouraged me to press for [antidiscrimination] legislation, and . . . assured me that if it would get to his desk he would sign it."[11] At the time he came to consider the Brown decision in 1953, Warren's views on race relations seem to have been relatively undeveloped, as were his views on many other issues he would be facing as a judge.

The Court, meanwhile, was deeply split on the Brown case, as revealed by its inner history in the 1952 term. Four justices—Black, Harold Burton, Douglas, and Sherman Minton—had in that term declared themselves personally opposed to racial segregation and in favor of overruling Plessy v. Ferguson,[12] the Court's 1896 precedent maintaining that "separate but equal" racially segregated public facilities did not violate the Fourteenth Amendment's Equal Protection Clause. Three justices—Tom Clark, Stanley Reed, and Chief Justice Fred Vinson—favored retaining Plessy, with varying degrees of enthusiasm, and two justices—Frankfurter and Jackson—were hard-pressed to find an adequate rationale for overruling Plessy, although personally unsympathetic to enforced racial segregation.

In response to repeated prodding by Frankfurter, the Court finally resolved to put the Brown case over for rehearing in the 1953 term. Frankfurter commissioned Alexander Bickel, at the time one of his law clerks, to draft a series of questions for reargument. The questions were largely intended as stalling devices. They gave comfort to both sides and raised issues, such as whether the history of the Fourteenth Amendment revealed that it was intended to embrace racial segregation in the public schools, that Frankfurter suspected were inconclusive. The major purpose of the reargument was to give the justices more time to congeal their positions on the segregation cases. Frankfurter and Jackson, especially, were fearful that an opinion that invalidated segregation but did so in a strident or unreasoned fashion would do greater harm than the continuance of the practice.

Those supporting reargument had not anticipated, of course, that Chief Justice Vinson would die of a heart attack in September, and that the man who replaced him would take a different position toward the segregation cases. From the beginning Warren saw *Brown* as a comparatively simple case. He felt that Stone and Vinson Court precedents had crippled the separate but equal doctrine. He believed that "separate but equal" systems rarely resulted in comparable educational facilities for whites and blacks and thought that such a showing would be comparatively easy to make. He also believed, despite his ambivalent experience with racism in California, that the injustice of an enforced separation of human beings based on their color was apparent. Unlike Vinson, who worried about Congress's reluctance to change segregated practices, the entrenched practices of segregation in the South, and the longstanding existence of the *Plessy* principle, Warren was mainly concerned with the problem of how a decision to eradicate segregation in the public schools could be effectively implemented.

Warren's forging of a unanimous majority for the *Brown* decision established his presence on the Court. The task was one especially suited to his skills: It involved convincing others of the necessity for an arm of government to act decisively and affirmatively where a moral issue was at stake. The eradication of segregation, in his mind, was comparable to the establishment of compulsory health insurance. Both were responses to an injustice; both sought to prevent humans from being disadvantaged through no fault of their own. A difference between the two responses was that Warren the judge did not need to rely upon another branch of government to make the response for him, as Warren the governor had had to rely upon the

California legislature. But Warren the judge faced two problems of comparable difficulty. He needed to convince persons affected by the Court's response that they should accept it, and he needed to enlist support for his position on the segregation cases within the Court itself.

In the *Brown* case the fortuity of Warren's being appointed to the chief justiceship rather than to an associate justiceship first assumed significance. The protocol of the Court is based on seniority: The Court's most junior member states his views last in conference debate and has no power to assign opinions for authorship. But the Chief Justice is treated differently: He is ranked first in protocol and in formal privileges notwithstanding his seniority. Thus new associate justices tend to receive insignificant opinions their first term, but not necessarily new Chief Justices, since the Chief Justice assigns opinions when he is in the majority. It would have been inconceivable for Warren to have written the opinion in *Brown v. Board of Education* if he had been an associate justice. Moreover, associate justices have no control over the internal management of cases: when they are argued, discussed in conference, and so on. That administrative task is reserved for the chief. Warren could not have carried out his strategy for bringing about a unanimous decision on the *Brown* case had he been an associate justice.

In addition, despite the convention that the Chief Justice has only one vote among nine justices and thus has no more power than any other justice, a Chief Justice has opportunities to exercise leadership not possessed by associate justices. The Chief Justice not only schedules the timing and determines the agenda for conference discussions, but he also controls the flow of discussion and has the conceivable option, in every case, of assigning the opinion. The Chief Justice also briefly states the facts of the cases to be discussed, and has an opportunity to advance his views on the merits of a case before anyone else has had a chance to speak. A person, such as Warren, who was accustomed to chairing meetings, managing agendas, and assigning office tasks, might find that the chief justiceship gave him ample opportunities to exercise power and to make his presence felt. If some of the tasks of a Chief Justice can be likened to those of a chairman of a small group called on regularly to make decisions, those were tasks that Warren had performed for much of his public life.

During the oral reargument of *Brown*, which took place in October, 1953, Warren remained largely silent, in keeping with the low profile of his first months as chief. He then delayed putting *Brown* on the conference agenda for two months, and by December 12, 1953, when the first Court conference on *Brown* was held, he was ready to declare his views. He began by stating that in his judgment one could not sustain *Plessy* unless one granted the premise that blacks were inferior to whites. He did not grant that premise, and consequently he was prepared to invalidate segregation and to insure equal treatment of all children in the public schools. But while he had no doubts about the principle of *Brown*, he had not resolved how it was to be implemented, and he suggested that the Court take some time and care in the framing of a decree, being sensitive, especially, to conditions in the Deep South. He suggested that no vote on the segregation cases be taken that day, but that the case be "talk[ed] over, from week to week . . . in groups, over lunches, in conferences."[13]

With this statement Warren communicated three messages to his colleagues. The first was that if there had been any doubt as to whether the Court would invalidate *Plessy*, that doubt was foreclosed. At least five justices would so vote, Warren being the fifth. The second was that Warren viewed the segregation cases as separable into two components: the framing of an opinion overruling or emasculating *Plessy* and the framing of a decree implementing the Court's decision. Warren's preliminary strategy on the segregation cases was, according to Justice Burton, to "direct discussion towards the decree—as probably the best chance of unanimity."[14] In this strategy he had the enthusiastic support of Felix Frankfurter, who had resolved to work behind the scenes for a decision invalidating segregation and may have suggested the strategy to Warren.

Warren's remarks communicated one other message. The message was that those on the Court who remained prepared to defend the "separate but equal" doctrine would have to confront Warren's assertion that segregation and the idea that blacks were inferior to whites were intimately linked. Without labeling defenders of the *Plessy* decision white supremacists, he conveyed that association. This was a familiar Warren technique, the argument to induce shame. Opponents of Warren in California had repeatedly had their positions labeled immoral, unethical, or unjust by Warren, and found that they were forced to defend themselves against such alleged polarizations of their views. Warren continued this practice in his oral questioning of counsel in arguments before the Supreme Court. In cases, for example, where the police from a given state had allegedly intimidated a person suspected of committing a crime, Warren would occasionally ask the lawyer arguing the case for the state why he "had treated [the suspect] this way." [15] The lawyer, of course, had not participated in the alleged intimidation and perhaps had not even met the state law enforcement officers whose conduct was being scrutinized. Warren, however, identified him with the practices and asked him to justify them.

Within a short time after Warren's remarks in conference, two additional justices on the Court revealed themselves as now prepared to support a majority opinion invalidating segregation. Clark, whom others in the 1952 term had thought to oppose a reversal of *Plessy*, [16] identified himself in the December conference as prepared to declare segregation unconstitutional, although he remained concerned about the mechanics of implementation. Frankfurter, who a year earlier had suggested that nothing in the Fourteenth Amendment prevented racial segregation in the states, now made an ambiguous statement in conference that stopped short of committing himself to any majority opinion but indicated that he favored overruling *Plessy*, and followed his remarks with a memorandum to the justices supporting Warren's position. [17]

Warren knew, before these events, that Frankfurter had resolved to work behind the scenes for unanimity; Warren was a willing partner in Frankfurter's efforts to conceal his true position from other justices on the Court. Apparently Frankfurter felt that he was supporting a result in the segregation cases that was inconsistent with his theories about judicial usurpation of legislative prerogatives and did not want that inconsistency to surface within the Court.

Warren next let the segregation cases simmer among his colleagues. On January 15, 1954, the day that Frankfurter circulated his memorandum on the cases, Warren scheduled a luncheon at which the cases were discussed. Throughout January and February the justices continued to discuss the case informally, all save Frankfurter and Jackson meeting regularly for lunch. In late February or March [18] a formal vote was taken. Eight justices voted to invalidate segregation; Reed voted to uphold it. Of the eight, Jackson had indicated that he would probably write a concurrence and had, as early as February 15, drafted a memorandum that was to be its basis. Warren assigned the majority opinion to himself and began to work on it in early April.

The Court had by now agreed on the strategy of separating its response to the cases into an opinion, which would invalidate segregation on principle, and a decree, which would implement the decision. The justices had also agreed to delay formulation of the decree another year, so that affected states could be invited to participate in its framing the following term. The hand of Frankfurter was visible here: His January memorandum had contained the phrase "all deliberate speed," which was later to be pivotal in the decree, and the Court had endorsed his and Warren's belief that implementation should be gradual and mindful of local conditions.[19] Warren had only two hurdles left—the achievement of unanimity, including the suppression of concurrences, and the production of an opinion that was, as he put it, "non-rhetorical, unemotional, and, above all, non-accusatory."[20]

In securing the first of these goals Warren was aided, it appears, by Jackson's heart attack on March 30, which left Jackson hospitalized and was to lead to his death. Warren produced an opinion on May 7, which shortly secured the consent of all the justices except Jackson and Reed. On May 10 Warren visited Jackson in the hospital and delivered his draft opinion. Jackson, whose condition had prevented him from working on any cases and who was therefore disinclined to develop his February memorandum into a concurrence, resolved to join the majority, making only minor suggestions. While Warren's opinion satisfied Jackson in its moderate tone and its absence of pretense, Jackson might well have written separately had he been in full health.

Meanwhile Warren had a conversation with Reed between the seventh and the twelfth of May.[21] One of Reed's law clerks, who was present at the conversation, reported that Warren said, "Stan, you're all by yourself in this now. You've got to decide whether it's really the best thing for the country."[22] The issues for Reed, Warren suggested, were issues of conscience and of the effect of a southern justice's dissent on a matter so pivotal to the South. Reed eventually capitulated. Formal unanimity was secured on May 15, with Warren making another visit to Jackson on that day.[23] The decision in the segregation cases was announced on May 17, with Jackson leaving the hospital to appear in the courtroom with the rest of the justices.

In two letters to Judge Learned Hand written after the *Brown* decision, Felix Frankfurter suggested that his covert partnership with Warren was the basis of forging unanimity in the segregation cases. The first letter, dated July 21, 1954, stated that "unanimity in the segregation cases . . . could not possibly have come to pass with Vinson" and that "the heart of the business was the wise and skillful ways by which the cases . . . were dealt with."[24] That letter was written at the time of Frankfurter's "honeymoon" with Warren, which will be discussed in a subsequent chapter.

In the second letter, written on October 12, 1957, after the honeymoon had ended, Frankfurter referred to "a lot of hogwash" that "has been written and uttered on [Warren's] share in the outcome [in *Brown*]—that he 'done' it, either by cracking the whip or by his winsomeness or both."[25] In this and other correspondence Frankfurter implied that while he could not talk about the internal details of the segregation cases, Warren's role was decisive only because of his amenability to the strategy suggested by Frankfurter, and that "if the 'great libertarians' had had their way we would have been in the soup."[26]

Frankfurter was correct in his judgment that unanimity in the segregation cases could not have occurred without Warren and also in his judgment that the decision was a political compromise, trading off the eradication of segregation on a constitutional basis for the implementation of the change "with all deliberate speed." But Frankfurter seriously overestimated his own role in forging unanimity. The plan of separating the opinion from the decree may have been Frankfurter's; the decision to reargue the case and avoid the deadlock forming on the Vinson Court was prompted by Frankfurter.

But the step-by-step process of converting recalcitrant justices to the majority position was formulated and carried out by Warren. Warren made the decision not to take a formal vote but to meet in informal lunches and conversations; Warren produced a draft opinion that won the consent of his colleagues without major alterations; Warren convinced Jackson and Reed of the importance of a unanimous opinion; Warren took special pains to conceal the decision from all but the participating justices until the moment it was announced.[27] No justice on the Warren Court that had witnessed the new Chief Justice's handling of the segregation cases could fail to sense his presence. James Reston, in a March, 1954 column, reported that other justices had found in the new chief "a sensible, friendly manner," a "self-command and natural dignity," a "capacity to do his homework," and "an ability to concentrate on the concrete."[28]

⚞

The *Brown* case, however, was more than an episode that conveyed to other justices the fact that Earl Warren was a person to be reckoned with. It was also a catalyst in helping Warren crystallize his thinking about the new office he held. There were two dimensions to *Brown*, and the case was susceptible to two different jurisprudential interpretations. One dimension of *Brown* was its short-range politics. Viewed from this perspective, it was a moderate, compromising decision, delaying action on its implementation, confining its impact to the public schools, avoiding emotional language or the stigmatization of segregationists, inviting representatives of the states to participate in the formulation of a forthcoming decree. Frankfurter saw the case in those terms. He was subsequently to break from the Court's ironclad unanimity on race cases by writing a concurring opinion in the 1958 case of *Cooper v. Aaron*[29] and to express doubts that *Brown* outlawed state laws forbidding interracial marriages.[30] In Frankfurter's view the decision in the segregation cases was a subtle and enlightened exercise in politics, made possible by his own ingenuity and the presence of a flexible and skillful politician as Chief Justice.

Another dimension of *Brown*, however, was its meaning as a philosophical statement. This dimension became apparent in the Court's treatment of racial segregation cases after *Brown*, in which segregation in other public facilities was summarily declared unconstitutional by unanimous per curiam opinions that cited *Brown*. As a philosophical statement—that racial segregation in public facilities was inherently unfair and unjust, whatever the context—*Brown* became an example of the use of the Constitution by the Supreme Court to compel action by other branches of government. The states and Congress (which had permitted segregation in the District of Columbia) were told by the Warren Court in *Brown* and its progeny that many of their existing practices were illegal, that they had to change those practices, and that if they did not, the Court would support those persons—whom everyone understood to be hitherto disadvantaged blacks—who pressed for change. Seen from this perspective,

*Brown* ushered in a new role for the Supreme Court in the twentieth century, that of an active enforcer of fairness and justice as embodied in the Constitution.

The feature of *Brown* that most clearly identified the emergence of the Court in this new role was the absence of conventional constitutional analysis in the *Brown* opinion. Warren's opinion for the Court invalidated segregation as a violation of the Equal Protection Clause not through any analysis of the historical meaning of that clause (the history was dismissed as inconclusive) or through a close analysis of precedents in the race relations area (they were treated summarily). Warren's opinion invalidated segregation on the basis of two findings. First, he found that "today" education was "perhaps the most important function of state and local governments," and that the Equal Protection Clause consequently required that "the opportunity of an education . . . be made available to all on equal terms."[31]

Second, he found that "[racially] separate educational facilities [were] inherently unequal," because "segregation of white and colored children in public schools has a detrimental effect on the colored children," being "interpreted as denoting the inferiority of the negro group."[32] The basis for Warren's first finding was his own conviction of the importance of education; the basis for his second finding was social science literature of the 1940s and fifties on racial prejudice and its effects. The opinion in *Brown*, in short, declared racial segregation unconstitutional through appeal to contemporary social perceptions rather than to constitutional doctrine.

The ideal behind *Brown*—the intuitive justice of equality of opportunity—was an ideal not explicitly codified in the Constitution. The force of the case was in the decisive, almost summary fashion with which Warren's opinion announced the Court's dedication to that ideal in the race relations area. When coupled with the per curiams after *Brown*, the decision conveyed a jurisprudential message: The Warren Court, at least in race relations cases, was going to fuse constitutional interpretation with a search for justice, reading constitutional language in a way that could make the Constitution harmonize with current perceptions of what justice required. And if the Court's readings compelled a change in the practices of other branches of government, the Court was not going to avoid insisting on that change merely because those branches were purportedly more democratic in composition than it.

In *Brown*, then, can be seen the seeds of an activist support for "liberal" policies, such as equality of opportunity, that was to constitute an important theme of the Warren Court. The *Brown* decision was a classic manifestation of mid-twentieth-century liberal theory in its effort, through the affirmation of principles such as equality of opportunity, to fuse the idea of affirmative, paternalistic governmental action with the idea of protection for civil rights and civil liberties. *Brown* also marked the origins of Earl Warren's stance as a liberal judge, a stance that will be subsequently explored. But my concern at this juncture is not with the nature of Warren's judicial liberalism. It is rather with *Brown* as a crucible in the development of Warren's activist role as a judge.

The two dimensions of *Brown* had suggested two possible roles for Warren the Chief Justice. One was as temporizer and compromiser, avoiding open confrontations on politically sensitive issues, burying internal differences beneath a hard-won statement of unanimity. Another was as activist promoter of the cause of justice, particularly justice for those who had been denied equal opportunities in American life. Both roles were consistent with Warren's California experience. He had been measured in his rhetoric and careful in the timing of his actions as a California public official; he had chosen, for the most part, not to polarize issues and to justify his decisions on the basis of propositions that had widespread public support. On the other hand he had been continually interested in having his offices take decisive action on issues, and he had not been deterred by the reluctance of other governmental institutions to be comparably activist. Warren had functioned, in California, as both the artfully "moderate" politician and the determinedly activist officeholder.

As Chief Justice of the United States, Warren was eventually to discard the role of temporizing politician for the role of activist judge. The subsequent course of his tenure revealed *Brown* to have been the seedbed of his activism. The momentum of the Court's mission in *Brown*, which came to be seen by Warren as a mission to vindicate ethical principles that were embodied in the Constitution, even if this meant dramatically expanding the Court's jurisdictional reach and power, eventually grew to the point where *Brown* now appears as the case where the Warren Court's ultimate character was first revealed.

But Warren did not immediately perceive the *Brown* decision in such terms. He remained cautious in his approach to the chief jus-

ticeship for at least two years after *Brown*. Warren's eventual aban-
donment of that cautious posture and his accompanying commitment
to activism were prompted by his conclusion that caution in a Supreme Court justice was an impractical, unnecessary, and ultimately
indefensible posture. That conclusion was largely precipitated by
Warren's exposure to the personality and jurisprudence of Felix
Frankfurter.

# 7

# Reacting to Felix Frankfurter

Many of Earl Warren's acquaintances did not adequately understand one feature of his temperament. They suspected that since Warren was gregarious, pleasant, modest, and cautious in his judgments, he was malleable or lacking in conviction. When such acquaintances were themselves eager to impose their views on others and capable of hasty and imprecise judgments about people, a serious misperception of Warren sometimes emerged. When Warren became aware that this misperception had taken the form of another's patronizing him or seeking to deceive him, he swiftly listed that person as one of his antagonists. And if the person had come to play a relatively important role in Warren's public career, Warren was even capable of framing his career goals in opposition to what he perceived to be the goals of that person.

We have seen this response surface in Warren in his relations with Culbert Olson; the response surfaced again in his relations with Felix Frankfurter. The relationship between Warren and Frankfurter was founded on a misconception, grew close for a time because neither realized the nature of the misconception, suffered as each began to engage in further reflections on the other, and eventually became antagonistic, sometimes openly so. The relationship was to prove decisive for Warren's career on the Supreme Court. Warren's eventual perception of Frankfurter as an untrustworthy antagonist who was "pressuring" him stimulated Warren to develop a theory of judging in stark opposition to Frankfurter's. Frankfurter's presence was, of course, not the sole basis for Warren's emergence as an activist judge. But Frankfurter's presence gave Warren a powerful impetus to assume that role.

᾿

By the time Warren was named Chief Justice, Frankfurter had settled into a position on the Supreme Court that, given his intellectual talents and experience, was one of ironic vulnerability. Frankfurter's career, prior to being named to the Court by President Franklin Roosevelt in 1939, had been one of conspicuous achievement, particularly in two areas. One area was as a link between academia and government: Frankfurter was the quintessential intellectual adviser to public servants. He had been an associate of Henry Stimson, United States Attorney in New York, and of Newton Baker, secretary of war in the Wilson Administration, a disciple of Oliver Wendell Holmes, an adviser and sometimes a paid agent of Louis Brandeis, and an intimate of Roosevelt.[1] He provided these politicians and public servants with a range of academic advice. Simultaneously, he made the academic world in which he functioned more aware of and accessible to politicians.

Frankfurter conceived of his function at Harvard Law School, where he taught from 1914 to 1939, as "bring[ing] public life, the elements of reality, in touch with the university, and, conversely, to help harness the law school to the needs of the fight outside."[2] His appointment to Harvard was to symbolize, he said at another time, "the part to be played by law schools and the law in the solutions of pressing public problems."[3] In the course of pursuing that function Frankfurter wrote books about the Sacco-Vanzetti case and the Supreme Court's use of the Commerce Clause, urged the appointment to the Court of persons, such as Learned Hand, whom he respected, and consulted with Roosevelt about the latter's 1937 plan to "pack" the Court. When he was appointed to the Court he felt, according to a recent biographer, "fully prepared for and capable of handling" the office of associate justice.[4]

The other area in which Frankfurter appeared to be conspicuously successful was what might be called personnel management. So bureaucratic a phrase for describing this quality of Frankfurter's would not have appealed to the possessor, who coined the word "personalia" to describe his ability to "guid[e] the clash of personalities."[5] He established networks of friends and acquaintances, recommended favorite students for clerkships with Holmes and Brandeis or for positions in the New Deal, and was active in faculty politics at Harvard. He prided himself on his awareness of the subtleties of personal relations, his wide circle of contacts in public life, and the success of his protégés. His theory of personnel management was hierarchical: He supported the "best" students, gravitated towards persons in positions of power, was impatient with mediocrity, and was captivated by promise. At the same time, he continually flattered and cultivated persons whose positions he coveted or respected: His published letters to Franklin Roosevelt were characterized by the editor of the volume as "sometimes . . . excessive and repugnant"[6] in their obsequiousness.

The position of associate justice on the Supreme Court seemed to be one ideally suited for a man of Frankfurter's specific talents. He was appointed to a Court whose jurisprudential character was apparently changing, moving it closer to positions Frankfurter had supported: Observers expected the Court of the 1940s to be sympathetic to the New Deal's efforts to increase federal power and to experiment with economic regulation, efforts that Frankfurter was known to favor. Indeed, Roosevelt regarded Frankfurter's appointment as providing him with a justice who could supply an intellectual basis for New Deal policies.[7] In addition, Frankfurter's talent for "personalia" seemed ideally suited for the "clash of personalities" common to decision making on the Court. In sum, Frankfurter's combination of intellectual ability and political experience made him appear to be the most promising of all Roosevelt's Court appointments; he was expected, by himself as well as others, to be the dominant justice on the Court.

By 1953, when Warren replaced Fred Vinson as Chief Justice, it was clear that Frankfurter's career on the Court had not fulfilled these expectations. Rather than becoming the Court's intellectual leader, he had painted himself into a jurisprudential corner, having staked out positions that made it difficult for him to respond to new issues with flexibility and open-mindedness. In addition, he had not emerged as a dominant personal force on the Court: His relationships with several other justices were charged with suspicion and rancor. "After . . . his first five years on the Court," Professor H.

N. Hirsch has written, "Frankfurter would mentally divide his colleagues into three categories—adversaries, allies, and potential allies."[8]

At the heart of the problem may have lain the fact that Frankfurter never fully transferred his allegiance from politics and academia to the Court. While he talked repeatedly about the obligations thrust upon him by his judicial role, and ostensibly divorced himself from both the academic and the political communities, he continued to try to function on the Court as he had in his pre-Court career— as an adviser and an academic, engaging in behind-the-scenes manipulations of his colleagues and keeping in close contact with his friends and disciples in the academic and political worlds. On the Supreme Court, however, he was no longer an academic or a political adviser: He was a judge, and the Court was a particularly inappropriate place for him to operate in his former ways. The results of this imperfect transformation were disastrous in terms of Frankfurter's credibility with his fellow justices; they also tended to undermine his jurisprudential posture.

By 1953, at least,[9] Frankfurter had identified himself with a restrictive version of judicial restraint, a theory that stressed the adverse consequences in a democracy of judges substituting their own judgments on constitutional issues for the judgments of more democratic or representative branches of government. This theory pictured Frankfurter as bound, except in extreme instances, by the policy judgments of others; he could make such judgments only in very limited situations. But while he was professing this constrained theory of judging, he was acting like a politician and policymaker within the Court: He repeatedly sought to cajole, flatter, or browbeat other justices into supporting the policy outcome he favored in a given case.

The gap between Frankfurter's public articulation of his role on the Court and his private behavior caused other justices to find him untrustworthy, tedious, or hypocritical. Frankfurter's acknowledged strengths in personal relations—his charm, his talent for flattery, his vivacity, his resonance—were diminished by the apparent purposiveness of his behavior. His talent for "personalia" was also diminished by the fact that his theory of personal relations was hierarchical. This feature of Frankfurter's public relations had emerged at Harvard Law School, where Frankfurter functioned as a link between powerful persons outside the law school and aspiring persons within it. In his role as a sponsor and an adviser he could both demand "promise" and loyalty from those whom he chose to sponsor and offer service and praise to those whom he sought to advise. While others perceived, not always charitably, that Frankfurter tended to divide the world into disciples and mentors, this characteristic was at least consistent with the principal role he had played at Harvard.[10]

On the Court Frankfurter's hierarchical theory of personal relations interacted in an unfortunate fashion with his tendency to see other justices as his allies or his antagonists. Having staked out a conspicuous jurisprudential position—the arch defender of self-restraint—Frankfurter then viewed other justices as either supporters of that theory or opponents of it. At the same time he tended to make judgments of the intellectual worth of his fellow justices that resembled the judgments he had made of students at Harvard: Some justices were "promising," others "mediocre." The perceived mediocrity of another justice's mind, for Frankfurter, made that justice a prime candidate to be won over to Frankfurterian jurisprudence, but at the same time it diminished Frankfurter's excitement at the prospect of having such a person as an ally. A "promising" justice, on the other hand, was also a potential ally, but if he turned out to resist Frankfurter's efforts he quickly became an antagonist.

Frankfurter consequently lectured, patronized, or ridiculed some of his fellow justices and engaged in open and frequent combat with others. Only rarely did he have an ally whose intellect he also respected—Robert Jackson and John Harlan, for instance—and in such cases his tendency to lecture or to patronize was reduced. When he had an antagonist whose intellect he respected—Hugo Black and William O. Douglas serve as examples—Frankfurter was also less expansive and more openly antagonistic.[11] But with respect to a far larger group of justices—"mediocrities" and potential allies—Frankfurter adopted the tone of an academic and political savant. Evidence suggests that with the bulk of those justices, which included Harold Burton, Tom Clark, Sherman Minton, Frank Murphy, Stanley Reed, Owen Roberts, and Fred Vinson, Frankfurter's method was counterproductive.[12]

When a new justice came on the Court, Frankfurter immediately conceived of him as a potential ally and began to attempt to curry favor with him. Warren's nomination had apparently not inspired Frankfurter: He had reportedly expressed concern about Warren's lack of legal experience and his conspicuously political background.[13] But the segregation cases awakened Frankfurter to the fact that Warren was not at all inexperienced in the management of delicate public issues, and Frankfurter began to see Warren not only as a potential ally but also as a "promising" colleague. At the same time, Warren was responsive to Frankfurter's support of his strategy to achieve unanimity in the cases and grateful for the intellectual aid of an experienced and sophisticated jurist.

Ɫ

The two years of the *Brown* opinion and the subsequent decree implementing that opinion[14] were the years of Warren and Frankfurter's brief honeymoon. Four months into the 1953 term Frankfurter was to write his longtime friend Charles C. Burlingham that Warren "has been discharging [the office of Chief Justice] . . . with true regard to its highest standards."[15] That letter, dated February 15, 1954, came after Warren and Frankfurter had struck a bargain in the segregation cases. In April of 1954, as the cases were at their critical stage, Frankfurter wrote Jackson that it was "a pleasure to do business" with Warren.[16] In early May, he wrote Burlingham that he was attempting to raise funds for a bust of Warren in the California State House in Sacramento,[17] and that same month, after the opinion in *Brown* had been handed down, he told Burlingham, in the context of discussing the segregation cases, that "one does not have to be a soothsayer to know that the new Chief Justice is one person and his predecessor was another."[18]

Warren, for his part, seemed to be equally solicitous of Frankfurter. After the 1953 term had ended, he corresponded frequently with Frankfurter over the summer, consulting him on ideas for a commencement address, the timing of the *Brown* decree arguments, and the scope of judicial review under the Due Process Clause. Frankfurter sent him a book of his essays, and Warren wrote that "I have been fascinated by its diversity as well as its profundity." In the same letter Warren spoke of "reading all [Frankfurter's] decisions on 'Due Process,'" which Frankfurter had sent him on Warren's request. "I am," Warren confessed, "still endeavoring to orient myself in that field."[19]

Throughout 1954 and into 1955 Warren sought Frankfurter's advice on a variety of subjects, ranging from the proper disposition of antitrust cases to speeches and addresses Warren had been asked to give. It was characteristic of Warren to seek help (especially when asked to write a speech) from persons toward whom he was favorably disposed; it was characteristic of Frankfurter to bestow advice upon anyone who sought it, especially if he thought well of the solicitor. Frankfurter's letters to Warren in 1954 and 1955 contain a bevy of suggestions and compliments. He instructed Warren that the Court should refrain from "pass[ing] on abstract constitutional questions."[20] He offered to lend Warren one of his law clerks, "those rather remarkable young men who have been sent to me, sight unseen, from year to year by my friends at Harvard," to help on the *Brown* decree.[21] He congratulated Warren on a speech Warren made before the American Bar Association, which "said important things that needed to be said" and "was wholly free from self-righteous complacency."[22] He advised Warren that Florida's practice of not responding to petitions to the Court for review of Florida judgments in criminal cases ought not to be tolerated by the justices.[23] And, at the risk of having Warren "bored by this discussion," he presented a lengthy exegesis on how to "invoke the sanctioned principle of avoiding constitutional adjudication by . . . construing . . . legislation so as to find a construction which avoids the constitutional [issue]."[24]

The letters between Warren and Frankfurter began to change their character in 1955. The easy informality of Warren's correspondence in the summer of 1954 was replaced with a certain stiffness, and Frankfurter's effusive asides ("I cannot close without saying again what a happy year this has been for me and I am sure for all the others"[25]) were increasingly absent, replaced by phrases such as "these are questions that bulk large in my concern."[26] A principal basis of Warren's attraction to Frankfurter, and Frankfurter's to Warren, had been Warren's neophyte status on the Court. As Warren felt more comfortable in his new position, he was less inclined to rely on others for advice; as Warren became less amenable to being "educated" by Frankfurter, some of Frankfurter's enthusiasm for the "new Chief" began to wane.

There were two additional reasons for Frankfurter's changing attitude toward Warren in this period. From the segregation cases on, Frankfurter was engaged in frequent correspondence with persons whose enthusiasm for Warren, particularly as a legal craftsman, was more restrained. Judge Learned Hand furnishes a vivid example. Hand gave an inkling of his attitude toward Warren in a June, 1954 letter, in which he "agree[d] [with Frankfurter] that Fred [Vinson] and Earl are to the last degree not 'fungibles,' with all that you imply," but indicated that Warren had "relatively a small capacity for verbal analysis."[27]

From that initial position Hand's discomfort with Warren increased. In a September 15, 1955 letter, he said that "I can't help it, [Warren] is not a distinguished party," and "allowing for your love affair [with Warren], I must in the end set it down as [an affair of the heart]." Warren might be "a soothing ameliorator" or "an influence" on the Court, but Hand longed for "the hard unflinching dialectic of [Holmes]."[28] A January 1, 1956 letter found "the remarks of one Earl Warren [at a judicial conference] . . . pretty tame stuff."[29] A letter on October 25 of that year said that "the more I get of your present Chief, the less do I admire him," and indicated that others had referred to Warren as " 'that Dumb Swede.' "[30] A July, 1959 letter said that "somebody is writing for Warren far better than at the beginning, though the results are worse."[31] Those comments were accompanied by exhortations to Frankfurter not to transmit his own views of Warren. "I *don't want* you to utter a syllable— I *beseech* you not to," Hand wrote in October, 1956;[32] that February he had written that "I neither expect, nor want you, to say anything. I recognize that you will, and must, be still as a Sphinx."[33]

The correspondence with Hand illustrates one feature of Warren's chief justiceship that was apparent early in his tenure. The basis of Warren's strength was repeatedly expressed in intangibles, qualities that were not easily captured in his written products. Persons, such as Hand, who did not experience Warren's presence and were compelled to evaluate him through his opinions often noted a "relatively small capacity for verbal analysis." To Frankfurter, who took pride in the academic standing of his own opinions, such criticism of Warren, when offered by persons whose intellectual judgments he respected, may have made an impression. "[S]top flagellating your soul about that Earl Warren," Frankfurter wrote Hand early in 1956. "[O]f course he ain't got no distinction nohow."[34]

But far more decisive in convincing Frankfurter that his initial enthusiasm for Warren was misplaced was the apparent shift in Warren's jurisprudential posture from a potential ally of Frankfurter to a potential antagonist. In an article published in December, 1956, political scientist Clyde Jacobs noted that Warren's voting pattern on the Court had begun to change. In the 1953 term, Jacobs claimed,

Warren had not "revealed any pronounced tendency to associate with the liberal bloc," which included Black and Douglas. He consistently supported the government in criminal law and procedures, and his "votes in cases arising out of loyalty-security considerations revealed a position opposed to that of Black and Douglas." And in the 1954 term, according to Jacobs, Warren occupied a "center position" on the Court. But by the 1955 term he "disclosed a definite pattern of alignment with the liberal bloc," and the Court had "gradually re-oriented itself in a more liberal direction." Jacobs concluded that in those three terms Warren had moved from "a center position, per-haps facing slightly in the conservative direction on matters of civil liberties" to a position that was "closely . . . align[ed] . . . with the Black-Douglas wing of the Court."[35]

Jacobs' analytical terminology was imprecise. He considered only certain cases, equated "liberal" with "libertarian," and defined "con-servative" as "support[ing] a restrictive interpretation of civil liber-ties."[36] Consequently his article cannot be taken as an authoritative general statement on Warren's jurisprudential stance. But Jacobs provided some documentation for a judgment that Frankfurter made about Warren's role on the Court during the first three years of his tenure. Warren, in Frankfurter's view, had become a member of what Frankfurter called, in a June, 1957 letter to Hand, the " 'hard core liberal' wing of the Court," whose "common denominator is a self-willed self-righteous power-lust," and whose members were "undis-ciplined by adequate professional learning and cultivated under-standing." In that same letter Frankfurter linked Warren with Black and Douglas, who made decisions on the basis of "their prejudices and their respective pasts and self-conscious desires to join Thomas Paine and T. Jefferson in the Valhalla of 'liberty' and in the mean-time to have the avant-garde of the Yale Law School . . . praise them!"[37]

Similar sentiments were expressed to other correspondents. Frankfurter wrote Justice John Harlan in September, 1958 that War-ren's "attitude toward the kinds of problems that confront him is more like that of a fighting politician than that of a judicial states-man."[38] He suggested to Philip Kurland in 1960 that Warren thought that the *Supreme Court Review*, which Kurland had founded, was a journal that Frankfurter had helped create as a means of criticizing liberal Warren Court majority opinions.[39] He reassured Alexander Bickel that "explicit analysis of the way the Court is doing its business really gets under the skin [of Warren and other 'liberals'], just as the praise of their constituencies, the so-called liberal journals, . . . fortifies them in the present result-oriented jurisprudence."[40]

Even after his retirement from the Court in 1962, Frankfurter went out of his way to criticize Warren to friends and disciples. He was sharply critical of Warren's handling of the Warren Commission, particularly his questioning of Jack Ruby, Lee Harvey Oswald's assassin, without a lawyer being present. "Warren is thin skinned," Frankfurter wrote Bickel, "but he certainly ought not to be allowed to escape the contrast between what he decides and what he does."[41] He suggested to Kurland that the Warren Commission had proceeded without attention to "due process of law," and stated that "Harlan nonplussed me the other day by telling me that Black told him the trouble with Warren is that he is no lawyer."[42]

﹍

Warren, for his part, passed from initial enthusiasm for Frankfurter as a colleague through a posture of annoyance to eventual sharp estrangement. While as late as December, 1958 Warren was capable of writing a warm sympathy letter to Frankfurter during one of the latter's illnesses,[43] he and Frankfurter had earlier abandoned all pretense of jurisprudential harmony. They had quarreled openly in Court in the 1958 case of *Caritativo v. California*.[44] In that case the Court sustained, in a per curiam opinion, the constitutionality of a practice whereby California prison wardens had the sole discretion to institute court proceedings to determine the sanity of condemned criminals in their custody. California law provided for the death penalty but prohibited the execution of insane criminals. If a warden, however, did not request a sanity proceeding, condemned prisoners could not themselves request it; their execution could only be blocked by a gubernatorial pardon.

Warren, under whose administrations the procedure had been initiated, voted with the majority in *Caritativo*. Frankfurter dissented, arguing that the system denied condemned prisoners due process of law. In his written dissent he called the California procedure "inherently arbitrary," and suggested that "the State of California should have on its conscience a single execution that would be barbaric because the victim . . . had no opportunity to show [that he was] mentally unfit to meet his destiny."[45] In addition, Frankfurter ex-

temporaneously condemned the California penal system while reading his dissent in court. According to Justice William Brennan, who
joined Frankfurter's dissent, Warren then responded to Frankfurter's
comments, defending the California system in words that "scarcely
obscured his resentment at Frankfurter's extemporaneous remarks."[46]

The previous term Frankfurter had professed an interest in Warren's opinion for the Court in *Watkins v. United States*,[47] a decision
striking down congressional power to engage in broad-ranging legislative investigations in the name of national security. The *Watkins*
case was a pivotal one in Warren's development as a judge. He concluded that wide-ranging legislative investigations were presumptively violative of individual privacy and dignity, a position he had
not taken in California. Warren was again eager to achieve unanimity
in the case, and wrote Frankfurter in May, 1957, while his opinion
was in preparation, "many thanks for the thought and attention you
gave to *Watkins* over the weekend."[48] But when Warren's final draft
circulated, Frankfurter disassociated himself from most of it in a concurrence. He told Hand that Warren "tried hard to persuade me not
to file the concurring opinion," but that he could not endorse "Warren's . . . excessive [holding] and [his] poor rhetoric."[49]

A decisive breach between Warren and Frankfurter came with
the case of *Cooper v. Aaron*,[50] a sequel to the *Brown* decision that
arose out of an atmosphere of massive resistance and violence in Little Rock, Arkansas. All the justices agreed on the result in *Cooper v.
Aaron*, and Frankfurter, mindful of his successful strategic partnership with Warren in *Brown*, proposed that each justice sign his name
individually to the opinion of the Court. Warren relished this kind
of open declaration of an individual's commitment to a principle and
heartily agreed. Then Frankfurter, after including his name among
the nine justices signing the Court's opinion, wrote a concurrence in
which he quarreled with portions of the opinion's language. Warren
regarded Frankfurter's concurrence as the equivalent of a betrayal.

Throughout his career, we have seen, Warren had openly broken
with persons he considered to have abused his trust or to have sought
to humiliate him. "The only thing that he would never forgive,"
William Sweigert said, "was anything that [he perceived] was done
selfishly or fraudulently. . . ."[51] After Frankfurter's actions in *Cooper
v. Aaron*, Warren regarded Frankfurter as his antagonist. In his
memoirs Warren said that "there was but one event that greatly dis-

turbed us during my tenure," and that was *Cooper v. Aaron*. "We were all of one mind in that case," Warren noted. "Mr. Justice Frankfurter however, [after proposing the individual signing procedure] informed us that he had many friends in the Southern states, and that he intended to reach them by writing and circulating a concurring opinion of his own." After Frankfurter's actions, Warren wrote, "some of the Justices stated that they would never permit a Court opinion in the future to be made public until it was certain that the views of all were announced simultaneously."[52]

The tone of Warren's account of the *Cooper v. Aaron* episode initially appears matter-of-fact, but on reflection two revealing features emerge. First, Warren stated that Frankfurter's concurrence was *the only event* that "greatly disturbed" the justices on the Warren Court. Surely that was not the case, given the number of controversial and heated Warren Court decisions. Second, *Cooper v. Aaron* was certainly not the only Warren Court case where members of a majority became aware of concurrences after the majority opinion had been completed. But perhaps *Cooper v. Aaron* was the only instance where Warren felt that his authority as Chief Justice had openly been threatened. From then on he regarded Frankfurter as his enemy.

Even before *Cooper v. Aaron* the estrangement of Warren and Frankfurter had become noticeable to observers; it continued until Frankfurter's retirement in 1962. In addition to the episode in *Caritativo v. California*, in a March, 1958 oral exchange over the case of *Trop v. Dulles*,[53] Frankfurter challenged Warren's listing of eighty-one cases where the Court had declared acts of Congress unconstitutional, pointing out that many of the cases had been overruled.[54] A last public contretemps occurred in April, 1961, when Warren reportedly responded to an oral Frankfurter dissent by "turn[ing] crimson, then purple," and snapping, "That was not an opinion. That was a lecture."[55]

&#x200B;

Frankfurter's experiences with Warren were a vivid and perhaps exaggerated microcosm of the surface and underside of Warren's personal relations. Associates and contemporaries initially confronted the Warren public persona: resonant, accommodating, seemingly malleable. Closer contact revealed the prideful and independent spirit underneath. Some associates, like Frankfurter, mistook the surface for the whole of Warren's presence, wounded Warren's pride, were perceived of as betraying his trust, and found themselves confronted by a stubborn and formidable opponent.

Other strong-minded justices who served with Warren, such as Harlan, Black, or Douglas, sensed or appreciated Warren's sensitivity to slights and either avoided estrangement, or, especially in the case of Black and Harlan, accepted Warren for who he was and fostered a warm, if largely unexpressed, relationship of mutual respect. Warren said in a tribute to Harlan that "[h]is dissents seared, but they never burned."[56] Black, who found Warren "more hard-headed and opinionated than I am,"[57] did not mistake Warren's affability for indecisiveness. Warren said that "having Hugo Black sitting at my side for sixteen years" was "one of the great privileges I enjoyed on the Supreme Court."[58]

Of all the justices on his Court, the one to whom Warren felt closest was William Brennan; Warren and Brennan's relationship can be seen as a foil to that of Warren and Frankfurter. The persons with whom Warren tended to grow close to during his many years of public service, held certain traits in common. Warren Olney and William Sweigert, for example, had both been younger than Warren by a decade or more, had been dedicated to Warren's interests, had been perceived by Warren as honest, wholesome, and likeable, and had been "kindred spirits" to Warren in their intuitive reactions to social issues. In 1956, when Brennan came to the Court, Warren was his chief and his chronological senior by fifteen years. Brennan's temperament, like Warren's, was gregarious, and, like Warren, he was capable of having his emotions deeply engaged. He shared Warren's enduring interest in politics; he witnessed, as did Warren, the increasing dissatisfaction of Eisenhower, who had appointed him, with his performance as a justice; he also witnessed the discomfort his increasingly activist judicial posture caused for Frankfurter, who called Brennan's work "shoddy" in a letter to Hand in 1961.[59]

In addition, Brennan shared with Sweigert and Olney an ability to put into practice policies to which Warren had given abstract support. Sweigert had been Warren's architect of executive policy; Olney his implementor of law enforcement decisions. Brennan was Warren's judicial technician. He was capable, in cases such as *Baker v. Carr*, or *New York Times v. Sullivan*,[60] of supplying doctrinal rationales for decisions in which Warren strongly believed.

Brennan also intuitively understood Warren's reactions to issues. There is a story told about one term of the Court when the justices were interested in making a more definitive statement about the constitutional status of obscenity and to that end had assembled samples of allegedly "obscene" literature in a room off of the justices' conference room. Warren, who was outraged by literature he found "indecent," refused to look over the samples, indicating that he would talk to Brennan instead.[61] While Brennan and Warren did not have identical views on obscenity,[62] Warren apparently felt that Brennan would understand his intuitive reaction to the material, and he Brennan's, well enough so that they could discuss the literature on the basis of Brennan's observations. The story may have been amplified with time, but it is consistent with the nature of Warren's and Brennan's relationship.

Meanwhile Frankfurter began to perceive Warren as an undistinguished craftsman and a potential jurisprudential antagonist; he sought first to cajole and then to lecture. The pattern of Warren's relations with Culbert Olson began to reappear. Frankfurter perceived Warren as susceptible to the "hard core liberal wing of the Court"; he consequently bombarded Warren with suggestions as to how he might avoid such "narrow minded prejudices." But the suggestions, now increasingly delivered from a posture that could have been taken as peremptory or patronizing, had the opposite effect: Warren felt suspicious of and eventually antagonistic toward Frankfurter.

As he had done with Olson, Warren began to define his goals in opposition to the goals of a perceived enemy. In the *Watkins* case he demonstrated a tendency to equate his responsibility to review the constitutionality of legislative acts with his intuitive sense of what justice required in a given case. The reach of congressional investigations had to be constitutionally checked, he felt, because investigative practices violated human dignity. Frankfurter, of course, deplored such a "prejudiced" view and disassociated himself from Warren's *Watkins* opinion; Warren saw Frankfurter's actions as that of an antagonist.

By the beginning of the Warren Court years, Frankfurter had rigidified his jurisprudential posture. He felt that judicial restraint was the only proper approach for an autocratic institution in a democratic society; he had developed a number of doctrinal devices to avoid judicial review of the constitutionality of legislation; he talked repeatedly of the obligations of judges to defer to more representative and democratic forums. Warren increasingly came to believe that Frankfurter's attitude regularly prevented justice from being done. He did not share Frankfurter's sanguinity about the capability of legislatures, Frankfurter's principal instrument of democracy, to formulate enlightened social policies. Much of Warren's public life had been as an opponent of special privilege in legislatures or as an activist official seeking to cajole or shame reluctant legislators into taking action for the benefit of persons without special privileges. As his personal relations with Frankfurter worsened after 1955, his intuitive opposition to Frankfurter's theory of judging increased. And as Warren began to associate himself with other, more expansive theories of the Court's constitutional review powers, Frankfurter began to see him as a prospective convert who had departed from the fold.

Thus the more steps Warren took to disassociate himself from Frankfurter's conception of enlightened judging, the more Frankfurter assumed the role of Warren's critic. And the more Frankfurter assumed that role, the more Warren characterized him, as he had characterized others in the past, as one of the numerous enemies Earl Warren had had to endure in his efforts to use the powers of his public offices in socially useful ways. Warren, in Frankfurter's mind, became a "martinet," hell-bent to impose his "liberal" views on the country; Frankfurter, in Warren's mind, became a foe of progress, aided in his efforts by a stable of academics and other critics whose motivation was to embarrass the Warren Court.

✒

It was typical of Warren to resolve to exercise the powers of his office in an expansive fashion after experiencing pressures from someone else to exercise those powers sparingly. Warren's activist stance as a judge can be said to have originated in his visceral reactions to two experiences on the Court in the 1950s. He had first experienced the crucible of *Brown*, an episode that had demonstrated for him that the Supreme Court could, in cases where high principles were involved, surmount jurisprudential and political difficulties and declare itself committed to equality and justice in American life. He had then experienced the realization that his first major ally on the Court, Frankfurter, was not committed to that point of view, but was rather pressuring him and others to recognize the technical, political, and theoretical problems inherent in the active promotion of social policies by a group of judges.

Warren identified Frankfurter as an opponent within three years of their partnership in *Brown*. He no longer welcomed Frankfurter's suggestions or felt he profited from Frankfurter's academic and judicial experience. He regarded Frankfurter as a potential threat to his authority, a manipulative force within the Court, and a person whose erudition and experience had not necessarily resulted in his having good judicial instincts. As Warren began to move away from the cautious posture of his first years on the Court he became inclined to trust his instincts, to emphasize the significance of getting "good" results as a judge. He came to see Frankfurter's purported subordination of personal preferences to a theory of judicial restraint as regularly resulting in decisions that were obfuscated by a flood of academic language, or unnecessarily self-conscious, or sometimes simply misguided. As Warren began to perceive Frankfurter's jurisprudence in such terms, he came to relegate Frankfurter to the status of, at best, a gadfly, and, at worst, an obstructionist. In either role Frankfurter became a person to be bypassed, not a person to be respected. Warren was embarking on a mission to do justice, and while he did not like criticism of the Court that proceeded from Frankfurterian assumptions, he resolved not to be influenced by that criticism.

Underneath Warren's hearty, pleasant persona, then, his peers on the Court confronted a Chief Justice who had confidence in his intuitive reactions, who had formed his own judgments about his peers, separating those he saw as potential supporters from those who appeared to be antagonists, who was used to getting his own way, and who was not afraid to speak his mind. They also confronted a Chief Justice willing to compromise on matters he regarded as relatively insignificant—technical language in a given opinion—to prevail on matters he thought important, such as decent and just results. This was the same Earl Warren contemporaries had confronted as governor, attorney general, and district attorney. If there was a continual discrepancy between Warren's persona and his composite personality, the discrepancy remained constant over time.

By 1962, the last year of Frankfurter's service on the Court, the same kind of self-fulfilling prophecy that haunted Culbert Olson had come to pass. Warren had become what Frankfurter feared he would become: one of the leaders of the opposition to Frankfurterian jurisprudence on the Court. In the final major case of Frankfurter's tenure, *Baker v. Carr*,[63] Frankfurter witnessed a humiliating defeat. One of his own opinions, *Colegrove v. Green*,[64] articulating one of his cherished doctrines, the doctrine that courts should not decide "political" questions, such as legislative reapportionment, that were the prerogatives of more democratic institutions, was scuttled.

Warren was an influential member of the *Baker v. Carr* majority, which included Frankfurter's old antagonists Black and Douglas; the opinion was written by Brennan, Warren's intimate. The opinion violated nearly every principle of Frankfurter's jurisprudence. It substituted the judgments of the undemocratic judiciary for those of the democratic legislature; it failed to draw a bright line between the judicial and political spheres; it seemed to be a direct response to contemporary political pressures to which the Court was supposed to be indifferent. Frankfurter deplored every facet of the majority's approach in *Baker v. Carr* and said so in a long and heated dissent. Warren reportedly said of the decision that if *Baker v. Carr* had been decided before 1954, *Brown v. Board of Education* would have been unnecessary.[65]

*Baker v. Carr* was yet another indication that Warren the Chief Justice would not necessarily hold to the substantive positions taken by Warren the California public official, who had opposed reapportionment of the California legislature. But that tendency in Warren was not remarkable by 1962. Warren's decision in *Watkins*, when compared with the stand he had taken on the loyalty oath controversy, had suggested that he would not feel bound on the Court by positions he had taken in California. The case was more remarkable for the stark opposition it showed existing between the jurisprudential rationale of the majority and that of Frankfurter's dissent. The majority brushed aside Frankfurter's doctrinal objections to judicial intervention in "political" questions as so many technicalities. When a legislative majority had perpetuated clear inequalities, resulting in some citizens' votes counting more than others, the "political question" doctrine simply did not prevent the Court from making a constitutional inquiry under the Equal Protection Clause. The issue, the majority in *Baker v. Carr* seemed to be saying, was one of equal justice versus academic refinements. Frankfurterian jurisprudence, in the face of serious issues of social justice, appeared as a cult of petty obstructionism.

This posture—impatience with obstructionist doctrine when jus-

tice called out to be done—was to become identified with Earl Warren's chief justiceship; it was to serve as a powerful alternative jurisprudence to that of Frankfurter. Most of Warren's energy on the
Court was directed toward achieving the "right" results. He did not
often agonize, as did Frankfurter, over an outcome in a case, nor did
he despair of finding an adequate constitutional basis for justifying
his intuitions, nor did he worry about being overly activist. He spent
his time on discerning results that seemed just and on marshaling
support for those results by attempting to convince others of their
inherent justice. He was, especially in his later years, the antithesis
of a Frankfurterian model judge, trusting his intuitions and his "prejudices" far more than the myriad doctrinal and institutional considerations to which Frankfurter asked judges to be sensitive. Warren
the activist Chief Justice was an opponent of Frankfurter; he had
become so, in large part, because Frankfurter had disappointed him,
patronized him, and generally "pressured" him into declaring his independence.

The relationship of Warren's interest in achieving just results to
his judicial activism is the subject of subsequent chapters. But one
consequence of Warren's interest in securing results that he thought
were just bears mentioning at this stage. In *Brown v. Board* Warren
had not been dissuaded in his efforts to secure a just result by the
potential opposition of others; in Warren's reaction to Frankfurter,
that opposition had often merely increased his stubbornness. To these
features of Warren the Chief Justice should be added his long experience in leadership positions, his dislike of dissension in offices that
he held, and his tendency to rely on his intuitive judgments.

This combination of factors produced a tendency in Warren to
dominate offices that he held, and to seek to equate the collective
views of those offices with his own views. Because of the skillful
manner in which he conducted personal relations on the Court, this
tendency rarely caused him difficulties. It did, however, hamper his
performance as chairman of the presidential commission that investigated the assassination of John F. Kennedy from December, 1963
to September, 1964. The consequences of Warren's presence on that
commission are the subject of the next chapter.

# 8

# The Warren Commission

There was only one other episode during Warren's tenure as Chief Justice that rivaled the decision in *Brown v. Board of Education* in its power to command the interest of the general public. In this second episode, as in *Brown*, Warren was the titular head of a group of persons who announced a unanimous decision; in both instances Earl Warren's presence was vital in securing that unanimity and in producing the eventual language that supported the decision. In both episodes the participants to the decision-making process were continually mindful of the political implications of their decision; in both episodes, none was more so than Earl Warren.

This other episode was not a Supreme Court case. It was the investigation of the assassination of President John F. Kennedy by a presidential commission of which Warren was the chairman. The activities of the so-called Warren Commission, its final report on the evidence surrounding Kennedy's assassination, and its relationship with other investigative bodies, notably the Federal Bureau of Investigation and the Central Intelligence Agency, have spawned as much controversy as any set of events in Warren's career. Chairing the Warren Commission was arguably the most visible act of Earl Warren's long career in public service.

In reviewing the facts of the assassination and the subsequent Report of the Warren Commission, many Americans have experienced a sense of not quite determining the truth. This reaction has

not abated with time: Sixteen years after the assassination, and fifteen years after the apparently authoritative explanation of its facts
by the Warren Commission, a Select Committee on Assassinations
of the House of Representatives reconsidered the commission's performance and concluded that its investigation of the assassination was
inadequate and that its explanations were "too definitive." [1]

The Select Committee's findings revived conspiratorial theories
of the assassination and brought to the edge of respectability a position hitherto reserved for polemicists: that there may have been a
"cover-up" of the true facts of President Kennedy's assassination by
the Warren Commission itself. The Select Committee itself did not
endorse this idea, concluding to the contrary that the Warren Commission conducted its investigation "with high integrity" and "arrived at its conclusions . . . in good faith." [2] But in an age when
skepticism about the true motives of high public officials has become
rife, the committee's faint-praise exoneration of the Warren Commission's conduct may not prove much of a palliative.

Many people, on reflection, still believe that there could have
been no "cover-up" of the Kennedy assassination for the simple reason that the commission's chairman, whom Lyndon Johnson called
the "personification of fairness and justice in this country," [3] would
never have participated in one. At the time of his appointment to
the commission in December, 1963, Warren had come to be known
as a champion of due process for accused persons, as an advocate of
equal justice for the affluent and the disadvantaged, and as a symbol
of the virtues of integrity, honesty, and decency in public life. The
idea of such a person "covering up" evidence of an assassination conspiracy remains to many implausible on its face.

On balance, however, Earl Warren's service on the Warren Commission did as much to foster conspiratorial theories of the assassination as it did to assuage them. The story of Warren's service on
the commission is the story of a man whose strong instincts for justice and candor collided with other strong elements in his nature—
his patriotism, his interest in making decisions swiftly, his awareness
of the political implications of his decisions, and his belief that matters of great public significance, if decided by a collective body,
should if at all possible be decided unanimously. It is the story of an
unhappy and unsatisfying experience in Earl Warren's life.

In discussing the Warren Commission as a feature of Earl Warren's chief justiceship one first needs to understand the extent of his participation. The commission was regarded by Warren, and has been regarded by commentators, as a sideline to his career as Chief Justice. Warren reluctantly accepted the post as commission chairman, spent only a small amount of his time on commission activities, and wrote only selected portions of the report that the commission issued in 1964. On the other hand, Warren was, according to commission staff members, "extremely well informed"[4] about the commission's inquiries, and made a number of personal decisions that had decisive effects on the eventual shape of the commission report. Moreover, Warren was perceived by President Lyndon Johnson as a symbolic figure whose chairmanship would make the commission credible to the public. Although it may be said that Warren's presence on the commission was a reluctant and distracted presence, it was not an insignificant one.

The appointment of Warren to the commission set the tone for his service. On learning that Johnson planned to appoint him, Warren resolved to decline the post and prepared arguments to support his refusal. Foremost among these arguments was that he thought extrajudicial services by Supreme Court justices tended to create conflicts within the Court and potential separation-of-power problems. Warren felt deeply about such issues. He rarely took part in any activity as Chief Justice that could remotely be conceived as raising potential conflicts of interest, and he regularly concerned himself with "the welfare of the Court" and the avoidance of "divisiveness and internal bitterness" among its justices.[5] His disinclination to become chairman of the commission was thus more than the standard disclaimer of a busy official.

Warren underestimated, however, the ability of Lyndon Johnson to find a chink in the armor of even the stubbornest of those with whom he had to deal. In a face-to-face talk the same day he heard of Warren's disinclination to serve, Johnson appealed to Warren's patriotism, stressing the possibility of global nuclear war should certain rumors surrounding the Kennedy assassination not be put to rest. "You were a soldier in World War I," Warren recalled Johnson saying, "but there was nothing you could do in that uniform com-

parable to what you can do for your country in this hour of trouble."[6] Johnson's version recalled the President's telling Warren that
"in World War I he had put a rifle to his shoulder and offered to
give his life, if necessary, to save his country," and that after Kennedy's assassination "the country [was] confronted with threatening
divisions and suspicions."[7] Whichever version is more faithful to the
exact words the president used, the message was clear. Warren capitulated.

Warren's conception of his service on the commission as that of
a soldier, and his conception of the commission's inquiry as that of
a patriotic mission, were to constitute one major theme of his tenure
as commission chairman. In Warren's initial speech to commission
staff members, he encouraged them to "determine the truth, whatever that might be"; on the other hand, he also "placed emphasis on
the importance of quenching rumors."[8] While the jurist in Warren
demanded a fair hearing, quenching rumors and allaying doubts was
part of the "patriotic" function of the commission, since it reinforced
within the American public the politically and socially beneficial pre-
sumption that the assassination was neither the result of a foreign
plot nor the natural consequence of a corrosive atmosphere of violence within American society. One commission staff member stated
that Warren believed that "the principal function of the Warren
Commission was to allay doubts, if possible." Warren, the staff
member recalled, "made no secret . . . among the staff that he
wanted doubts la[id] to rest."[9]

The theory of the assassination that harmonized most closely with
a patriotic function for the commission was that Lee Harvey Oswald, a demented individual, had acted alone. A demented, nonideological, nonconspiratorial assassin reinforced a public belief that in
America only "nuts" assassinate presidents. Warren, for his part,
strongly believed that Lee Harvey Oswald had acted alone in the
assassination and that Oswald's actions were those of a disturbed
fanatic, not a hired gun or a political terrorist. "If the sole responsibility of the Commission had been to determine who shot and killed
President Kennedy," Warren wrote in his memoirs, "it would have
taken very little work." If he had been a prosecutor prosecuting Oswald for murder, Warren felt, he could have obtained a conviction
"in two or three days."[10]

Oswald in Warren's view "had been a misfit all his life," was "a
complete malcontent," and was "disoriented" and "violence-prone."[11]
He was "without friends and associates, an absolute 'loner' wherever

he worked." He was "incapable of working or living satisfactorily with anyone"; he had "a disposition and orientation that would not enable him to plan, counsel with or take orders from anyone." These qualities "b[ore] directly on his capability as the alleged focal point of a conspiracy of the magnitude dreamed up by some people." In short, Oswald fit perfectly into "the history of American presidential assassinations," where "in all but one of those instances, the assassin was the sole perpetrator," and "in each of them, the killer was [a] misfit in society." [12]

Warren's characterization of presidential assassins was a convenient one. There had been three presidents previously assassinated, two of whom, Garfield and McKinley, had been shot by single individuals who could be called "misfits," and the third, Lincoln, who had been shot by an assassin who was very possibly part of a conspiracy. In characterizing Oswald as a prototypical presidential assassin, Warren was intertwining a belief that Oswald was the kind of person who was a "loner" with a belief that Oswald was the kind of person in American society who assassinated presidents. Convicting Oswald would have been "easy" for Warren the prosecutor not only because of affirmative evidence that Oswald fired the shots but also because Oswald so perfectly fit the prototype.

By dwelling on Oswald's "misfit" status, Warren saw himself as helping "lay doubts to rest." He was reassuring the American public that a familiar, if shocking, pattern had repeated itself: the surfacing of a demented loner who assassinates an American president. He was also ignoring some deviations from that pattern, such as the group of persons who fired at President Truman outside Blair House in 1950. Warren was attempting to further a principal purpose of the Warren Commission: "the . . . job [of] running down wild rumors." [13] Lyndon Johnson was to say that the Commission "was born . . . out of the Nation's suspicions;" [14] Warren the soldier and patriot was to allay them.

The theme of patriotism also surfaced in Warren's relationship with other investigative agencies during the commission's existence. The Select Committee on Assassinations concluded that the Warren Commission "failed to investigate adequately the possibility of a conspiracy to assassinate the President," and that this failure was partially attributable to "the failure of the Commission to receive all the relevant information that was in the possession of other agencies and departments of the government." Specifically, the Select Committee found that the Federal Bureau of Investigation and the Central Intelligence Agency were "deficient in [their] sharing of information" with the commission.[15]

To understand Warren's relationship with the FBI and CIA one has to recall not only his intense patriotism but his relative lack of sophistication about internal Washington politics and international affairs. We have seen that Warren was not, at the time of his appointment as Chief Justice, a "national" politician in the sense of having spent long years of service in Washington or having traveled extensively throughout America. Despite his involvement in national Republican politics, he was essentially a Californian, familiar with the patterns of government and politics in his own state but not particularly concerned with matters outside its borders, with the conspicuous exception of law enforcement and national defense matters, which touched close to home. Although by 1964 Warren had been Chief Justice for ten years and was a visible and controversial national figure, he was not a Washington "insider": His connections with other agencies of the federal government had been essentially limited to matters before his Court.

Warren's provincialism, his patriotism, and his relative lack of contact with official Washington reflected themselves in some decisions pertaining to the commission that in retrospect appear startling. Warren's first choice for general counsel to the commission was Warren Olney, his staff intimate, counselor, and friend during his years as a California public official. Olney, who had come to Washington ten years earlier as an assistant in the Justice Department's Criminal Division on Warren's recommendation and had been named director of the Administrative Office of the United States Courts by Warren in 1958, was thought by several commissioners to be too close to Warren to serve as general counsel. Commissioner Gerald Ford, articulating the sentiments of this group, opposed Olney's prospective appointment as "moving in the direction of a one-man commission,"[16] and Warren withdrew Olney's name. Warren's next choice, J. Lee Rankin, a former solicitor general of the United States who had contributed to the government's brief in *Brown v. Board of Education*, was regarded as a more credible candidate, and was unanimously approved.

Another of Warren's early decisions resulted in the commission's relying on the investigative reports of existing agencies, notably the FBI and CIA, rather than conducting its own independent investigation of the assassination. In his first session as chairman, Warren defined the commission's role as "evaluat[ing] evidence as distinguished from . . . gathering evidence."[17] He also stated in that session that he did not think that the commission needed subpoena powers to carry on its activities,[18] and in another session, the next day, said that he had not been in touch with the CIA about the investigation because he "did not want to put the CIA into this thing unless they put themselves in."[19]

Certain events caused Warren to modify these original determinations. An FBI report on the assassination to the commission, received in December, 1963, was widely condemned by the commissioners as inadequate: Warren said that he had found nothing in it "that has not been in the press."[20] And on January 22, 1964, a rumor surfaced, through Waggoner Carr, the attorney general of Texas, that Lee Harvey Oswald had been an informant for the FBI. Warren convened a special unpublicized meeting of the commissioners that evening; one commissioner, Gerald Ford, described the meeting's atmosphere as "tense and hushed."[21]

Warren concluded that if the rumor were true, "nothing the Commission or anybody [did] could dissipate" a conspiracy theory on the assassination. He found the implications of the rumor "terrific" and suggested that the commission investigate the matter "independently," as well as attempting to verify the rumor through a conversation with Director J. Edgar Hoover of the FBI.[22] Warren had earlier expressed concern about relationships between government investigative agencies and had suggested that "perhaps we ought to have a thorough investigation . . . as to the relationship between the FBI and the Secret Service and the CIA. . . ."[23]

Despite his misgivings, Warren placed very little pressure on investigative agencies to cooperate with the commission. He indicated that he did not want to be "unfriendly or unfair" to Hoover;[24] he did not press for any investigation of the FBI, and no investigation materialized; he relied on the CIA's assurances that it would provide the commission with all relevant information. In fact, the FBI apparently regarded the commission "more as an adversary than a partner in a search for the facts of the assassination,"[25] thought that the commission was "seeking to criticize the FBI,"[26] deliberately withheld information from the commission, and attempted to limit the scope of the commission's investigation of FBI activities by leaking some information to the press in the hope of ending inquiries prematurely.[27]

The CIA, for its part, apparently did not volunteer information to the commission but waited for specific requests, withheld information from the commission about possible CIA plots to assassinate Fidel Castro, with whom Oswald had been rumored to be linked, and allowed its representatives to give misleading testimony before the commission.[28] As a consequence the Warren Commission was placed in the awkward position of relying on the fact-gathering powers of investigative agencies that were not prepared to use their powers fully for the commission's behalf.

The commission's difficulties with investigative agencies cannot solely be traced to Warren's initial decision not to develop an independent investigative staff. The image of the FBI in the early 1960s was that of an impressive and incorruptible organization. As J. Lee Rankin put it in 1978, "I never thought the Bureau was capable of [lying to other government agencies]. . . . I was rather sanguine about that and I don't think the country believed the FBI would do such things. . . . It was a time when I am sure all the Commissioners and I certainly believed that Mr. Hoover would not [make statements] unless [they were] the truth."[29]

Similarly the CIA, while identified with the ill-fated Bay of Pigs invasion of Cuba in the early days of the Kennedy Administration, had not taken on the image, as it was to during the "Watergate" years of the 1970s, of an organization that often concealed or misrepresented its activities even to agencies with whom it cooperated. One staff member of the Warren Commission said that he "trusted" the CIA;[30] another said he had been "favorably impressed"[31] by the agency while in college; a third stated that he was "favorably disposed" towards the CIA because "they had seemed to have high caliber people out of my college."[32] Warren's decision to rely on the agencies for investigative expertise was made in an atmosphere in which the credibility of the FBI and CIA was assumed. It nonetheless proved to be a decision that was naive, foreclosed investigative possibilities, and has undermined the definitiveness of the commission's report.

Warren also took a cautious approach, as commission chairman, toward the disclosure of information affecting "national security." He refused to have the commission examine the FBI's file on Oswald on the grounds that it probably contained information affecting national security, which the commission could not make public. He attempted, on similar grounds, to prevent records of testimony presented to the commission from being made public, and announced that the commission was not planning to publish its report, citing national security and expense as his rationales.

These acts seemed to rest on a distinction between controversial issues of a domestic nature, such as criticism of government or its agencies (Warren insisted on disclosing criticism of the FBI and the Secret Service in the commission's report), and controversial matters affecting national security, which he regarded as nondisclosable. The same man who was outraged, in *Watkins v. United States*, at the House Un-American Activities Committee's "broad-scale intrusion into the lives and affairs" of persons who had been critical of governmental policies[33] was willing to protect the confidentiality of an FBI investigation into the sexual habits of the Oswalds.

Being a patriot and protecting the security of the nation did not seem to Warren to be inconsistent with being a civil libertarian. While he believed that government could not unduly infringe rights of citizenship, he also believed that government had a proper role in protecting its citizens from outside threats, and in the course of securing this protection could engage in covert activities, such as the undercover operations of his Alameda County District Attorney's Office in its campaign to investigate Communist activity on the Oakland waterfront in the 1930s. While Warren was generally suspicious of secret activities by governmental officials—in his California campaigns, for example, he repeatedly claimed that the open disclosure of connections between lobbyists and legislators would check the power of special interests—he tolerated covert activities when national security was at stake.

In some respects Warren's posture toward the FBI, the CIA, and "national security" in general during the Warren Commission's tenure resembled his posture toward the military architects of the Japanese relocation policy in World War II. The defense of the nation was at stake, he had reasoned in endorsing that policy. Certain agencies, including his own attorney general's office, were dedicated to preserving that defense. Those agencies necessarily operated covertly and arbitrarily; that they did so in a sincere attempt to serve patriotism and national defense made all the difference.

The theme of patriotism united with a second major theme of Warren's service on the commission, the theme of family privacy, to produce his most controversial decision as chairman. Warren ruled that the commissioners need not examine the photographs and x-rays taken

of President Kennedy's wounds during his autopsy. In the last pages
of his memoirs Warren defended that decision, largely on the basis
of the privacy of intimate family relationships. He had seen the pictures himself and recalled, "they were so horrible that I could not
sleep well for nights." Persons "with ghoulish minds" were clamor-
ing for artifacts of the assassination, hoping to "prey . . . on the
morbid sentiments of people."

Warren, whose own father had been murdered by a blow to the
head, found this "sensation monger[ing]"[34] an affront to the sensibilities of the Kennedy family. He instructed that the pictures be
removed from the commission's file of documents and be made available only with the consent of the Kennedys. No one else involved in
the production of the Warren Commission Report saw the pictures.[35] One commission staff member recalled that "the sensitivity
of the [Kennedy] family" was "the only reason which was ever alluded to for not seeing x-rays and photographs."[36]

In his memoirs Warren stated that he was "solely responsible"[37]
for the decision to withhold the pictures and x-rays from the commission. In testimony made in 1978 before the House of Representatives' Select Committee on Assassinations, however, J. Lee Rankin, referred to a "communication" between Robert Kennedy and
Warren in which Kennedy expressed "hope" or "something stronger"
that "the x-rays would not become a part of the official record of the
Commission."[38] Robert Kennedy apparently stressed the Kennedy
family's concern for privacy in making this request.

Rankin testified that he had discussed the matter with Warren
and that he agreed with Warren's decision to withhold the x-rays
and photographs from the other commissioners and staff. In April
and May, 1964, Arlen Specter, one of the commission's staff members, wrote two memoranda to Rankin arguing that it was "indispensable" that the photographs and x-rays be made available to the
commissioners and the staff,[39] but Warren, through Rankin, denied
Specter's request. Specter recalled in 1978 that "an attitude on the
part of the Kennedy family that . . . the photographs and x-rays
would get into the public domain" was essential to Warren's decision.[40]

The controversialism of Warren's decision barring those writing
the commission report from access to the x-rays and pictures has
stemmed from a furor over the "single-bullet" theory of the assassination. Under this theory one bullet fired from Oswald's gun struck

both Kennedy and Governor John Connally of Texas, wounding Kennedy in the throat and Connally in the chest, thigh, and wrist. The numerous wounds caused by this bullet made many observers, including Warren,[41] originally skeptical of the theory, but ballistic tests subsequently convinced the commission sufficiently to include the theory in its report.[42]

The x-rays and photographs taken at Kennedy's autopsy, critics of Warren's decision have argued, might have revealed whether the "single" bullet actually entered Kennedy's body at an angle sufficient to wound both him and Governor Connally. (Kennedy was sitting in the left rear seat of an open-top presidential limousine, Connally in the right "jump" seat.) They might also have revealed whether the second bullet, which entered Kennedy's brain, entered from the rear (the direction from which Oswald fired) or the front. The doctors attending Kennedy at the autopsy, who had also seen the x-rays and photographs, testified to the Commission that both bullets entered from the rear, but they were not required to produce the x-rays or photographs during their testimony.

Fifteen years after the publication of the Warren Commission Report, the House's Select Committee on Assassinations concluded that "scientific acoustical evidence establishes a high probability that two gunmen fired at President John F. Kennedy" and that Kennedy "was probably assassinated as a result of a conspiracy."[43] The Select Committee, found, however, that the unknown second gunman missed Kennedy. It agreed with the Warren Commission that the two bullets that struck the president were fired from the rear by Oswald, and that the commission's single-bullet theory of the wounding of Kennedy and Connally was supported by the available evidence.[44] Thus had Warren made the x-rays and photographs available, it seems, the commission would not have altered its conclusions. But the skepticism of critics of the report about those conclusions might well have been allayed.

Warren's decision was one that, in retrospect, seems excessively protective of the Kennedy family. It is hard to imagine Earl Warren the prosecutor suppressing evidence at the request of relatives of a murder victim. In deciding not even to make the x-rays and pictures available to the commissioners and the staff in private, he had made at least one out of three assumptions. He had either assumed that the sensitive material would inevitably circulate beyond the commission, or that the access of even that small circle of persons to the x-rays and photographs would somehow "defile the memory and image of President Kennedy as a vibrant young leader,"[45] or that the commission was bound to make all material that it examined part of the public domain.

The first two assumptions seem unlikely. The staff was relatively small and handpicked by Rankin, who supported Warren's concern for the Kennedy's family privacy. Individual members and staff would have viewed the x-rays in the presence of each other. The pictures may well have been horrible, but the loss of a few nights sleep seems rather far from a judgment that an assassinated president's "image" had thus been "defiled."

The third assumption seems most plausible. It was consistent with the fact that Warren, as one staff member recalled, "really wanted everything that was going to be viewed by the Commission to be part of the record," and therefore felt that he could not show the pictures to staff without including them in the report. In the April 30, 1964 meeting of the commission, Warren, the other commissioners, and Rankin discussed the pictures and the single-bullet theory. Rankin noted that "we have a very serious problem in the record." The single-bullet theory assumed that one of the bullets that hit Kennedy passed through his body "and then through Governor Connally," but Connally had testified "that that couldn't have happened, [and] that the bullet that struck him [was] one that did not hit the President."

Rankin then suggested that the x-rays and photographs of Kennedy's head might clarify the situation, but that he "thought we could avoid having those pictures being a part of our record, because the family has a strong feeling about them." He suggested that "a doctor and some member of the Commission" examine the pictures, so that "we can avoid any question that we have passed anything up that the Commission should know." Warren then interrupted and declared that if such a suggestion were to be entertained, it should be done "without putting those pictures in our record." "We don't want those in our record," he said. "It would make it a morbid thing for all time to come."[46] The commission took no action on Rankin's proposal, and it drew conclusions about Kennedy's murder that were supported by visual evidence that only Earl Warren, out of the entire commission staff, had seen.

Throughout the investigation Warren showed a striking concern for the sensibilities of persons directly affected by Kennedy's assassination and manifested this concern by sparing them from rigorous public questioning. After Marina Oswald's initial testimony before the commission, she said of Warren that "like all really great persons, he made me feel *kak doma* [at home]" and that he was "not just a gentleman but a gentle man."[47] Norman Redlich, a staff member who had been asked to prepare Rankin for a "very extensive" examination of Marina Oswald,[48] wrote a memorandum to Rankin implying that the questioning of Marina had been far too lenient and suggesting "that Marina Oswald has repeatedly lied to the Secret Service, the FBI, and this Commission."[49]

Warren also expressed a reluctance to question Mrs. Jacqueline Kennedy, whose testimony the commission regarded as important but whom some commissioners did not want to interview. In the December 16 meeting of the commission, John McCloy raised the question of Mrs. Kennedy's testimony and stated that he was opposed to her being questioned. Senator Richard Russell said that he "wouldn't like for the whole Commission to do it," but felt that "someone" ought to, and Gerald Ford suggested that it could be done "most informally" by someone who knew her well. Warren attempted to avoid questioning Mrs. Kennedy at all, stating that "I don't think we know [enough] about this thing to question witnesses" and that he did not see how the commission could question Mrs. Kennedy formally "because it doesn't seem an honorable way."[50] Later, in the meeting of January 21, 1964, he told Russell that "we have never bothered Mrs. Kennedy."[51] Eventually Warren questioned Mrs. Kennedy briefly and informally in her apartment, a decision that at least one commission staff member strongly opposed.[52]

Warren had been deeply moved by Kennedy's assassination. Earl Warren, Jr., said that his father and Kennedy "were always tremendously cordial to each other" and that Kennedy would "call [Warren] up in regard to certain judicial appointments, and asked his counsel and so forth."[53] When asked to speak at a memorial service two days after Kennedy had been killed, Warren departed from his usual tone on ceremonial occasions. He called the assassination a "horrible deed," the assassin a "misguided wretch," and the event itself one "stimulated by forces of hatred and malevolence such as today are eating their way into the bloodstream of American life." "What a price," he exclaimed, "we pay for this fanaticism!"[54] These statements, of course, suggested that Warren was anything but objective about the Kennedy assassination.

Thus it appears that a number of forces—his strong sense of family intimacy, his respect for Kennedy and his solicitude toward those immediately affected by Kennedy's assassination, the recollection of an unsolved murder in his own family—combined to make Warren extremely protective of the "private" features of Kennedy's assassination. At first glance this protectiveness might seem inconsistent with Warren's confidence in the judgments of the public at large, his expansive reading of First Amendment "newsworthiness" privileges on the Court,[55] and his concern for openness in government. But Warren's belief in a general principle of public access to the governmental processes of public officials—embodied in his insistence that all evidence to which the commissioners were exposed appear in the public record—did not extend to support for the public's "right to know" the details of events affecting national security or the intimate details of family life.

Warren, as noted, made a preliminary decision not to publish any report of the Warren Commission's findings; only pressure from other commissioners induced him to change his mind. The commission, in his view, was dealing with materials sensitive to national security and to the Kennedy family.[56] He also believed that the public had no right to know details about his father's murder, or his daughter Virginia's struggle with polio, or the Warren family's home life unless he chose to communicate that information. Obviously members of the public had no right to "relive the assassination of President Kennedy with the aid of pictures of his disintegrated head."[57]

♪

One also sees the impact of Warren on the Warren Commission in the time frame in which the commission's report was prepared. Warren's history as a public official, we have seen, had been marked by a concern for promptness in the dispatch of business, whether the business consisted of cases in court, legislative proposals, deadlines for speeches, or the Supreme Court's docket. While Warren himself tended to procrastinate, especially with respect to tasks he did not relish, he placed a great deal of pressure on staff and subordinates to perform work quickly and to meet deadlines.

When Warren was appointed chairman of the commission he was apparently given the impression that the commission's tenure would last about three months. That expectation, at any rate, was what he conveyed to staff. Rankin, for example, was told by Warren that his job as general counsel "would take only 3 months at the outside."[58] The commission staff had been hired by the end of January, 1964, and the commission's hearings began on February 3 of that year. Five months later the hearings were still going on, with the commission's report yet to be written. In May, Rankin instructed the staff to "complete their summary of the case" by June 1, but only one staff member, Arlen Specter, who was investigating the "basic facts" of the assassination, submitted his report by that date.[59]

The date of June 1, 1964 as a termination date for the commission's work, which was unpublicized outside the commission staff, had been formulated by Warren at the commission's meeting on January 21, 1964.[60] When one staff member, Howard Willens, met with Warren in June and indicated that he would not be able to meet that deadline—as was the case with all the staff members except Specter—Warren reportedly lost his temper.[61] On June 27 the commission announced that its report would not be released until after the Republican National Convention on July 13; the deadline for termination was then extended to August. In August the deadline was extended to September. The report eventually was submitted to President Johnson on September 24 and released on September 28.

In an oral history interview about his service with the commission, Warren stated that "there was no deadline of any kind for us."[62] He also initially expressed doubts that the commission was affected by the fact that 1964 was an election year, since the commission was formed in 1963. When reminded that most of the commission's business took place in 1964, Warren said that the presidential election was "no factor at all."[63] But in a January 21, 1964 meeting of the commission, Warren said, in proposing a June 1 target date for completion of the report, that "if this should go along too far and get into the middle of a campaign year . . . it would be very bad for the country."[64] In addition, three commission staff members stated in 1978 that "[Warren's] main motivation in wanting the work done, which he repeated several times to different members of the staff, was that he wanted the truth known and stated to the public before the Presidential election of 1964 because he didn't want the assassination in any way to affect the elections."[65] Willens, recalling Warren's angry response to hearing about delays in June of 1964, speculated that "the Chief Justice was very discouraged . . . that the deadlines he had hoped could be met were not any longer realistic ones." There was "certainly concern," Willens said, "about whether the report would come out in advance of the 1964 election."[66]

Several staff members, in 1978 testimony before the Select Committee on Assassinations, indicated that Warren kept regular pressure on the staff to complete their work, one indicating that "[t]he Chief Justice was a very strong force on the Commission and was interested in receiving periodic reports or documents showing the progress of the Commission's work."[67] The staff members, however, did not claim that Warren asked them to "sacrifice thoroughness for speed."[68] In fact, in the summer of 1964, when staff reports began to flow into the commission, Warren himself functioned to delay the release of the report. The reason for the delay underscores a final theme of Warren's service on the commission, his desire that the final report be unanimously adopted by all the commissioners.

✍

Commission staff members testified that their drafts of sections of the report were rewritten before they appeared in print,[69] although they were not consulted about the rewriting. The rewriting procedure was evidence that as the report neared completion issues remained on which the commissioners were divided. Two issues, in particular, were troublesome: the number of shots that hit Kennedy and the question of evidence of a conspiracy. With respect to the first issue, John Sherman Cooper, one of the commissioners, recalled a "debate" within the commission "as to whether there were two shots or three shots or whether the same shot that entered President Kennedy's neck penetrated the body of Governor Connally." Cooper himself held to the view "that there had been three shots and that a separate shot struck Governor Connally."[70]

Warren, who at this point believed in the single-bullet theory, proposed resolving the issue with ambiguous language, which stated the conclusion as follows: "Although it is not necessary to any essential findings of the Commission to determine first which shot hit

Governor Connally, there is very persuasive evidence from the experts to indicate that the same bullet which pierced the President's throat also caused Governor Connally's wounds."[71] This language was eventually accepted by all the commissioners. A 1978 House committee report referred to Warren's tactic as "resolv[ing] the issue . . . by the use of agreeable adjectives, rather than by further investigation."[72]

The conspiracy issue was potentially more serious. Commissioners Richard Russell and Hale Boggs remained unpersuaded that no conspiracy surrounded the assassination, and Russell reportedly intended to make public his objections in a separate statement in the commission's report. At this point Warren rewrote the conspiracy finding, stressing the language "the Commisson has *found no evidence* that either Oswald or Jack Ruby was part of any conspiracy," conceding that "the possibility of others being involved with either Oswald or Ruby cannot be established categorically," but indicating that evidence of a conspiracy "has been beyond the reach of all the investigative agencies and resources of the United States."[73] On reading that language, as Rankin recalled, "the Commissioners . . . were all willing to accept [it] . . . and [agreed to] not dissent from or want any minority report and they so voted."[74] Warren had secured unanimity again, and had been willing to pay for it even at the price of a short delay.

One staff member, recalling "changes in tone" between the staff drafts and the commission's report, identified Warren with the changes. "Earl Warren was adamant," he testified, "that the Commission should make up its mind on what it thought was the truth and then would state it." This desire for unanimity and forthrightness, where possible, "was consistent with [Warren's] philosophy as a Judge. The *Brown v. Board of Education* decision . . . was unanimous. I think [Warren] was at great pains to make sure it was." On the commission, as well, "a great deal of time was spent on . . . getting a unanimous opinion from all the Commissioners . . . because Earl Warren felt it was best that they make up their mind as to what they thought the truth was and then try to settle it."[75]

᚜

The Warren Commission and *Brown v. Board* can be seen, in this view, as of a piece. In both episodes Warren was a "tremendous presence"—an "overriding strong force."[76] In both episodes he came to conclusions early, pressed continually for a resolution of the issues, was sensitive to potential dissenting views, adopted broad and ambiguous language to satisfy objections, and pressed hard for unanimity. In both episodes his central concern, notwithstanding his own disclaimers, was with reaching a just, sound, and politically sensible result. In *Brown* he desired a result that was forceful yet strategically presented; in the Warren Commission he wanted to identify Oswald as a misfit loner assassin to allay doubts, as far as possible, about conspiracies and to avoid drawing the assassination into the 1964 election campaign. In both episodes his sympathies were deeply engaged. In *Brown* he had been moved by NAACP counsel Thurgood Marshall's vivid pictures of black and white children playing together before and after school but being forcibly separated by the school system.[77] In the Kennedy assassination he had been moved by the "horrible deed" itself and the consequent bereavement of the Kennedy family.

But the consequences of Warren's presence on the Warren Commission invite contrast to the consequences of his presence in *Brown*. In the segregation cases he galvanized a reluctant Court into action on behalf of a simple but highly significant moral proposition, that the separation of one race from another is necessarily stigmatic to the race being separated. If Warren had not been on the Court, the *Brown* decision might not have been unanimous and might not have generated a moral groundswell that was to contribute to the emergence of the civil rights movement of the 1960s. Despite all our current difficulties with the law of race relations, it is hard to deny that the *Brown* decision has led to a greater appreciation in American society of the moral, as well as the legal, insignificance of skin color as a guide to human conduct. It is also hard to deny that American society is better off for that enhanced appreciation.

Whether contemporary American society is comparably better off because of Earl Warren's presence in the Warren Commission is a much more problematic issue. To be sure, many commission staff members have pointed out in discussing their experiences that there is something cosmic about presidential assassinations that gives them an air of unreality, and simple "lone-assassin" theories do not seem sufficient to match the vast significance of the event. Moreover, the Kennedy assassination was the first to take place in the age of visual mass media. The enormous impact of a presidential assassination being rendered on television has served to leapfrog the events in Dallas over the more "forgettable" assassinations of McKinley and Garfield to rival the assassination of Lincoln in historic significance. Such events have a quality of constantly fascinating and bedeviling the public, and purportedly authoritative commentators on those events, such as the Warren Commission, have a way of being under constant and very severe scrutiny.

But there remain some decisions made by Warren on behalf of the commission that seem to have diminished the effectiveness of the report which consequently failed to "allay doubts" about the Kennedy assassination. The most striking of these decisions, the decision not to allow others to view the x-rays and photographs, deprived the commission of evidence that might have supported the least credible portion of its theoretical reconstruction of the assassination, the single-bullet theory. Other decisions—Warren's tentative resolution not to publish the commission's findings, his pressure on the staff to assemble and report on their findings in less than six months, his brief and gentle interrogations of Jacqueline Kennedy and Marina Oswald, his papering over of differences among the commissioners in the final text of the report—helped to create an impression among polemical critics that the report was a "white-wash" or a "cover-up."[78] There was a fair amount of irony in that impression, since the most serious of Warren's errors stemmed from his refusal to create a distinction between "private" information through which the commission would come to its findings and "public" information through which others might assess those findings. Nonetheless his failure to release information allegedly affecting national security or the privacy of persons intimately connected with the assassination did not aid the commission's credibility.

Warren, we have seen, relied too extensively on the investigative apparatus of the FBI and CIA and was deliberately misled by both agencies. But, as noted, it does seem unfair to criticize him or the commission for such reliance. The commissioners regarded themselves as a temporary agency with limited resources and a short term of operation. They had no initial reason not to trust the FBI and CIA, who were, after all, agencies of the United States government helping to investigate the murder of a United States president. When evidence that the FBI's investigations might have been deficient came to Warren's attention, he, in contrast to several other commissioners, expressed an inclination to pursue an independent investigation of the FBI itself.[79] He also called at one point, as noted, for "a thorough investigation . . . as to the relationship between the FBI and the Secret Service and the CIA in connection . . . with [the assassination]."[80] These investigations never materialized, and as a result the commission's credibility has been severely damaged. But in an atmosphere of strong public support for the CIA and FBI, and those agencies' hostility to the commission's investigations, there was not much the commission with its relatively skeletal staff, could have done to counter its dependence on the agencies.

On balance, however, the themes that marked Warren's service on the commission combined to diminish the commission's status as an authoritative source on the Kennedy assassination. Warren's instinctive ability to make decisions swiftly and to devote his energies to convincing others of the rightness of his choice had, in the case of the murder of John F. Kennedy, resulted in his seizing upon a theory—the lone, demented assassin—that seemed hard to reconcile with ballistic evidence. Moreover, the commission had access to no other evidence, Warren having taken a narrow view of its investigative powers. Having immediately embraced the single-assassin theory, and then later the single-bullet theory that supported it, Warren became an advocate for their unanimous adoption by his fellow commissioners.

When Warren was faced with opposition, he resorted to using ambiguous language in the report. This language undermined the authoritativeness of the commission's findings. In addition, he continually pressed for promptness in the issuance of the report so as to avoid drawing the commission into politics. While he succeeded in keeping the report from becoming an election issue in 1964, he also insured a short-term life for the commission that limited the staff's and the commissioners' ability to produce a polished document. The result was that while Warren achieved the short-term goals that he sought, he did not achieve his primary purpose. The report that his commission produced has not served, over time, to quench rumors and allay doubts. Indeed the inadequacy of the Warren Commission's investigation may be the single most disquieting feature of the Kennedy assassination.

Warren's performance on the Warren Commission can ultimately be traced to his temperament and to his conception of the commission's role. He did not want to serve on the commission, and he refused to reduce his work as Chief Justice during the period of the commission's existence. His decision set a pattern: the other commissioners came to see their service on the commission as no more than part-time jobs. Only Rankin and the staff worked full-time on the commission's investigation; the commissioners met infrequently; and even then some members missed several meetings.

At the same time Warren was accustomed to "taking charge" of an office, making unilateral decisions, keeping close tabs on staff, and conducting office business with promptness and dispatch. Given Warren's approach to officeholding, the occasional participation of other commissioners, and Warren's status as chairman, he had sufficient opportunity to see that the commission's performance was consistent with his beliefs about its proper function.

Warren's conception of the role of the commission, we have noted, was that of a mechanism for allaying public doubts about the circumstances of Kennedy's murder. He wanted the commission to get at the truth, of course, and to disclose most of the information that it obtained, but above all he wanted to assuage rumors and alleviate fears of conspiracies. By focusing on a lone-assassin theory of the assassination, by relying on the investigative reports of the FBI and the CIA, by accepting the single-bullet hypothesis, by keeping the commission's report from becoming a political issue, and by compromising on language in the report to avoid the surfacing of internal dissension, Warren did his best to "allay doubts" by conforming the Kennedy assassination to what he hoped would be seen as a familiar pattern of presidential assassinations in America.

The problem with Warren's approach was its context. He sought to portray a sensitive and consuming subject, the assassination of an American president, in a simple and vulnerable fashion. Had Kennedy's assassination not been so cosmic an experience for many Americans, and therefore the object of such continuous public scrutiny, the vulnerability of Warren's approach might not have become apparent. But subsequent investigations of the Kennedy assassination have revealed that all the evidentiary lynch pins of the Warren Commission Report were shaky at best.

The single-bullet theory was repudiated by John Connally, one of the persons whom Oswald shot, and has been challenged in subsequent literature.[81] The FBI and CIA did not report evidence to the commission of apparent connections between Oswald and persons outside the United States. Not all the commissioners who signed the report believed in all its conclusions. And a subsequent congressional investigation of the assassination advanced the hypothesis that more than one gunman fired at President Kennedy.

The apparent weakness of the evidence on which the commission's report was based, the paucity of the commission's investigative apparatus, and the lack of support that the commission received from the FBI and the CIA have invited critics to suggest that the commission was engaging in a "cover-up." While these suggestions have yet to be based on any significant evidence, the commission did exhibit tendencies of fastening on a convenient and "safe" explanatory hypothesis—the lone-assassin theory—and taking pains to minimize the weaknesses or internal difficulties with that hypothesis.

Earl Warren can fairly be identified with those tendencies of the Warren Commission. That this is so attests to the singlemindedness with which he would pursue goals when he thought those goals were worthwhile. Warren's singlemindedness on the Warren Commission can be said to have an opposite effect from his singlemindedness in *Brown*. Whereas *Brown*, despite the academic difficulties its opinion caused, can arguably be seen as a "simple case," capable of a direct yet profound moral resolution, the assassination of John Kennedy was not, despite Warren's claims in his memoirs, a simple murder case. The Kennedy assassination was not "simple" because of a combination of fuzzy and contradictory evidence and the timeless significance of presidential assassinations.

Warren's swift conclusions about the assassination and his vigorous insistence on unanimity in the commission report exposed the commission to subsequent criticism. The felicitous interaction of Warren's personality and a momentous public decision, so apparent in the segregation cases, was not repeated in his service on the Warren Commission. The experience of chairing the commission "was obviously a terrible, terrible strain on him," Earl Warren, Jr., said of his father.[82] The commission episode was "terribly exhausting and most unpleasant" for Warren, one of his law clerks recalled; Warren "didn't like to reminisce" about it.[83]

These initial chapters on Warren the Chief Justice have stressed those of Warren's features identified with his "presence": interaction with colleagues, personal convictions, techniques of decision making, "essences" and "tones." Such features are not often associated with analyses of the careers of judges; they are not easily extracted from judicial opinions. More common in studies of judges are personal traits relevant to their performance as jurists. Most prominent among those is usually the intellectual bent of a judge's approach to his office: the jurisprudential perspective that gives the mass of his decisions a consistency and a scholarly identity. Warren has regularly been regarded by critics as having failed to supply a universally applicable jurisprudential basis for his opinions. He is said to have been a judge for whom instinctive results were far more important than analytical reasons. The next portion of this study seeks to explore the meaning and assess the accuracy of that criticism.

Warren as a cadet in Army Officer Training School. Camp Lee, Virginia. The photograph was taken in the spring of 1918.

Warren, U.S. Webb (the incumbent Attorney General of California), and William H. Waste, Chief Justice of the California Supreme Court, at a dinner launching Warren's candidacy for Attorney General in 1938. Webb and Waste are writing contributions to Warren's nonpartisan campaign, which had a budget of $35,000.

The Warren family enters the Governor's Mansion on Warren's succession to the Governorship of California in January, 1943.

Vice-Presidential candidate Richard Nixon, Presidential candidate Dwight Eisenhower, and Warren in a publicity photo in October, 1952, after Nixon's "Checkers" speech of September 23.

Warren assumes a magisterial pose on his first day as Chief Justice of the
United States, October 5, 1953.

A photograph of Warren released to the press on his seventy-fifth birthday,
March 19, 1966.

# THREE

## THE CHIEF JUSTICE: JURIST

# 9

# Warren's Theory of Judging

In an instructive commentary on Earl Warren as Chief Justice, Anthony Lewis suggested that Warren "made no attempt, in opinions or otherwise, to propound a consistent theory of how a judge interpreting the Constitution should approach his task." Warren "evidently felt unconfined by precedent or by a particular view of the judicial function"; his opinions were "difficult to analyze because they [were] likely to be unanalytical." Warren seemed to be asking students of his Court to "put aside . . . [s]ome of the qualities valued in the judicial process—stability, intellectuality, craftsmanship"—and to search for "the just result." For Lewis, a Warren opinion was "a morn made new—a bland, square presentation of the particular problem in that case almost as if it were unencumbered by precedents or conflicting theories."[1]

Lewis's commentary described Warren as a jurist and juxtaposed Warren's performance as a judge against an ideal of enlightened judging. As a description, it was not inaccurate, although I shall argue that Warren's theory of judging was considerably more complex than his critics have indicated. As an evaluative statement, identifying effective judging with qualities such as "intellectuality," "craftsmanship," "stability," and analytical skill, Lewis's comments raise the fundamental issues of what constitutes valuable judging in

American society and who should make that determination. Earl
Warren's career as a jurist compels examination of those issues.

᪥

My analysis of Warren as a jurist begins with some preliminary assertions. The Constitution spoke directly to Warren as a person, a
judge, and an American citizen. He conceived of the Constitution as
an embodiment of values that he believed in and as a basis for granting him, as a judge, power to protect those values. Warren repeatedly emphasized that he had a "duty" under the Constitution to see
that his understanding of its imperatives was implemented, and he
saw the Constitution's imperatives as ethical imperatives.[2] The ethical imperatives that Warren read in the Constitution were so clear to
him, and his duty to implement them so apparent, that matters of
doctrinal interpretation were made simple and matters of institutional power became nearly irrelevant. If a branch of government
had engaged in or tolerated a practice that Warren found inconsistent
with his conception of American citizenship under the Constitution,
the power of that branch was thereby undermined. Warren's perception of constitutional values overwhelmed institutional values in his
decision making calculus.

The ethical imperatives that guided Warren as a judge reflected
his personal morality in that Warren held a set of values that he
believed represented moral truths about decent, civilized life. It was
inconceivable to Warren that these values would not be embodied in
constitutional principles, since he believed that they formed the essence of American democracy. Indeed, Warren felt *bound*, as a judge,
to consider ethical imperatives in his adjudication; in his view, they
deserved as much consideration as explicit constitutional language,
and perhaps more. His principal concerns as a jurist were to discern
the underlying ethical structure of the Constitution and to apply rig-
orously its ethical imperatives, even if such an application resulted
in a failure to achieve orthodox doctrinal consistency.[3]

Warren's juristic perspective cannot completely be divorced from
the intellectual context in which he and the Warren Court functioned. I have discussed that context in some detail elsewhere[4] and
shall here attempt only the barest recapitulation. Warren's tenure on
the Supreme Court was marked by the dominance of a particular
twentieth-century academic theory of judicial performance. The the-

ory, which had been influential in academic and in some judicial circles since the early years of the century, assumed that judges were lawmakers and worried about the consequences of unrestrained judicial lawmaking in a democratic society.

The chief constraints on judges as lawmakers, the theory assumed, could no longer come from mystical notions of law as a brooding omnipresence in the sky or of judges as oracles charged with discovering the meaning of legal principles that were unintelligible to laymen. The chief constraints had to come from judges' own good sense. Good sense took two principal forms: institutional good sense, the disinclination to usurp the proper powers of other branches of government, and doctrinal good sense, the good sense to base judicial decisions on properly articulated reasons. If judges, especially Supreme Court justices interpreting the Constitution, did not exercise good sense, the public would perceive judicial lawmakers as autocrats, substituting their judgments for those of more democratically elected branches of government or equating legal principles with their personal convictions.

Various versions of this theory were articulated by judges and scholars throughout the twentieth century. In 1905 Justice Holmes, dissenting in *Lochner v. New York*,[5] said that "the accident of [judges] finding certain opinions natural and familiar or novel and even shocking ought not to conclude [their] judgment upon the question whether statutes embodying them conflict with the Constitution of the United States."[6] To Holmes's statement, intended to exemplify institutional good sense, was added, especially after the 1930s, statements by academicians of doctrinal good sense. Lon Fuller distinguished between decisions based on "fiat" and decisions based on "reason";[7] Henry Hart and Albert Sacks emphasized the requirement that judging be an exercise in the "reasoned elaboration" of the bases for a decision;[8] Herbert Wechsler asked Supreme Court justices to ground their constitutional law decisions in "neutral principles" that transcended the immediate social implications of the result.[9]

In crude summary form, the theory equated effective judging with "judicial restraint," either as manifested in a cautious exercise of the judiciary's power vis-à-vis other lawmaking branches, or as manifested in judicial efforts to suspend bias and to justify decisions on a "principled" doctrinal basis, "principled" denoting the qualities of rationality, impartiality, and intelligibility. Lewis's description of

Warren as a jurist began by assuming the correctness of "judicial restraint." Since Warren's approach to judging minimized stability, consistency, intellectuality, and craftsmanship, Lewis suggested, his decisions were "unanalytical," and he deemphasized "qualities valued in the judicial process."

Lewis did not suggest that Warren's lack of judicial restraint necessarily diminished his stature as a Chief Justice, but he clearly felt that Warren's posture diminished his stature as a jurist. Other advocates of judicial restraint, in assessing Warren's performance, have been more critical. Philip Kurland, reviewing Warren's memoirs, suggested that "[h]istory may yet decide" that Warren's chief justiceship was essentially one where "political power, including judicial power, was . . . exercised for the advancement of what was 'right and good' as [Warren's] personal ideolog[y] defined those . . . terms." Given this characterization, Kurland doubted that "Warren's place in history is one in a triumvirate with John Marshall and Charles Evans Hughes" and invited equation of Warren's stature with that of his "immediate predecessor and immediate successor: Fred M. Vinson and Warren E. Burger." [10]

In characterizing Warren as a jurist who failed to follow the canons of judicial restraint, critics have not distorted Warren's stance. Warren specifically rejected both institutional and doctrinal good sense. In his memoirs, for example, he called "the so-called doctrine of 'neutral principles' " a "fantasy," stating that "[a]s the defender of the Constitution, the Court cannot be neutral." He also said, in the same paragraph, that "[t]he Court sits to decide cases, not to avoid decision, and while it must recognize the constitutional powers of the branches of Government involved, it must also decide every issue properly placed before it." [11]

Warren's perspective as a jurist was formulated outside the dominant academic theoretical perspective of the twentieth century. To the extent that he took notice of various articulations of judicial restraint, whether by judges such as Felix Frankfurter or John Harlan or academicians such as Hart, Wechsler, or Kurland, he was either unconcerned about them or antagonistic to them. And despite what Warren's critics have concluded, rejection of institutional and doctrinal good sense does not entail the acceptance of "result-oriented jurisprudence." It only entails acceptance of result-oriented jurisprudence as defined by the dominant twentieth-century academic theory of judicial performance. Warren rejected that theory, but he did not endorse a theory that judges were free to do as they liked.

⬿

We have seen that well before he became Chief Justice, Warren developed the idea that certain ethical principles were both a manifestation of fundamental values in American society and a necessary foundation of enlightened government. The development of this belief during Warren's earlier career had a profound influence on him as Chief Justice.

Warren's concern for the ethical roots of conduct became a characteristic of his service as a California public official. As district attorney of Alameda County he replaced a traditional style of staff member—the part-timer whose law practice might present him with conflicts of interests—with dedicated, hardworking, nonpartisan staff of his own choosing. Warren served notice that his office would not make deals with political bosses or countenance corruption in public officials: Under his tenure the district attorney's office became known as a "Boy Scout organization." Warren's deputies, in their memoirs, remembered his insistence that they not break the law to enforce the law.[12]

Casting issues in ethical terms and making others aware of the importance he attached to moral principles was a familiar pattern in Warren's California career. He refused to have a permanent campaign organization because he thought such groups tended to become interested in maintaining their own patronage. He summarily fired his publicity director for giving out unauthorized and misleading information. As governor he declined to "log-roll" with legislators. He did not deviate from his promise to put statesmanship ahead of partisan politics, refusing to campaign for other Republican candidates.

Throughout his career in California Warren identified himself with issues that he thought basic to a decent and humane society. In championing those issues he was opposed consistently by persons whose resistance was directed not only to the merits of the issues but also to Warren's purported usurpation of their institutional powers. Local sheriffs resented District Attorney Warren's invasions of their power to investigate gambling, prostitution, and organized crime, but Warren, in his thirteen years as district attorney of Alameda County, perceived the local sheriffs as corrupt and therefore countenanced the invasions. Governor Culbert Olson sought, in his four-year term between 1938 and 1942, to retain civil defense powers in his office rather than that of Attorney General Warren, but Warren doubted Olson's administrative effectiveness and the patriotism of some Olson supporters. Warren consistently challenged Olson's powers and eventually ran against Olson in 1942.

Legislators attempted to demarcate the limits of Governor Warren's institutional authority by inviting him to bargain with them on pieces of legislation; Warren declined to acknowledge any limits by refusing to bargain, feeling that if his causes were truly just, the public would support his incursions on legislative prerogatives. In some instances, such as the issue of compulsory health insurance that surfaced after World War II, Warren was not able to bypass the legislature despite his conviction that his cause was just; these setbacks only made him more suspicious of the ways in which opposing branches of government used their powers.

Thus Warren came to the Supreme Court with a long history of pitting his perceptions of decent and humane behavior against those of recalcitrant, competitive lawmakers. From his earliest days as a district attorney he had seen himself as a crusader for fundamental standards of morality, doing battle against the special interests. He was to retain that conception of himself as Chief Justice. That Warren had compromised some of his ethical principles in the context of politics did not mean that he adopted a relativistic attitude toward them. His ethical principles were absolutes, but sometimes they could only be partially vindicated. Warren did not see his ethical principles as mere personal likes or dislikes; he had become convinced that the principles to which he adhered were necessary features of a "decent" life. Given this conviction, it was obvious to Warren, if not to others, that such principles were embodied in the United States Constitution.

✍

In the 1938 gubernatorial campaign letter to Robert Kenny, Warren had called the protection of civil liberties "the most fundamental and important of all our governmental problems" and had defined "the American concept of civil rights" as encompassing "not only an observance of our constitutional Bill of Rights, but also the absence of arbitrary action by government in every field and the existence of a spirit of fair play on the part of public officials." Protection for civil liberties, for Warren, was linked to the "prevent[ion] [of] government from using ever-present opportunities to abuse power through harassment of the individual." [13] In 1955, his second year as Chief Justice, Warren expressed comparable sentiments in an essay in *Fortune* magazine. "Our legal system," he maintained, "is woven around the freedom and dignity of the individual. . . . [T]he Constitution exists for the individual as well as for the nation. . . . The American constitutional system . . . places the fundamental law above the will of the government." [14]

The Bill of Rights, in Warren's jurisprudence, was a means by which governmental usurpations of individual freedom were checked. The rights enumerated in the Constitution were simply "the natural rights of man" or the "common-law rights of Englishmen." [15] The American legal system was "a mature and sophisticated attempt . . . to institutionalize" a "sense of justice" that was inherent in "the nature of man" and was manifested in a desire for freedom from "the terror and unpredictability of arbitrary force." [16] The Bill of Rights, by codifying this sense of justice, provided a basis by which American law could be brought "more and more into harmony with moral principles." [17] Application of the Bill of Rights protections to new situations was a necessary part of the "pursuit of justice," with which Warren equated "the capacity to generalize and make objective one's private sense of wrong, thus turning it to public account." [18]

The pursuit of justice, as Warren defined it, was thus "a continuing direction for our daily conduct." The Bill of Rights needed revision with time. "We will pass on," Warren said, a "document [that] will not have exactly the same meaning it had when we received it from our fathers." For the Bill of Rights protections to be real, Warren argued, they needed constant application by the judiciary. A "better" Bill of Rights was "burnished by growing use"; a "worse" one was "tarnished by neglect." [19]

Two themes present in Warren's 1955 interpretation of the Bill of Rights were central to his stance as a jurist. The first was his identification of the language in the Bill of Rights with protection of the natural rights of man against the arbitrary actions of government. Warren's language in making this identification was characteristically broad and abstract. He equated the "natural rights of man" with the "common-law rights of Englishmen," and he associated the "rights" with the much more specific and potentially restrictive language of constitutional amendments. He set no doctrinal limits on the protection of individual rights against the state, and he mentioned no institutional limits on the power of the judiciary to protect those rights. Warren read the Bill of Rights as an invitation to the judiciary to "generalize and make objective" a "private sense of wrong," turning it to "public account" in "the pursuit of justice."[20]

The second theme of Warren's 1955 essay was the need for "constant and imaginative application" of the language of the Bill of Rights to new situations by the judiciary. Certain assumptions were implicit in Warren's articulation of this theme. He assumed that the meaning of the Bill of Rights necessarily changed with time; the interpretations of the framers would necessarily differ from those of him and his contemporaries. He assumed that continued and active application of the Bill of Rights was necessary to make its protections "real" and that active application would necessarily alter the meaning of the words in the Bill of Rights. He also assumed that the principal applier of the Bill of Rights would be the judiciary because "an independent judiciary" was "[t]he sign" of "the tradition which places the fundamental law above the will of the government."[21] One could expect considerable judicial activism in "the pursuit of justice."

Warren's 1955 essay, like his 1938 statement on civil rights, was not given much notice by contemporaries because in both cases Warren's language was sufficiently broad and opaque to invest his thoughts with the tenor of a Fourth of July oration. Two features of the essay, however, should have stood out. First, Warren said that "[t]he pursuit of justice is not the vain pursuit of a remote abstraction";[22] he followed that sentence with a discussion of the changing nature of the Bill of Rights and the need for specific application of its provisions.

Second, Warren recognized that the persons whose rights were most often threatened in American society were minorities; such groups were "quicker to band together in their own defense than in the defense of other minorities." They did not always "react to the truism," Warren wrote, "that when the rights of any individual or group are chipped away, the freedom of all erodes." Warren suggested that if "each minority, each professional group, and each citizen would imagine himself in the other's shoes, everybody's rights would have firmer support." He then equated "[t]he beginning of

justice" with the aforementioned capacity to turn one's private sense
of wrong to public account, making it "general" and "objective."[23]

Warren's conception of himself as a protector of "objective" moral
values and his identification of lawmaking with the pursuit of justice
combined to make his perspective as a jurist primarily an ethical one.
In a 1962 address Warren characterized law as "float[ing] in a sea of
Ethics" and claimed that "without ethical consciousness in most people, lawlessness would be rampant." In "civilized society," Warren
argued, "law . . . presupposes the existence of a broad area of human conduct controlled only by ethical norms. There is thus a 'Law
beyond the Law.'"[24]

Applying the dictates of ethics to difficult decisions, Warren said
in his 1962 address, required training "in the discernment of right
from wrong and in the will to accept the right without the slightest
duress." An enlightened ethical perspective, for Warren, was associated with the ability to "discern the right in the midst of great
confusion and to pursue it." Persons having this ability "kn[e]w how
to discover the ethical path in the maze of possible behavior."[25] There
were three imperatives in Warren's calculus of judging. The judge
needed to search for the "Law beyond the Law," to discern right
from wrong "in the midst of great confusion," and to discover the
ethical path. The judge needed to do these things because American
society was founded on basic distinctions between right and wrong,
justice and injustice, freedom and arbitrariness. The Constitution
embodied those distinctions. The judge had a duty to articulate them.

Warren's view of judging did not define imperatives in conventional institutional or doctrinal terms, and thus he gave far less weight
to institutional or doctrinal considerations than many of his cele-
brated colleagues on the Warren Court. Warren could join a Frankfurter opinion in a 1961 birth control case[26] that had as its premises
a narrow institutional theory of the Court's power to decide controversial political matters and a cautious doctrinal approach to novel
constitutional issues, and then endorse, in another birth control case
decided four years later,[27] a view of the Court's review powers that
was too institutionally activist for Hugo Black and too doctrinally
novel for Harlan. Warren had simply determined that ethical principles dictated a different result.

Warren's interest in the consistent application of ethical principles that he thought were consonant with the pursuit of justice also

resulted in his regularly joining opinions, the language of which he did not necessarily endorse, and sometimes paring down his own language to retain a majority for his position. This trade-off between doctrine and results is not uncommon in Supreme Court decision-making, but Warren made it more readily than many justices because the moral values that he sought to apply could frequently be linked with a number of different constitutional justifications. An example of his relative indifference to the doctrinal basis of his opinions can be found in a comparison of his first draft of the opinion in *Bolling v. Sharpe*,[28] outlawing segregation in the District of Columbia, with his final product.

Professor Dennis Hutchinson has discovered, in the papers of Justice Harold Burton in the Library of Congress, a memorandum from Warren to the other justices dated May 7, 1954, ten days before the opinions in *Brown v. Board of Education*[29] and *Bolling v. Sharpe* were handed down.[30] In the memorandum, intended as a draft opinion in *Bolling*, Warren outlawed segregation in the District of Columbia on the doctrinal ground that the practice constituted an "arbitrary deprivation of liberty in violation of the Due Process Clause" of the Fifth Amendment.[31] The "liberty" being deprived, he argued, was the "fundamental liberty" of "access to the education which the government itself provides."[32] His argument was thus a substantive reading of the Fifth Amendment's Due Process Clause, with the term "liberty" being read as a fundamental right to receive a public education.

Warren believed that education was a "fundamental right" in the sense that he equated it with basic precepts embedded in American civilization. Such a right, Warren reasoned, must certainly be reflected in the values that formed the basis of that civilization. He had worked as governor to enhance the prestige and keep down the tuition costs of the University of California system; in his companion opinion in *Brown* he called education "the very foundation of good citizenship" and "perhaps the most important function of state and local governments."[33] Warren's conception of constitutional adjudication assured that he would find a duty in the Supreme Court to uphold that right.

Other justices, however, found no such "right" in the Due Process Clause of the Fifth Amendment. To have found such would have been to pour substantive content into the Fifth Amendment's Due Process Clause, a judicial practice that influential justices on the

Warren Court, especially Black and Frankfurter, had gone on record as deploring as an unauthorized exercise of power.[34] No evidence has yet surfaced as to what exchanges took place between Warren and other justices after the circulation of his draft opinion in *Bolling* on May 7, but Warren's final draft, circulated to the justices on May 15 and made public on May 17, entirely omitted the "fundamental liberty" analysis.

The opinion now justified invalidating segregation in the District of Columbia only on the ground that "[l]iberty under law extends to the full range of conduct which the individual is free to pursue" and that liberty could not "be restricted except for a proper governmental objective." Segregation in public schools was "not reasonably related to any proper governmental objective," for the reasons advanced by the Court in *Brown*. Since the states could no longer maintain segregated schools, Warren asserted that "it would be unthinkable that the . . . Constitution would impose a lesser duty on the Federal Government."[35]

Warren's final draft in *Bolling* stripped the opinion of much of its doctrinal significance. The *Bolling* case was different from *Brown* because the District of Columbia was an arm of the federal government, not a state, and there was no equal protection clause in the Fifth Amendment to apply against the federal government. Warren's final draft, however, suggested that the absence of an equal protection clause was inconsequential. His analysis of the Fifth Amendment's Due Process Clause indicated that segregation was an "arbitrary deprivation of [the] liberty"[36] of black children, but the analysis did not discuss what that "liberty" was or why it could not be restricted "except for a proper governmental objective."

Warren gave no indication, in short, of the doctrinal basis on which the Court was reviewing and invalidating the District of Columbia's educational practices. The analysis of educational rights made in his first draft had supplied him with a doctrinal basis. That analysis was out of current favor but had some doctrinal support in early Court cases.[37] By omitting the analysis Warren had essentially reduced the *Bolling* opinion to saying that segregation in the District of Columbia was being outlawed because its continuance was "unthinkable" after the outlawing of segregation in the states in *Brown*. That argument made sense primarily as an ethical proposition.

Warren omitted the fundamental liberty analysis in *Bolling* because he wanted to maintain unanimity in his Court. For Warren the segregation cases were simple exercises in the pursuit of justice. The practice of racial segregation in the public schools, being based on assumptions of racial superiority, was immoral and therefore wrong; the Court's duty was to eradicate it. The precise doctrinal steps that the Court took to justify the eradication through constitutional analysis were far less important to Warren than the Court's reaching the result of eradication unequivocally and unanimously. He did not want any of the justices equivocating about the result because of doctrinal technicalities; if the doctrinal analysis offended, he would modify it or delete it. He himself saw no equivocation possible on the ethical issues raised by *Brown* and *Bolling*, but he did not want to present potential wafflers with a technical way out.

*Bolling v. Sharpe* was a microcosm of Warren's technique of judging. In the decision of cases before the Supreme Court, Warren began his pursuit of justice by putting himself "in the other's shoes." One of his law clerks has said that Warren would "put himself right in the position of the person who was affected"[38] by a legal practice or rule under consideration and thereby "try to get to the essence of the case."[39] Often this technique revealed the "innate justice" of a case. In *Bolling* innate justice required treating black school-children in the District of Columbia, who, like their counterparts in the states, had a "right" to equal educational opportunities, the same way that those counterparts were to be treated after *Brown*.

Warren had said in his 1955 essay that "[a]ny child expresses . . . [a sense of justice and a desire for justice] . . . with his first judgment that this or that 'isn't fair'."[40] Numerous onlookers have recorded[41] Warren making similar "was it fair" queries of litigators before the Court. That question, in fact, has been so linked with Warren that he is remembered as putting it constantly to counsel: He actually put it only rarely. His sense of the fairness or justice of a case, however, was crucial to his ultimate decision. Fairness and justice were inquiries through which Warren determined for himself how a case squared with the ethical imperatives of the Constitution.

Warren's search as a Supreme Court justice was thus for the "essence" of a case that revealed to him the proper ethical stance "in the midst of great confusion." As noted, he regularly grasped the essence of a case by identifying with the claimant whose freedoms were being restricted. Often this identification yielded the discovery that an ethical principle dictated the vindication of a claimant's constitutionally protected moral right. School children could not be denied educational opportunities simply because of their skin color.[42] A suburban voter's vote was worth every bit as much as that of a voter on a farm.[43] A person detained for questioning by the police could not be deprived of legal representation simply because he or she was too poor to afford it.[44]

On occasion, Warren considered an individual's claim to be stripped of merit by the ethical principles of the Constitution. Gamblers were not deserving of protection against self-incrimination.[45] Dealers in obscene materials were ordinary criminals; their alleged First Amendment rights did not warrant serious consideration.[46] In reaching these disparate results, Warren "discover[ed] the ethical path . . . without the slightest duress."[47] Existing legal doctrine for him did not always reflect the "law beyond the law."

Once Warren had satisfied himself about the essence of a case and its appropriate ethical outcome, he became an advocate for the "proper" result. He went into conference, for the most part, with his mind already made up on results.[48] Before oral argument Warren instructed his clerks to write him bench memoranda, without disclosing how he stood on the disposition of a case. The bench memos were intended "to help flesh out the case"[49] by suggesting questions that the briefs submitted to the Court did not fully address. After hearing oral argument and reflecting on the case, Warren had in most instances come to a decision; he regarded conferences as an opportunity to see where others stood.

᭡

Warren's "craftsmanship" as a jurist was thus of a different order from that identified with enlightened judging by proponents of judicial restraint. Warren saw his craft as discovering ethical imperatives in a maze of confusion, pursuing those imperatives vigorously and self-confidently, urging others to do likewise, and making technical concessions, if necessary, to secure support. In believing his concessions on matters of doctrine to be "technical," Warren was defining his own role as a craftsman. It was a role in which one's sense of where justice lay and one's confidence in the certainty of finding it were elevated to positions of prominence in constitutional adjudication, and where craftsmanship consisted of knowing what results best harmonized with the ethical imperatives of the Constitution and how best to encourage other justices to reach those results.

Warren's definition of craftsmanship thereby reversed the relative weight assigned by influential twentieth-century theorists to the rightness of results and the doctrinal integrity of reasoning in constitutional adjudication. For proponents of judicial restraint the "rightness" of a result depended on the doctrinal integrity of the reasoning used to justify it; for Warren the vindication of moral principles provided its own justification. Warren's results were consistent with the ethical perspective from which he viewed social issues, which he thought to be embodied in the Constitution and which he believed to be basic to civilized life. His perspective did not give great weight to traditionally "legal" arguments where they were barriers to an application of the proper "ethical norm" and thus often appeared to eschew conventional techniques of legal reasoning, such as close analysis of a judicial precedent, the language of a statute, or the text of the Constitution. Since Warren's justifications for a result were often conclusory statements of what he perceived to be ethical imperatives, his reasoning as a jurist was regularly opaque. But opaque or unconventional reasoning is not the same as no reasoning. It merely invites one to analyze Warren's jurisprudence at a different level.

I now seek to develop the assertions made in this chapter through an analysis of four substantive areas of law to which Warren applied his theory of judging. The analysis of Warren's decisions made in each of these areas is itself unconventional in terms of standard academic criticism. The analysis does not begin, for example, by considering the doctrinal framework from which Warren viewed a particular area of law, for Warren very rarely approached decisions with a doctrinal framework. Nor does it consider seriously Warren's perception of the institutional constraints on judges.

The analysis instead focuses on Warren's discovery of ethical imperatives in the Constitution that imposed certain obligations on him as a judge. The analysis also addresses the relationship between Warren's personal convictions and this discovery process and examines the rhetorical techniques, including conventional doctrinal analysis, by which Warren attempted to communicate the rightness of his choice. My attempt to elucidate Warren's jurisprudence should not be taken as an attempt to defend it in any particular; chapter sixteen will reconsider the significance of Warren's approach.

# 10

# *Lessons in Civics*

A set of cases that were of prime importance to Warren were those that he saw as yielding fundamental lessons in civics. The term "civics" has no precise legal meaning, is not mentioned in the Constitution, and conveys associations of the elementary or secondary school classroom. Warren was not deterred by such associations: He took his civics seriously, writing a small book on the subject immediately after retiring from the Court.[1] While Chief Justice, he had seen potential civics lessons in three widely different kinds of cases: those defining the constitutional status of American citizenship, those determining standards for citizen participation in the electoral process, and those asking under what circumstances the government could invade the dignity or privacy of individual citizens.

☙

The 1958 companion cases, *Perez v. Brownell*[2] and *Trop v. Dulles*,[3] gave Warren an opportunity to examine the Constitution's protection of the status of being an American citizen. Both cases involved the expatriation of an American for doing an act forbidden by the Nationality Act of 1940. One act, the basis for expatriation in *Perez*, was voting in a foreign election; the other, the basis in *Trop*, was desertion of the armed forces in time of war. The Court split on the cases, sustaining expatriation in *Perez* and invalidating it in *Trop*. Warren dissented in *Perez*, joined by Black, Douglas, and Charles Whittaker, and wrote for the same three justices in a plurality opinion in *Trop*. Brennan, who had joined the majority in *Perez*, concurred with Warren's result in *Trop* but did not join his opinion.

The striking feature of Warren's opinions in the cases was his assertion in both that American citizenship was "not subject to the general powers of the National Government and therefore cannot be divested in the exercise of those powers."[4] Citizenship for Warren was "man's basic right"; it was "nothing less than the right to have rights."[5] Warren based this claim not on any language in the Bill of Rights but on a more general conception of the relationship between citizens and government. He maintained that the American government was "born of its citizens" and that it existed "[t]o secure the inalienable rights of the individual."[6] Since in the American system of government sovereignty rested in the citizenry, giving the government power to deprive citizens of their citizenship status was impermissibly transferring sovereignty to the government.

Warren's argument, with its emphasis on fundamental principles of social organization rather than on language in the Constitution, was reminiscent of Chief Justice Marshall's appeals beyond the constitutional text to "general principles which are common to our free institutions."[7] But Warren was operating in a period in the Court's history when such appeals were regarded by most constitutional theorists as inappropriate. The constitutional question in *Perez* was whether a congressional statute providing for expatriation as a consequence of certain specified activities violated a provision in the Constitution. Warren's response was that the statute violated the basic spirit and political theory of the Constitution.

The above analysis, emphasizing an ethical structure to the Constitution, was sufficient for Warren. Citizenship could only be voluntarily relinquished by the sovereign citizenry. But in the *Trop* case Warren sought to maintain a majority, and in *Perez* five justices had rejected his view of citizenship. He thus advanced a doctrinal argument to justify the result in *Trop*: Expatriation was a "cruel and unusual punishment" under the Eighth Amendment.

If Warren's theoretical approach to citizenship issues was unusual for his time in not resting on explicit constitutional language, his construction of the Eighth Amendment in *Trop* was equally unusual in that it propounded an institutional theory of judicial review of congressional legislation that was strikingly unconventional. First, Warren claimed that even though Congress had characterized the Nationality Act of 1940 as other than a penal law, expatriation was a "punishment." This claim assumed that the Court could conclude for itself what the purpose of a congressional statute was despite Congress's statement to the contrary.

Warren next maintained that expatriation was a "cruel and unusual punishment," because it was "offensive to cardinal principles for which the Constitution stands."[8] The source of those principles was Warren's perception that expatriation "destroy[ed] for the individual the political existence that was centuries in the development."[9] Destruction of that "political existence" was the equivalent of "the total destruction of the individual's status in organized society."[10] But nothing in the Constitution suggested that the right to political existence was absolute. The Constitution's own language suggested that a political existence could be restricted to qualified persons.

Warren's test to ascertain the meaning of "cruel and unusual" in the Eighth Amendment was "the evolving standards of decency that mark the progress of a maturing society."[11] Under that test expatriation was clearly indecent, but the death penalty, he conceded, was not. The destruction of a person's political existence was thus more indecent than the destruction of his or her life. While Warren's concession about the death penalty was merely a bow to established precedent, his own interpretative standard for the Eighth Amendment suggested that its meaning could be made the equivalent of what a group of judges thought to offend standards of decency at any given time. Warren, in short, had made no serious effort in his Eighth Amendment argument to analyze the doctrinal history or the current doctrinal status of the Eighth Amendment. He had simply asserted that since citizenship was man's most basic right, involuntary expatriation was necessarily a cruel and unusual punishment.

Warren then justified his conclusion in *Trop* by declaring that the judiciary had a "duty of implementing the constitutional safeguards that protect individual rights."[12] Since citizenship was a "fundamental right," the Nationality Act's restrictions on it conflicted with a provision of the Constitution, and therefore the judiciary had "no choice but to enforce the paramount commands of the Constitution."[13] But there was no indication in the Constitution that citizenship was a specially protected right that could not be divested, let alone that it was a fundamental right. The "conflict" between the Eighth Amendment and the Nationality Act had been created by Warren, with no support from any other justice. The "duty" of the Court to invalidate the Nationality Act on Eighth Amendment grounds was hardly obligatory.

The *Perez* and *Trop* opinions were characteristic of Warren's jurisprudence in that they violated existing academic canons of judicial craftsmanship and employed Warren's own canons. If one catalogs the standard analytical steps of a judicial opinion reviewing the constitutional limits of congressional legislation—articulation of the constitutional right purportedly affected, investigation of the purposes of objectives of Congress in passing the legislation, formulation of the doctrinal standard of judicial review operative in the case, resolution of the controversy through textual interpretation of the Constitution—Warren had largely ignored existing wisdom about how justices were to undertake those steps. He had called citizenship a "fundamental" right without support in the case law for that proposition. He had substituted his own judgment about the purposes and objectives of the Nationality Act for that of Congress. He had found a judicial duty to enforce the Constitution where it conflicted with an act of Congress without elaborating the basis of the conflict. He had failed to interpret the Eighth Amendment in a way consistent with established readings of its meaning.

Warren did not see any of these features of his expatriation opinions as significant. What was significant to him was the essence of the *Perez* and *Trop* cases, the basic ethical principles involved, and the relationship between the Constitution, the judiciary, and those ethical principles. Under Warren's canons of craftsmanship his *Perez* and *Trop* opinions were enlightened performances. The opinions exposed the fundamentally defective character of the Nationality Act of 1940. The act, under the guise of laying down conditions for nationality or administering the armed services, placed restrictions on the exercise of American citizenship that were simply inconsistent with the meaning of being a citizen. Citizenship, being "man's most basic right," could not be restricted in that fashion unless government in America did not exist for the consent of the governed.

Thus one "essence" of the expatriation cases was which fundamental values were at stake; another was the practical meaning of what Congress was seeking to achieve in passing the legislation. While expatriation laws had long been regarded as not penal, and while

Congress had disclaimed the punitive features of the law being tested, the Nationality Act, as applied to the petitioners in *Perez* and *Trop*, was in fact a penal law. American citizens who committed acts inconsistent with Congress's conception of national allegiance were punished by being expatriated. In operation the Nationality Act attempted to condition American citizenship on "proper" voting behavior or on "proper" performance in wartime. The conditioning process was in the form of a deterrent: If citizens behaved improperly, they lost their citizenship. As a matter of elementary justice and morality, Congress's efforts were unprincipled.

The ethics of the *Perez* and *Trop* cases thus dictated for Warren that the Court invalidate the penal sections of the Nationality Act. Since significant ethical principles—the "cardinal principles" of a civilized and decent society—were at stake, Warren, as a justice, had no presumptive reponsibility to defer to Congress if he felt Congress's behavior to be unconstitutional. He especially had no responsibility to accept Congress's own characterization of the purposes of the Nationality Act: That characterization ignored reality and was potentially self-serving. His "duty" lay elsewhere: to see that the proper ethical resolution of the basic issues raised by *Perez* and *Trop* was made. In achieving that resolution he relied on the Constitution because in American jurisprudence the Constitution prevailed over acts of Congress that conflicted with it and because the Constitution embodied "the evolving standards of decency" in America.

As Chief Justice, Warren was charged with the responsibility of resolving conflicts between Congress and the Constitution. He discovered what was at stake in *Perez* and *Trop*, and he concluded what the decent, humane, and just resolution of the cases should be. He believed that the Constitution stood for decency, humanity, and justice, and thus he expected that its ethical imperatives—its "paramount commands"—would require invalidation of the penal sections of the Nationality Act. He found such imperatives either in the "cardinal principles," embodied in the entire document, that "citizenship is man's most basic right," or in the "decency" requirements of the Eighth Amendment. In the expatriation cases, he did what he was supposed to do as a judge—pursue morality and justice and achieve the proper ethical outcome. That was his craft.

One senses in the expatriation cases that Warren's stance as a jurist differed from the stances of his academic critics in a manner that their evaluations of his performance have not often emphasized.

Warren did not accept but then imperfectly put into practice the conventional canons of judicial craftsmanship; rather, he bypassed them altogether. The fundamental issue raised by Warren's performance as a jurist is not, as has regularly been suggested, whether "good" results reached by the Court are justifiable if the reasoning justifying them is "bad." [14]

The issue is whether ignoring the established jurisprudential canons of judicial performance that Warren ignored constitutes "bad" judicial reasoning. If Warren's alternative set of canons constitutes a credible blueprint for judicial performance, then, even though Warren's reasoning was unconventional, critics of his performance need first to ascertain whether his reasoning was "bad." The pursuit of that inquiry requires a fuller understanding of the way in which Warren's reasoning operated in specific circumstances.

The next set of Warren's "civics lessons" cases involved citizen participation in the affairs of government, as exemplified by voting in state elections. Here Warren's views were quite explicit. Since the American government existed for its citizens, since the views of the people were "often in advance of their leaders," and since the people were expected by the framers "always to be in control" of the government, their rights as citizens to make their views known through the electoral process could not be infringed. The Constitution, Warren believed, "endowe[d] all citizens with the right to participate in their government." [15]

Like Warren's assertions about the constitutional status of citizenship in the expatriation cases, his assertions about a constitutional right to participate in government were not obvious on their face. Although the Constitution assumed a republican form of government, with a degree of citizen participation, it did not envisage an unrestricted right in each citizen to make his or her views known through the electoral process. In fact, the right to vote was sharply qualified at the time of the framing of the Constitution by considerations of gender, of race, of servitude, and of economic condition. If the "right" of citizen participation in governmental affairs was principally embodied by participation in the electoral process, the Constitution simply did not grant all citizens that right. It merely assumed that the republican form of government envisaged some regular, if limited, citizen participation.

Warren's opinion in *Reynolds v. Sims*,[16] the second of the major voting rights cases of the Warren Court, represented his principal juristic statement on citizen participation. The proposition that Warren advanced in *Reynolds v. Sims*—that electoral districts in both houses in a state legislature could be apportioned only on the basis of population—was one that he had been somewhat slow to endorse. We have noted that as late as 1948, while governor of California, Warren had defended the then-existing apportionment of the California state senate, which gave each county one representative regardless of its population, thereby insuring that Los Angeles County could cast no more votes in the senate than mountain counties with less than one hundred thousand residents.

By 1958, however, Warren showed signs of rethinking his position on reapportionment issues. As so frequently happened in his new role as Chief Justice, he began to conceive of an issue as raising principles that did not have to be compromised by the political process. A 1958 case from Georgia, *Hartsfield v. Sloan*,[17] challenged the constitutionality of the county unit system, by which voters from rural counties were given power disproportionate to their numbers. The Court denied a motion for leave to file a petition to compel a district judge to consider the constitutional validity of the county unit system. Warren dissented, saying that he thought a rule to show cause should issue.[18]

*Baker v. Carr*[19] came four years later, reversing the Court's well-established practice of treating reapportionment cases as raising exclusively "political" questions that the federal judiciary should decline to entertain. Behind Warren's vote with the majority in that case was a conviction that judicial deference to legislators in reapportionment matters was a sham because those possessing disproportionate power in a legislature had no incentives to surrender it. As governor it was to Warren's advantage to know which mountain senators were power brokers in the California legislature; the system had the virtue, from his point of view, of predictability. After coming on the Court, however, Warren "was led to the conclusion that it was unconstitutional to overweigh the value of some voters and underweigh the value of others."[20]

Warren reached that conclusion by his familiar recourse to "cardinal principles." The Constitution assumed that "ours is . . . a representative form of government through which the rights and responsibilities of . . . all the people" were protected. That protection needed to be effectuated through "representatives who are responsible to all the people, not just those with special interests to serve."[21] Accordingly, when Warren came to find a constitutional basis for the principle of "one man, one vote" he declared that a constitutional "right to vote freely for the candidate of one's choice" existed, and that "a debasement or dilution of the weight of a citizen's vote"[22] unduly infringed upon that right.

As in the expatriation cases, Warren's argument equated broad principles of political theory with specific constitutional rights. While it is a principle of republican government, representativeness is not necessarily offended by the votes of some citizens counting more than the votes of others. If the states were not prohibited by the Constitution from restricting the right to vote itself—and nowhere in the original Bill of Rights are there any prohibitions on state restrictions of suffrage—it would seem to suggest that "the right to vote freely" was not initially regarded as something essential to a representative form of government. Later constitutional amendments contained specific prohibitions against basing infringement of the right to vote on grounds of race or color, but the draftsmen of those amendments apparently assumed that voting rights could otherwise be restricted.[23]

Warren's opinion in *Reynolds v. Sims*, however, announced that a constitutional right to have one's vote counted equally existed and that legislative attempts to assign different weight to different votes were unconstitutional because they violated "the essence of a democratic society."[24] The consequence of *Reynolds v. Sims* was that when a state, through a popular referendum of its citizens, voted not to apportion one of its legislative houses on the basis of "one man, one vote," its action was constitutionally invalid.[25] In that example the constitutional right to have one's vote counted equally seemed to prevail even if it violated the principle of representativeness.

Warren invariably referred to *Baker v. Carr* as the most significant case of his tenure, a choice that followed from his sense of the fundamental importance of lessons in civics. Warren saw *Baker* and its progeny as striking at three evils that he considered to be obstacles to enlightened government: the presence of special interests, the selfishness of public servants, and the imperfect vindication of participatory rights held equally by all citizens. The presence of unequal apportionment systems confirmed his longtime concern that lobby groups and other particular "interests" would seek to control the legislative process. Such systems were simply efforts on the part of special interests to insure that their views would be more fully represented than those of other sectors of the public. Moreover, Warren reasoned, one could not expect those public officials benefiting from such systems to prefer the public interest over their own prerogatives, even though they had been elected to serve the public. Reform of malapportioned legislatures could not be expected to emanate from special interests or those legislators who profited by associations with them.

Reform of malapportioned legislatures was a task especially suited for the judiciary to perform, Warren believed. First, the principle on which reapportionment rested—equal participation for all citizens—would necessarily become compromised in the legislative forum as interest groups and legislators distinguished their particular goals from the general goal of equal voting representation. Second, since a cardinal principle of the Constitution was equality of citizen participation, the right to have one's vote counted equally was a constitutional right. Consequently the judiciary had a "duty" to afford that right the full measure of protection. A "denial of constitutionally protected rights," Warren wrote in *Reynolds v. Sims*, "demands judicial protection; our oath and our office require no less of us."[26]

One can readily note, as in the expatriation cases, a sharp divergence between Warren and his critics on the "craftsmanship" of *Reynolds v. Sims*. To apostles of judicial restraint, Warren's opinion was institutionally and doctrinally unsound. He had brought the Court into an area historically reserved for the political branches of government, overriding established decisions of his own Court that had cautioned against judicial resolution of "political questions."[27] In so doing he had put the Court in the apparently embarrassing position of compelling the voters of a state to apportion their legislature on one basis even when they had signaled a preference for another basis.

Warren had based his purported usurpation of legislative prerogatives on an interpretation of the Constitution that was neither faithful to its literal text nor consistent with the context in which it had been framed. He had substituted homilies such as "[c]itizens, not history or economic interests, cast votes" for doctrinal analysis.[28] He had apparently invited the public to think that "every major social ill in this country can find its cure in some constitutional 'principle' " and that the Court was "a general haven for reform movements."[29]

Warren considered all these criticisms to be unresponsive to the issues of principle at stake in the reapportionment cases. If there had been a long tradition of the votes of some American citizens counting less than others, that tradition was not made any more justifiable by its longevity. There had been an equally long, and equally unjust, tradition of treating citizens of one race as inferior to citizens of another. The Court had intervened in the political process to correct the latter injustice because its members suspected—correctly if the reaction of segregationist states to *Brown v. Board of Education* was any indication—that there were no incentives for legislators to correct the injustice themselves. The Court was intervening in the reapportionment cases for the same reason. Malapportionment benefited the very persons who could remedy it. Human nature being what it was, the existing prerogatives of legislators were a barrier to their sensing the full justice of a principle.

Warren himself had been slow to sense the essence of the reapportionment cases; he assigned *Reynolds v. Sims* to himself "knowing that this would create much comment in California."[30] On the Court he felt freer to probe "political" issues in search of their underlying principles; he no longer profited from political compromises. By putting himself in the "shoes" of voters discriminated against by malapportionment, Warren gauged the essential justice of their position. The American system of government "required fair representation"; fair representation "meant equal representation in which one man's vote had the same value as every other."[31]

In grasping this principle, as in grasping the principle that citizenship was man's most basic right because it was the right to have rights, Warren believed that he had learned a valuable lesson in civics. The American republic was founded on citizen participation; citizen participation was most fully realized in the electoral process. The electoral process could not justify counting one citizen's vote less than another if the American republic was also founded on principles of elementary fairness and justice.

The reapportionment cases were thus much more than a passing "reform." They were declarations that in the American system of government one citizen's vote was as worthy as another's. Warren did not see how such a declaration could be opposed on principle. He could understand the fact that reapportionment had been opposed as a practice, for the parochial angle of one's vision often prevented one from seeing the bedrock principles at stake in an issue. But once the principle of reapportionment was clearly understood, it could hardly be seen as other than a vindication of the spirit of American democracy.

Warren's judgment about the reapportionment cases has thus far been accurate. Despite the fact that the Court decided an issue traditionally reserved for legislatures, despite the weakness of textual support in the Constitution for the principle of one person, one vote, and despite strident and formidable opposition on his own Court to the opinions, Warren's ethical arguments on reapportionment have penetrated the public's consciousness. Framed in Warren's terms, the reapportionment cases appear as a struggle between the average voter, seeking not to have the fortuity of his residence affect the weight of his vote, and special interests advantaged by a fortuitous weighting of votes.

Such a struggle pits the values of equality and fairness against the value of entrenched privilege. Throughout American history, when rhetoric championing the values of equality and fairness has struck a chord in the public, entrenched privilege has been hard to justify. It does little good, in such an instance, to say that the issues are more complicated than the rhetoric makes them seem or that champions of equality and fairness are usurping their powers. Warren's craft in the reapportionment cases seems to have been an ability to convey to the public the civics lesson that he had learned.

It has been suggested that the concept of "liberty" in Western political thought embraces two strikingly different ideas, the idea of "affirmative" liberties, liberties that follow from being a member of a state that exists for the benefit of its citizens, and the idea of "negative" liberties, liberties that amount to private rights against the state.[32] The expatriation and reapportionment cases, from Warren's perspective, involved "affirmative" liberties, citizenship and citizen participation being benefits that individuals receive as a consequence of creating an apparatus that governs them. Warren also felt strongly, however, about "negative" liberties. Foremost among such liberties he ranked the right of individual citizens not to have their dignity or privacy invaded by agencies of the state.

The theme of negative liberties most clearly manifests itself in Warren's decisions in the area of criminal procedure, which will subsequently be considered. Warren once described his controversial opinion in *Miranda v. Arizona* [33] as having "simply said that when the law puts upon a man by putting him in restraint and taking him away from his home and his family and friends and starts to put him behind bars . . . he's then . . . entitled to have representation of counsel." [34] In seeking to ensure various constitutional protections for persons accused of crimes, Warren saw himself as vindicating an innate right of citizens to be presumptively free from the coercive mechanisms of government. He had developed this approach in a handful of diverse cases testing the limits of legislative investigative powers.

Warren's posture in the loyalty oath controversy at the University of California from 1949 to 1952 had revealed that, even in instances where he supported a broad governmental policy that had the effect of restricting individual rights against the state, he might oppose means chosen to effectuate the policy if he thought them to be unduly coercive. As a member of the University of California's Board of Regents, Warren had consistently voted against the imposition of a loyalty oath on university faculty and staff, but as governor he had called a special session of the California legislature in September, 1950 to institute an oath for all state employees, knowing that this oath would be applied against University of California personnel.

Warren was genuinely divided in the loyalty oath controversy between his deep patriotism and his fear of governmental intrusiveness. His early decisions on the Court indicated that he had not yet discovered an "essence" to investigative cases. In *Barsky v. Board of Regents*, [35] for example, decided in his first term on the Court, he tolerated the suspension of a physician's license to practice that resulted from the physician's unwillingness to disclose the records of an allegedly "subversive" organization to the House Un-American Activities Committee. And in *Peters v. Hobby*, [36] written a year later, he concluded that governmental employees accused of disloyalty had no constitutional right to confront their accusers.

The case that seems to have clarified Warren's thinking on gov-

ernment investigations was *Watkins v. United States*,[37] decided in 1957.
After *Watkins*, Warren saw efforts on the part of governmental bodies to obtain information about potentially subversive persons as
"broad-scale intrusion[s] into the lives and affairs of private citizens."[38] For the first time Warren's ruminations on internal security
issues had yielded a clear ethical principle: Since a citizen has a right
to live his or her life free from the intrusion of government, governmental agencies could not restrict that freedom without a specific
and compelling reason.

In *Watkins* a former official of the Farm Equipment Workers International Union, in testimony before the House Un-American Activities Committee, freely revealed his earlier cooperation with the
Communist party, but refused to reveal the names of persons who
were allegedly members of the party. He claimed that such questions were not "relevant to the work" of the committee and denied
that the committee had a "right to undertake the public exposure of
persons because of their past activities."[39] Congress subsequently indicted him for "contempt of Congress," citing his refusal to answer
the specified questions. He appealed his conviction, questioning the
authority of the committee to compel disclosure.

Warren found the "critical element" in *Watkins* to be "the weight
to be ascribed to the interest of the Congress in demanding disclo-
sures from an unwilling witness."[40] He asserted that "the mere sum-
moning of a witness and compelling him to testify, against his will,
. . . is a measure of governmental interference";[41] that "there is no
congressional power to expose for the sake of exposure";[42] that the
House Un-American Activities Committee had been "allowed, in essence, to define its own authority";[43] and that congressional investigations of the kind undertaken in *Watkins* could "lead to ruthless
exposure of private lives in order to gather data that is neither desired by the Congress nor useful to it."[44]

"[E]xcessively broad charter[s], like that of the House Un-
American Activities Committee," Warren argued, placed the courts
"in an untenable position" in their efforts to "strike a balance between the public need for a particular interrogation and the right of
citizens to carry on their affairs free from unnecessary governmental
interference."[45] If the courts simply assume that "every congressional investigation is justified," they would be "abdicat[ing] the responsibility placed by the Constitution upon the judiciary to insure
that the Congress does not unjustifiably encroach upon an indi-

vidual's right to privacy nor abridge his liberty of speech, press, religion or assembly."[46]

Characteristically, Warren's *Watkins* opinion blurred together ethical principles and more traditional constitutional arguments. The precise constitutional claim made by the witness in *Watkins* was that he had not received an adequate explanation of the pertinency of the questions that he had been criminally punished for failing to answer. Congress had constructed an explanation after the fact, but the explanation given at the time was inadequate. Hence, the petitioner in *Watkins* argued, he had been denied due process under the Fifth Amendment because he could not have known with sufficient particularity that his failure to answer these questions would lead to possible criminal prosecution for contempt. Warren accepted that argument completely, holding that the petitioner "was . . . not accorded a fair opportunity to determine whether he was within his rights in refusing to answer, and his conviction is necessarily invalid under the Due Process Clause of the Fifth Amendment."[47]

Warren's discussion of the precise constitutional issue being decided occupied only five pages of his thirty-five-page opinion. The rest reviewed the history of legislative investigations, revealed the presence of "a new kind of congressional inquiry unknown in prior periods of American history,"[48] noted that this "new phase of legislative inquiry" raised issues concerning "the application of the Bill of Rights as a restraint upon the assertion of governmental power,"[49] and engaged in the sweeping assertions about the dangers of unlimited congressional investigative powers previously quoted. The consequence was that the *Watkins* opinion appeared to be a broad condemnation of "those investigations that are conducted by use of compulsory process [and thus] give rise to a need to protect the rights of individuals against illegal encroachment."[50] This reading of *Watkins* is consistent with Warren's civics perspective. While the constitutional basis of *Watkins* can be reduced to a narrow claim of pertinency under the Due Process Clause, Warren suggested in the opinion that unfocused congressional investigations potentially violated rights of privacy, speech, press, religion, assembly, and most significant of all, the "right to carry on affairs . . . free from unnecessary governmental interference."

After *Watkins* there was little doubt about where Warren stood on cases in which he perceived that an agency of government was unduly interfering with an individual's presumptive right to carry on his affairs. In *Sweezy v. New Hampshire*, [51] decided the same term as *Watkins*, Warren invalidated the contempt conviction imposed on a university professor by the New Hampshire legislature for failing to respond to questions put to him by the state attorney general, a "one-man legislative committee" empowered to investigate "subversive activities."[52] The professor, Paul M. Sweezy, declined to discuss, among other things, the content of an ostensibly "subversive" lecture he had given to students at the University of New Hampshire.

"There is no doubt," Warren said, "that legislative investigations . . . are capable of encroaching upon the constitutional liberties of individuals. It is particularly important that the exercise of the power of compulsory process be carefully circumscribed . . . particularly in the academic community."[53] In *Sweezy* "there unquestionably was an invasion of petitioner's liberties in the areas of academic freedom and political expression—areas in which government should be extremely reticent to tread."[54] Indeed, Warren could not "now conceive of any circumstance wherein a state interest would justify infringement of rights in these fields."[55]

Warren sought to buttress the broad presumptive freedom to pursue one's affairs without governmental interference with a constitutional right to confront agents of government who attempted to place limits on that freedom. In *Greene v. McElroy*,[56] a 1959 decision, an aeronautical engineer employed by a private engineering corporation challenged the denial of his security clearance by the Army-Navy–Air Force Personnel Security Board. In the several administrative proceedings in which the engineer undertook to challenge the board's action, he was never given access to "confidential" information obtained about him from FBI investigations. The information, which was the basis of the denial of his security clearance, asserted that the engineer had associated with alleged Communists, had had Communist literature in his home, and had associated with officials of the Russian embassy. After exhausting his administrative remedies, the engineer filed suit in federal district court, and the Supreme Court granted certiorari to examine several issues, among them the narrow question whether Congress or the president had authorized the summary procedures used in the denial of security clearances.

Warren's opinion for the Court did not confine itself to that question, although Warren said that the Court was holding only that the procedures had not been explictly authorized.[57] Warren maintained, in the course of his opinion, that there was a constitutional "right to hold specific private employment and to follow a chosen profession free from unreasonable governmental interference,"[58] and that "principles . . . immutable . . . in our jurisprudence" required that "where governmental action seriously injures an individual" he had a right to the safeguards of "confrontation and cross-examination."[59] For Warren, *Greene* was a case where "substantial restraints on employment opportunities of numerous persons [had been ] imposed in a manner which is in conflict with our long-accepted notions of fair procedures."[60]

The above statements were not explicit justifications for the decisions in *Greene*, nor were they accurate summaries of established constitutional doctrine. None of the cases that Warren cited for the proposition that there was a constitutional right to hold a job or profession free from unreasonable governmental interference had declared such a right.[61] A right to confront and cross-examine witnesses had not been extended to administrative hearings.[62] The "immutable principles" of confrontation and cross-examination that Warren invoked in *Greene* had been formulated principally for criminal trials, not for "all types of cases where administrative and regulatory actions were under scrutiny."[63] Justice Harlan, in his concurrence in *Greene*, said that his "unwillingness to subscribe" to Warren's opinion was "due to the fact that it unnecessarily deals with the very issue it disclaims deciding."[64]

Taken together, *Watkins*, *Sweezy*, and *Greene* reveal that Warren's principal concern in cases involving governmental investigations was with the intrusiveness of the investigative process itself. He simply thought that it was wrong for the House Un-American Activities Committee to expose for the sake of exposure, for the New Hampshire legislature to interfere with academic freedom, or for the Personnel Security Board of the armed forces to limit, on nebulous and unchallenged grounds, an engineer's employment opportunities. The source of the wrong was the general evil of undue governmental restrictions on individual freedom. Warren attempted to oppose that general evil by invocation of specific constitutional protections, but his invocations were vague and not well substantiated; his majority opinions were subscribed to on much narrower grounds.

The reluctance of Warren's colleagues to endorse his doctrinal explanations did not deter him from adhering to the principle that he espoused in *Watkins*. He never failed, in the remainder of his career, to vindicate a claim by an individual that an overbroad government investigation had seriously infringed upon his rights. It did not matter whether the investigative context was the House Un-American Activities Committee[65] or a state investigative body,[66] whether the issue involved academic freedom[67] or the compulsory registration of American Communists,[68] or whether the effect of the investigation was to restrict travel,[69] service in a labor union,[70] or employment in a defense plant.[71] Warren had grasped the "essence" of cases arising out of governmental investigations: The citizenry had a presumptive right to conduct their affairs free from governmental interference.

Once Warren had discerned the validity of that moral principle, the process to judgment became easy. Almost any constitutional argument—the language of "liberty" in the Due Process Clause of the Fifth Amendment, the Sixth Amendment's guarantee of confrontation in a criminal proceeding, the First Amendment's protection of speech and assembly, even the Bill of Attainder Clause of Article I[72]—could serve to justify his conviction that the state had unnecessarily interfered with a citizen's negative liberties. Once again doctrine had become the handservant of ethics in Warren's jurisprudence.

Negative and affirmative liberties were ultimately linked in Warren's theory of civics. Because government in America existed for the benefit of American citizens, the status of being a citizen presupposed certain affirmative benefits that government was compelled to bestow fairly and equally (participation in the electoral process) and could not summarily remove (the status of citizenship itself). And because government in America existed for the benefit of American citizens, the government could not use its powers to infringe upon rights that it had been created to secure. The idea of government in America thus assumed both governmental responsibilities toward the citizenry and restrictions upon the impact of government on its citizens.

The civics lessons that Warren attempted to teach in the governmental investigation cases were most probably the source of the "Impeach Earl Warren" billboards that came to dot portions of America in the later years of the Warren Court. Almost all the cases where the Court tested the scope of governmental investigative powers involved persons suspected of being Communists or Communist sympathizers. Warren's caution in the first such cases he considered can very likely be traced to his approval of governmental efforts to identify potentially "subversive" persons, efforts he had undertaken as a California law enforcement official. After *Watkins*, however, Warren concluded that such governmental activities seriously frustrated the exercise of liberties of citizenship, and thus he became less moved by the fact that fear of subversion was the rationale for the intrusive activities.

In his autobiography Warren recalled Senator Joseph McCarthy having once said, on the floor of the Senate, "I will not say that Earl Warren is a Communist, but I will say he is the best friend of Communism in the United States." [73] He also recalled the John Birch Society's campaign to impeach him; the American Bar Association's condemnation of the *Watkins* decision and others as "aiding the Communist cause"; [74] and his subsequent resignation from the American Bar Association. In addition, Warren told a story about President Dwight Eisenhower's reaction to the investigative cases. In 1965 Warren and Eisenhower were flying on a presidential plane to London to attend the funeral of Winston Churchill. Eisenhower confessed to Warren that he had been "disappointed" in Warren's performance on the Court: He had anticipated that Warren would be a "moderate" as a judge, but Warren had not been. Specifically, Eisenhower referred to "those Communist cases"; although he had not read them, he claimed to "know what was in them." Warren then asked Eisenhower what he "would do with Communists in America," and Eisenhower replied, "I would kill the S.O.B.s." In commenting on Eisenhower's reaction, Warren said that he "was sure [the last] remark was merely petulant rather than definitive." [75]

Although Warren was surprised at Eisenhower's assessment, because "I had always considered myself a moderate," [76] Warren's view of judging could not be called an exercise in moderation. He believed that

> in the Supreme Court, the basic ingredient of decision is principle, and it should not be compromised and parceled out a little in one case, a little more in another. . . . If the principle is sound and constitutional, it is the birthright of every American, not to be accorded begrudgingly or piecemeal or to special groups only, but to everyone in its entirety whenever it is brought into play. [77]

His civics lessons decisions were consistent with that view. Once he had discerned a "sound and constitutional" principle he made it "the birthright of every American." Citizens could only relinquish their citizenship voluntarily: It did not matter that they were citizens who had left the country to evade military service or citizens who had allegedly deserted the armed forces. One citizen's vote could not count more than another's, even if a popular referendum had decided otherwise. Government agencies could not expose the private activities of citizens "for the sake of exposure," even if some of those activities would be thought by many persons to be subversive of national security. There was to be no compromising and parceling out of the cardinal principles for which the Constitution stood, even if textual support for those principles was stronger in some instances than in others.

The lessons in civics Warren learned aided him in more than the decision of the cases considered in this chapter. They formed the bedrock of his jurisprudence by linking his stance as an ethicist to his stance as a judge. They were, for Warren, fundamental insights of political theory: They revealed the basic structure of American government and therefore were cardinal principles for which the Constitution stood. With the civics cases Warren's 1938 statement about civil rights was resurrected. They had reminded him of his earlier belief that the Constitution protected citizens against "arbitrary action by government in every field" and insured "the existence of a spirit of fair play on the part of public officials." These were beliefs—now juristic convictions—that Warren was to put into practice again and again.

# I I

# "A Right to Maintain a Decent Society"

On June 25, 1969, two days after retiring from the Supreme Court, Earl Warren granted an interview to a California chain of broadcasting stations. The interview briefly discussed Warren's public career, touching on his political campaigns, his relationship with the University of California, and his decisions on segregation, reapportionment, and criminal procedure. Midway through the interview Morris Landsberg, the interviewer, recalled that Warren had "said awhile back that pornography was the court's most difficult area," and asked Warren to elaborate. "It's the most difficult area," Warren said, "[because] we have to balance two constitutional rights with each other." The "state and national government [have] a right to have a decent society," but "[o]n the other hand, we have the First Amendment." Obscenity cases raised the question of "how far people can go under the First Amendment . . . without offending the right of the government to maintain a decent society."[1]

Decency, we have seen, was one of the values that Warren believed was basic to civilized existence. His test for what constituted decent behavior was straightforward and conventional: what Warren thought that people of average sensibility would recognize as wholesome and upright conduct. Under this definition a number of activities could be seen as indecent because they appealed to mankind's baser instincts and discouraged people from living wholesome and honorable lives. Gambling, prostitution, bootlegging, and dealing in drugs were examples of such activities. Warren's characterization of gambling as preying on the poor, "taking the paycheck out of the hands of the worker," and affecting a family's food, shelter, and education has previously been quoted.[2]

Trafficking in pornographic materials was another indecent activity. The indecency of pornography, for Warren, came not so much from its inherent salaciousness as from its calculated distribution to a susceptible public. "[P]roducers and writers," he noted in 1972,

> continue to justify [the distribution of pornographic materials] by saying that they are merely satisfying the public demand. . . . There is a considerable measure of truth in the pornographers' estimate of the marketplace for filth since they would hardly be producing in this vein if people were to withhold their patronage.[3]

Warren believed that human beings were inherently susceptible to temptations of the flesh and the spirit, but he also believed that such temptations were often destructive and should be suppressed or channeled. He conceded the inevitability of a public market for obscene literature; he also sought to regulate that market. His ideal solution to the problem of traffic in pornographic materials was to "eliminate the profit": If that were achieved, "the traffic in such shoddy merchandise [would] vanish."[4]

Warren was hampered, however, in his attack on pornography by the language of the Constitution and by the reluctance of law enforcement officials. He sought to treat traffic in pornographic materials as unprotected indecent conduct and to suppress it: Most obscenity cases were simply "general criminal cases."[5] But obscene literature presented Warren with problems. The suppression of "indecent" literary materials did not merely involve regulation of a distribution process, it involved censorship of the printed word. Warren conceded that such obscenity cases involved "expression," and their standards could be "applied to the arts and sciences and freedom of communication generally."[6] Thus, while in some instances "commercial exploitation of the morbid and shameful craving for materials with prurient effect"[7] could be separated from the literary expressions in the materials and punished, in other cases such a separation could not easily be made.

Where judicial efforts to suppress pornography were mere pretexts for the regulation of literary expression, courts, in Warren's view, were engaging in "government censorship." Censorship was "a withering function"; it had been "demonstrated throughout history to be dangerously oppressive."[8] Warren did not like boards of censorship, he did not like the Supreme Court acting as a board of censorship in obscenity cases, and he especially did not like having to wade through masses of allegedly obscene publications in his capacity as a member of the Court.

In its 1966 term, the Court, having unsuccessfully attempted various formulations of the constitutional status of obscene literature, sought to arrive at a more definitive formulation. The result of that effort was the amusing decision in *Redrup v. New York*,[9] where the Court declared that "obscenity" for constitutional purposes meant whatever five justices of the Court currently thought was obscene. In the earlier stages of *Redrup*, however, the Court had been more optimistic about the possibility of reaching a definitive solution, and to that end a number of sexually explicit publications had been assembled in a room off the main conference room. Justices from time to time would wander into the room and examine the publications, but Warren refused even to look at the material.

Warren told his clerks that he continued to believe that obscenity cases should be treated as ordinary criminal cases, with an emphasis on the conduct of the defendant, and that Justice Brennan could review these materials for him.[10] He had no doubt that "obscenity [was] not protected under the free speech clause of the Constitution," but he found it "very difficult," as did his fellow justices, "to write a verbal definition of what obscenity is."[11] His solution was to advocate "the reestablishment of a moral tone for our nation," which would "restore public and private language to generally accepted norms of decency."[12] Such a reawakening, he hoped, would take the Court out of the censorship business and dissipate the market for pornography.

Warren admitted that, despite all his Court's efforts to clarify the unprotected status of obscene expressions and to define what was "obscene," an "increasing decadence" in "our theater, our films and the fiction we read" had occurred.[13] "Some of the things that go through the mail" were "just unspeakable";[14] families were "ashamed to share [visual arts] together."[15] Continued and flourishing traffic in pornographic materials was principally a result, he felt, of inadequate law enforcement. No one "has been able to write the definition for obscenity that juries can follow"; policemen "don't like to enforce [obscenity laws]"; prosecutors "don't like to prosecute them"; judges "don't like to determine what is obscene and what isn't obscene"; there was "a general inclination for everybody to do nothing and blame somebody else." [16]

The deteriorating stance of the nation toward "indecent" acts distressed Warren, who believed in the maintenance of decency through protection of potentially susceptible persons and punishment of those seeking to play on their susceptibilities. Warren's obscenity decisions stand out as the one instance in his tenure as Chief Justice where his doctrinal stance on a constitutional issue partially isolated him from the rest of his Court. Not only was Warren's solution to the obscenity problem never endorsed by a majority of his colleagues, but the Warren Court moved, without explicitly acknowledging it, toward a far narrower definition of "obscenity" for constitutional purposes than Warren's approach would have permitted.

✍

Warren is reported to have once said, after considering an allegedly pornographic book, that "[i]f anyone showed that book to my daughters, I'd have strangled him with my own hands." [17] The language does not sound like Warren's, but the sentiment rings true. We have seen that the Warren family was an emotionally close unit, with Warren occupying the role of a patriarch. Gender roles were clearly defined and traditional: None of Warren's daughters pursued a professional career, and all his sons at some point in their lives considered becoming lawyers. We have also seen that a dominant theme of the Warren family's life, which became accentuated as Warren came to occupy more prominent public positions, was protection of the family's privacy from the scrutiny of outsiders. Protectiveness was a familiar stance for Earl Warren, reflected in his paternalistic attitude toward disadvantaged persons, his affection for children, his fatherly attitude toward many of his law clerks, and his strong interest in separating his private and public lives. A sense of responsibility about protecting his daughters from exposure to obscene literature was consistent with Warren's paternalism.

Obscenity also raised for Warren the aforementioned issue of individual susceptibility to degenerative forces. Warren, we have noted, was an orthodox early twentieth-century Progressive in his moral attitudes. He felt that most people were easily tempted by corruption or lust but that such persons, if properly encouraged by the state, were capable of resisting such temptations. His standard response to a social problem with moral overtones, such as prostitution or gambling, was to seek the advice of independent, nonpartisan "experts," secure recommendations that habitually defined the problem as related to environmental pressures and called for affirmative governmental redress, and seek to implement the recommendations by such means as well-publicized closings of dog tracks or vigorous raids on brothels. In using government as a paternalistic moral force, Warren felt he was helping susceptible persons resist temptation.

Finally, we have seen that Warren's role as a "crusading" district attorney in California was consistent with his conception of himself as a champion of the "public" interest pitted against "special" interests. Pornographic literature presented Warren with another potential crusade. The "special interests" were peddlers in smut; the public was personified by susceptible men or vulnerable women; the Court was given an opportunity to protect such persons from themselves. Restricting the distribution of pornography was the standard response of a law enforcement official. Joined with these curbs on special interests were exhortations to the citizenry to forgo decadence and to invest American society with a more wholesome moral tone.

As a crusading district attorney, Warren had only to concern himself with the intended opposition of those adversely affected by his crusades: He could count on abstract public support for decency and wholesomeness. As Chief Justice, however, he could strip pornographers of constitutional protection only on the basis that decency, understood by Warren to exclude pornographic literature, was one of the ethical values embedded in the structure of the Constitution. He was able to discern a constitutional "right to maintain a decent society," but problems remained. First, as a Supreme Court justice, Warren was no longer part of the branch of government charged with enforcing laws, and those charged with the responsibility of eradicating pornography seemed to have no particular enthusiasm for their task. Moreover, large segments of the public did not seem to care whether traffic in pornographic materials was vigorously suppressed or not. To Warren, it was curious why the declining moral tone of society was something that many Americans did not take very seriously.

↰

With hindsight, one can piece together the general approach Warren sought to have the Court take in obscenity cases. Warren's approach was designed to accomplish three goals. First, he wanted to achieve a separation of the illicit use of potentially obscene materials from the content of the materials themselves. Second, he wanted to delegate determinations of what constituted "obscene" expression to local units of government. Third, he wanted to prohibit the censorship of literary and artistic expression.

Warren sought to achieve these goals and to reconcile his belief that pornographic literature was not protected by the Constitution with his concern for First Amendment values by conceptualizing the obscenity cases as conduct cases, not speech cases. The gravamen of obscenity was the distribution of offensive materials. Being ordinary criminal cases, obscenity cases could be treated as prostitution cases were treated, by suppression of the illicit traffic and exhortations to citizens to avoid the illicit marketplace. Moreover, as conduct cases, they did not raise issues of censorship because what was being suppressed was commercial activity, not artistic or literary expression.

*Kingsley Books, Inc. v. Brown*[18] and *Roth v. United States*,[19] two cases from the 1956 term, show Warren's early efforts to separate conduct from expression in obscenity cases. In *Kingsley*, the state of New York, after an adjudication that certain books were obscene, sought to confiscate and destroy them. A majority of the Court held that this procedure was not a prior restraint on speech, given the obscene status of the books, and that the state could proceed. Warren dissented, distinguishing between the destruction of materials that had been in the possession of a person criminally prosecuted for obscenity and the destruction of the materials where no such prosecution had taken place. The procedure in *Kingsley*, Warren maintained, "place[d] the book on trial. . . . The personal element basic to the criminal law [was] entirely absent." Obscenity, Warren felt, should be determined by the manner of use of a publication. "It is the conduct of the individual" that was to be judged, he said, "not the quality of art or literature." New York's procedure "savors too much of book burning."[20]

In *Roth*, Warren affirmed the convictions of two bookdealers, one in New York and one in California, for distributing obscene literature. *Roth* was different from *Kingsley*, Warren believed, in that the defendants had been afforded the protections of a criminal trial, had been found to have possessed the requisite knowledge of their acts, and had purveyed "textual or graphic matter openly advertised to appeal to the erotic interest of their customers." They were "plainly engaged," Warren found, "in the commercial exploitation of the morbid and shameful craving for materials with prurient effect."[21] In obscenity trials, Warren repeated, "[i]t is not the book that is on trial; it is a person. The conduct of the defendant is the central issue."[22]

The *Kingsley-Roth* sequence demonstrated Warren's interest in achieving a stark separation between individual conduct and artistic expression. The defendants in *Roth* had known that they were catering to a "morbid and shameful" craving in the public; that had been their central purpose in distributing the literature. Warren would punish them for appealing to a susceptible public's baser tastes and thus violating canons of decency. No such justification existed for the punishment in *Kingsley*, since the defendant had not yet been convicted of a comparably indecent act. The separation between conduct and expression thus avoided "[m]istakes of the past" in the area of obscenity, for where purveyors had acted indecently, "the obscenity of a book or picture" was not truly at stake.[23]

The dichotomy Warren saw between conduct and expression seems hard to sustain upon analysis. The conduct of the defendants in *Roth* would hardly have been indecent if the materials they had distributed had been inoffensive. Materials with no "prurient effect" did not cater to any "morbid and shameful" public craving; it was the prurience of the materials that encouraged purveyors. Thus to say that the obscenity of a book or picture was not at issue in cases like *Kingsley* or *Roth* was to assume away difficulties. It made no sense to try booksellers for knowingly distributing salacious or pornographic literature without a preliminary judgment that the literature merited such adjectives. Indeed Warren conceded that "[t]he nature of the materials" was "relevant as an attribute of the defendant's conduct."[24]

The distinction between constitutionally protected expression and unprotected conduct can best be understood as an effort by Warren to reconcile the conflicting values he perceived to be at stake in obscenity cases. Although willing to deny First Amendment protection to obscenity, Warren was concerned about "broad language" in obscenity cases that could "eventually be applied to the arts and sciences and freedom of communication generally."[25] Resort to the "color and character" of the setting of certain literary expressions was a way to buttress their suppression:[26] If materials containing the expressions were being purveyed by pornographers, suppression was made easier.

While Warren could seize upon the "indecent" conduct of purveyors as a means of resolving his doubts about restrictions on literary expression, he remained troubled about censorship in obscenity cases. He was hostile to the idea of censorship itself, as his lengthy dissent in the 1961 case, *Times Film Corp. v. Chicago*[27] indicated. That case validated a Chicago ordinance that required submission of all motion pictures to an examining board for licensing prior to public exhibition. The owners of a film, "Don Juan," refused to submit the film for examination, claiming that the ordinance constituted a prior restraint on free speech. Warren agreed, stating that the Chicago ordinance gave "formal sanction to censorship in its purest and most far-reaching form"[28] and noting that the Chicago licensers had banned a scene depicting the birth of a buffalo, a movie that used the words "rape" and "contraceptive," a scene where a girl was slapped, and films portraying life in Nazi Germany.[29] In Warren's view censorship "endangered the First and Fourteenth Amendment rights of all others engaged in the dissemination of ideas."[30]

Warren realized that if the Court were to adopt a position that obscene literature was not entitled to First Amendment protection it might have to choose between permitting blanket restrictions on artistic or literary expression and making individual determinations of what was obscene in a given case. The latter would put the Court in the awkward position of being a board of censors, a position Warren sought to avoid. In an exchange with counsel during the oral argument of an obscenity case in the 1965 term, Warren asked whether the Court "had to read all [the literature under question] to determine if [it had] social importance." "I'm sure," he said, "this Court doesn't want to read all the prurient material in the country to determine if it has social value. If the final burden depends on this Court, it looks to me as though we're in trouble."[31]

Warren had a hand in a number of efforts the Court made during his tenure to develop a workable but unburdensome approach to the law of obscenity. In *Roth*, Warren joined the Court majority's definition of obscenity as material that "appeals to [a] prurient inter-

est,"[32] a definition that was suggested would eventually restrict obscenity prosecutions to hard-core pornography.[33] In *Jacobellis v.
Ohio*,[34] a 1964 decision, Warren retained the *Roth* definition "until a
more satisfactory [one] is evolved," but suggested, in dissent, that
the enforcement of obscenity laws be localized. There was no "prov-
able 'national standard' " for what was obscene, Warren argued, "and
perhaps there should be none." He was willing to countenance "material being proscribed as obscene in one community but not in another" so as to avoid "this Court [establishing] itself as an ultimate
censor . . . making an independent *de novo* judgment on the question of obscenity." Delegation to local communities was "the only
reasonable way . . . to obviate the necessity of this Court's sitting
as the Super Censor of all the obscenity purveyed throughout the
Nation."[35]

No one on the Court accepted Warren's "localizing" solution; by
*Jacobellis* only he and two other justices retained their support of the
*Roth* test. Warren also repeated in *Jacobellis* his belief that "the use to
which various materials are put—not just the words and pictures
themselves"—was to be considered in determining whether a work
was obscene.[36] That distinction had never been endorsed by another
justice, but in the next significant Warren Court obscenity case,
*Ginzburg v. United States*,[37] it was made the basis of the Court's opinion. In "close" cases, Justice Brennan wrote for the Court, "evidence
of pandering may be probative with respect to the nature of the material in question" in an instance when "the publications . . . cannot
themselves be adjudged obscene."[38]

The remarkable feature of the *Ginzburg* opinion was its suggestion that somehow the use to which allegedly obscene materials were
put affected their constitutional status. Under this test not only would
different defendants (say panderers and librarians) have different degrees of constitutional protection, but the same publication might be
"obscene" or not depending on how it was presented. Consequences
such as these had plagued Warren's approach to obscenity from its
first formulation in *Roth*. The Court seemingly endorsed Warren's
approach in *Ginzburg* simply to allow itself to restate a stringent definition for obscenity[39] and to convict a pornographer.

It was during the oral arguments in *Ginzburg* and its companion
cases that Warren commented that "we're in trouble" if the Court
had to be "the final censor to read all the prurient material in the
country." He had hoped that his localizing of enforcement standards

in *Jacobellis* might help relieve the Court of that burden, but no other justice had endorsed that strategem; Justice Brennan specifically rejected it for the majority in *Jacobellis*.[40] In the 1966 term Warren joined another majority's efforts to minimize the Court's role as a "Super Censor." In *Redrup v. New York*, as noted, the Court concluded that it would employ summary per curiam opinions in obscenity cases whenever a majority of five justices, for whatever reason, found the material in question not to be obscene. Given the composition of the Court at the time, the *Redrup* procedure had the effect of equating obscenity with hard-core pornography. Two justices, Black and Douglas, opposed any bans on expression, "obscene" or not; one justice, Potter Stewart, would ban only hard-core pornography; and two justices, Brennan and Warren, continued to endorse the *Roth* test, which now insisted that literature be *utterly* without redeeming social value.

Warren's acquiescence in *Redrup* indicated how eager he was to free the Court of its putative censor role. The *Redrup* solution allowed individual justices to ignore elements of pandering, to substitute their own unarticulated preferences for the *Roth* test, and to avoid the issue of national versus local standards, all in order to save the Court from having to read masses of "prurient" literature. This solution seemed to indicate a certain amount of fatalism on Warren's part. He remained concerned, even after his retirement from the Court, about "increas[ing] decadence" and a deteriorating moral climate. He continued to be outraged by pornographers and hopeful that someday the market for pornographic materials would dry up, and he took the presence of offensive literature as a serious attack on standards of decency. But he had evidently become reconciled to the fact that obscenity prosecutions were something that local authorities had little interest in obtaining. Rather than encourage censorship in others, he had reluctantly taken it upon himself. After ten years of being confronted with smut, he had had enough.

The Warren Court made one more effort to address obscenity, in the remarkable 1969 case of *Stanley v. Georgia*,[41] and here again Warren joined the majority. In *Stanley*, the state of Georgia, in the process of conducting a lawful search of a private home for evidence of bookmaking activities, seized and confiscated three reels of concededly obscene films. Georgia argued that the state interest in protecting its citizens from exposure to obscene literature justified regulating what one read or viewed in one's home. The Court disagreed, holding that the state's power to regulate obscenity "simply does not extend to mere possession by the individual in the privacy of his own home."[42] After *Stanley* it appeared that allegedly obscene materials lost their quality of obscenity once they entered a private home, again suggesting that the evil of obscenity lay primarily in its commercial distribution and thereby implicitly resurrecting Warren's dichotomy between conduct and expression.

The joining of the *Stanley* majority by Warren marked the end of a painful journey through the briar patch of obscenity law. From an initial concern in *Roth* to exclude certain "prurient" literature from the ambit of First Amendment protection and to punish purveyors of smut, Warren had moved to embrace, first, a test for obscenity that restricted it to hard-core pornography unless the materials in question had been grossly "pandered"; second, an acknowledgment that the line for protection of obscenity lay wherever five justices of the Court subjectively thought it should; and third, absolute constitutional protection for obscene material read or viewed in one's home. By the end of his tenure, Warren, whose views on the right of the states to "maintain a decent society" had not changed since he came on the Court, found himself in the awkward position of having contributed to a dramatic and far-flung expansion of constitutional protection for tasteless and vulgar expression. No wonder he found pornography the Court's "most difficult area." No wonder he wistfully called for a reawakening of the nation's moral tone.

Warren's itinerary in the obscenity cases, however, did not mean that he simply shifted his opinion on obscenity issues. Although Warren's derivation of constitutional principles was unorthodox, it was not naive. He recognized that values embedded in the Constitution might often conflict, and he was not always certain about how much weight to attach to competing values in given situations. An obscenity case could conceivably involve three constitutional values prized by Warren: decency, privacy, and the protection of expression. For Warren, *Roth* was a case that involved only decency, whereas *Kingsley Books*, given its procedural posture, was a case that involved only protection for literary expressions. *Stanley*, in contrast, involved all three values; on balance, Warren decided that the privacy value, ultimately linked in *Stanley* to First Amendment values, outweighed the decency value.

Moral issues invariably brought out Warren's paternalism. The profit taking of the gambler, the pimp, and the pornographer outraged him, stirring his impulses to protect susceptible persons from themselves. His philosophy of government assumed that such protection could be achieved and that with the aid of a paternalistic state humans could rise above their baser drives. Warren approached obscenity with strong paternalistic and regulatory tendencies: Pornographers were common criminals, and their activities warranted vigorous prosecution.

As Warren confronted obscenity issues as a judge, however, he began to see that government regulation invited government censorship. Decency, in the clutches of the censor, could become repression; the state in the guise of protectiveness could seek to control people's thoughts. The people who shared Warren's robust interest in maintaining decency too often emerged as the people for whom the birth of a buffalo was obscene. A regular reader of history and political thought, Warren knew how little it took for petty censors to use their limited powers to stifle great art and literature. However much he believed in decency, he was unwilling to delegate the definition of decent conduct to parochial bluenoses.

Moreover, Warren was unwilling to occupy the role of censor himself. He was more than bored with allegedly obscene literature; he was deeply offended by it. He did not want to search for some social value in works he deplored. He hoped that enlightened law enforcement officials and a concerned citizenry would help him stamp out obscenity, but that did not come to pass. Grudgingly, always excepting the panderers and the dealers in smut, he came to endorse enhanced protection for vulgar expressions, because to do otherwise was to give government too much power to dictate what was written and what was read. Government, after all, existed to secure human rights, not to stifle them. Even so worthwhile a value as decency could be made a means for restricting one's freedoms. As unfortunate as were the drives that attracted certain persons to the salacious and the lewd aspects of life, those drives could not be smothered in a regime of censorship.

Warren's ideal model of obscenity regulation rested heavily on the notion that decent citizens and law enforcement officials would readily decide what constituted decent and indecent modes of literary expression and rapidly suppress the latter. Under that model, pornographers would quickly be put out of business, courts would not have to act as censors, and the moral tone of the nation would remain healthy. Unfortunately Warren's model broke down at its initial stages. Not only were the public and the law enforcement community incapable of swiftly demarcating the decent from the prurient, they seemed disinclined to do so. Rather than delegate that task to a board of censors, Warren reluctantly took it on himself. The consequence was that he and his peers could not find an easy basis for separating the obscene from the nonobscene and ended up facilitating traffic in "indecent" materials.

It is a measure of Warren's broad-mindedness that when he saw the First Amendment issues that obscenity cases raised, he took them seriously, even though such a stance permitted a fairly broad circulation of materials that he considered "unspeakable." It is a measure of his absolutist approach to moral issues that he never seriously questioned the strict maintenance of decency in society or perceived the extent to which conventions about decency change and are affected by dissenting "indecent" expressions. Warren simultaneously believed in the freedom of the individual from government oppression and in the primacy of the state on moral questions. He and his fellow Progressives saw no contradiction between moral paternalism and support for civil liberties: By being protected from exposure to the debauched aspects of life, individuals were being liberated from their own destructive tendencies. A free citizen, for Warren, was a citizen dedicated to the maintenance of a decent society.

# I 2

# Law Enforcement
# from the Other Side

Law enforcement, we have seen, was one of the significant themes
of Earl Warren's career. Law enforcement launched Warren's entry
into public service; law enforcement provided him with a base to
enter politics; Warren's early attitudes on social issues reflected his
law enforcement training. Of all the departures from existing law
made by Warren and his Court, revision of established rules in the
area of criminal procedure might have seemed, to those who had
known Warren as district attorney and attorney general, the least
likely for him to have supported. He had been a tough prosecutor
who had once made political capital of his conviction rate;[1] he had
been a persistent foe of gamblers, bootleggers, and confidence men;
he had kept a watchful eye on suspected Communists in labor organizations; he had sought to expand the powers of the law enforcement agencies in which he had served. The idea that Earl Warren,
whom Raymond Moley called in 1931 the "most intelligent and politically independent district attorney in the United States,"[2] would
be charged with being "soft on crime" as a judge would have been
regarded as highly improbable by those who knew his California
background.

Inconsistencies between Warren's career as a California public
official and his performance on the Court have, of course, been a
persistent commentator's theme. Warren himself, when confronted

with apparent inconsistencies, either minimized them, denied them, or distinguished between his roles as a politician and as a judge. In the case of his criminal procedure decisions, his response took the form of a denial. He said in 1969:

> If anybody could show me anything that we have done in the time that I have been on the Court other than to insist that a man is entitled to counsel at all times after he has been put upon by the Government in a criminal case, and entitled to fair treatment, to due process in the trial of his case, I would concede that we had perhaps done something wrong. I can't think of any such thing.[3]

Warren had said a year after he joined the Court that American criminal justice was "pockmarked with . . . procedural flaws and anachronisms"; persons sometimes were "arrested, tried and convicted without being adequately informed of their right to counsel," or were denied such a right because they "[could] not afford to exercise it."[4] He had simply put those observations into practice as Chief Justice: He saw nothing striking about his efforts. "If there's been change in my attitude," he said in 1968, "I'm stronger for law enforcement than even then."[5]

Others saw Warren's judicial stance toward criminal procedure differently. Lloyd Jester, who had worked for Warren in the Alameda County District Attorney's Office, said that Warren's protective attitude toward criminals as a judge was in direct response to his awareness of the coercive powers of an ambitious prosecutor. When Warren became Chief Justice, Jester claimed, "he just plain assumed that every other district attorney, every other prosecutor in this land of ours acted and was acting in the same manner in which he himself had acted in the past. . . . [H]e just turned turtle, 'cause he was going to correct all of these ills."[6]

In attempting to explain the relationship between Warren's views of law enforcement as a California public official and as Chief Justice, this chapter focuses on Warren's changing perception of what "law enforcement" meant. Warren said in 1969 that although "enormous crime problems" existed in the years that he was a district attorney, the crime of the 1950s and sixties was "a different kind of crime." Law enforcement when Warren was a prosecutor had meant fighting "the bootlegging, the high-jacking, the rum-running and all of the crimes that surrounded that liquor business"; law enforcement in the years of the Warren Court dealt with "the robberies and the burglaries and the muggings and the rapes and all these other individual crimes that largely emanate from the slums in our cities."[7] Law enforcement officials in the latter period were not so much crusaders against corruption and vice as they were protectors of the public against violence.

The transformation of the archetypal criminal from a member of an organized gang of conscious lawbreakers to a disadvantaged resident of the inner-city ghettos affected the way Warren viewed law enforcement. Whereas lawbreakers in Warren's prosecutorial years had principally been profiteers consciously seeking to trade in illicit materials, lawbreakers in his judicial years were regularly "people in our big cities who [were] living in ghettos, without any employment of any kind." Fighting crime in the fifties and sixties meant more than vigorous prosecution of criminals; it meant "get[ting] rid of the ghettos" so that people with "no schooling [and] no skills" were no longer "easy prey to all kinds of bad influences."[8]

The professional criminal of the Prohibition era, Warren believed, knew what he was doing and took his chances; the average Warren Court criminal, he thought, might have turned to crime because of disadvantage or out of degradation. One could even see "street crime" in the fifties and sixties as a protest against inequality and disadvantage in American life: a desperate plea to be able to participate fully in American society. Although the Court could do nothing about the deplorable conditions of urban America, it could at least ensure that the process of criminal justice did not add to the degraded status of those participating in it.

Warren Court criminal procedure cases thus embodied, for Warren, principles of fairness and equality that were part of the ethical structure of the Constitution. He saw the criminal justice system as a personification of the encroaching hand of government; he saw the average Warren Court criminal as being "put upon" by that system, stripped of humanity and dignity and opportunities for fair treatment. He put himself in the "shoes" of the average criminal and noted how features of criminal justice acted to condition an incarcerated person's opportunity to defend himself on constitutionally irrelevant grounds such as wealth or race. From the vantage point of the incarcerated, he looked upon police officials and prosecutors with a skeptical eye. Warren's altered perspective produced a resolution: If government was going to deprive some of its citizens of their liberty and their humanity, it was at least going to effectuate that deprivation fairly. Otherwise Warren would reveal himself to be as vigorous a defender of disadvantaged criminals as he had been a prosecutor.

Warren's experience with criminal procedure cases began, as had his experience with cases involving legislative investigations, with a decision in which his mature views were suppressed. *Irvine v. California*,[9] like *Barsky v. Board of Regents*[10] and a deportation case, *Galvan v. Press*,[11] came at a time when Warren, "still groping around in the field of due process,"[12] seemed disinclined to depart from Court precedents that restricted individual liberties. The California police in *Irvine* had illegally implanted microphones in a gambler's bedroom and used the evidence to obtain a conviction. Warren had no sympathy for gamblers, but he was shocked by the police actions. He alone joined Justice Jackson in ordering the record of the case to be sent to the attorney general for investigation and possible prosecution of the offending police.[13]

Warren's vote in the *Irvine* case, however, was to uphold the conviction. A 1949 Court precedent, *Wolf v. Colorado*,[14] had held that illegally procured evidence did not need to be excluded from a state criminal trial, and Warren "agreed not to overrule it at that particular moment."[15] He hoped that Attorney General Brownell would institute an investigation into the "terrible abuse of power on the part of the police,"[16] but no investigation was ever forthcoming. Warren repeatedly told the story of the *Irvine* case later in his career: Among the conclusions he drew was that the Court could not rely on other branches of government to remedy abuses that came to the Court's attention.[17] After *Irvine*, Warren rarely sustained a conviction that he thought had been unfairly obtained and never relied on law enforcement bodies to discipline themselves. Seven years after *Irvine* he voted to overrule *Wolf v. Colorado*.[18]

The feature of the criminal justice system that aroused Warren's strongest emotions was the confession made during incarceration. Confessions, for Warren, were consequences of the pressures that the system placed on detained individuals. They were the primary goal of zealous police, the key to a successful prosecution, the capitulation of a free person to authority. As a district attorney, Warren had taken pains to secure confessions and had walked the line of unethical official conduct.

In the 1937 *Point Lobos* murder case, previously discussed, Warren's men had questioned one suspect, Frank Conner, continuously for five and one half hours without an attorney present, denied him access to any attorney, misrepresented the whereabouts of one attorney he had retained, and kept him in continuous custody for twenty-one hours without sleep. All this had taken place before Conner had been formally booked and arraigned. His confession, later repudiated, was the principal basis on which he was convicted. After serving four years in prison, Conner was paroled, by then in a deteriorated mental condition.

While strongly denying any illegal or unethical conduct in the *Point Lobos* case, Warren soon revealed, on the Court, an interest in scrutinizing activities of law enforcement officials that had resulted in confessions. In 1958 he dissented from two cases in which the Court validated convictions obtained from suspects who had been denied access to an attorney during interrogation.[19] Justice Douglas wrote for himself and Warren in one of the cases that "[t]he mischief and abuse of the third degree will continue as long as an accused can be denied the right to counsel at this the most critical period of his ordeal."[20]

In 1959, writing for the Court, Warren reversed a conviction in a confession case, *Spano v. New York*,[21] the facts of which resembled those in the *Point Lobos* murder. In *Spano* the defendant, who had been indicted for murder, surrendered himself for questioning in the company of his attorney. Having been advised by his attorney not to answer questions, he was questioned for eight hours without his attorney being present. At a point in the interrogation, Spano's friend, a police officer whom Spano had notified of his intention to surrender, joined the interrogation. The friend falsely informed Spano that he was in trouble because Spano had called him, that he had lost credibility as a result of Spano's failure to confess, and that consequently his job on the force was in jeopardy. Spano then confessed. In invalidating the confession, Warrren maintained that Spano's "will was overborne by official pressure, fatigue and sympathy falsely aroused."[22]

The ethical principle of *Spano*, for Warren, was that "the police must obey the law while enforcing the law." He referred to "the abhorrence of society to the use of involuntary confessions" and suggested that this attitude did "not turn alone on their inherent untrustworthiness." There was also a "deep-rooted feeling" that "in the end life and liberty can be as much endangered from illegal methods used to convict those thought to be criminals as from the actual criminals themselves."[23] Involuntary confession cases symbolized the coercive powers of the state and the helplessness of the incarcerated individual, a state of affairs that aroused Warren's fears of oppressive government. Police illegality gave him the opportunity to strike out at those fears.

After *Spano* the course to Warren's famous opinion in *Miranda v. Arizona*[24] was straightforward and swift. Five years after *Spano*, in *Escobedo v. Illinois*,[25] denial of access to an attorney when a police investigation had focused upon one suspect was held constitutionally impermissible by a five-man majority, including Warren. The defendant in *Escobedo* had confessed, but he had not been advised of his right to keep silent, and he had not been allowed to have his lawyer present while he was being questioned. As Warren put it,

> [h]e called for his lawyer, his lawyer called for him and the police told both of them that they could see each other only when they got through with Escobedo. And then they went through him and against his protests they took his confession and convicted him on his confession.[26]

Warren thought this official coercive behavior no more tolerable than that in *Spano*.

With *Escobedo* decided, Warren said, "the question arises: if he's entitled to a lawyer when his lawyer is present, when is he *first* entitled to a lawyer?"[27] That question was answered—in great detail— by Warren for the Court in *Miranda*. Four cases in which defendants had been questioned outside the presence of their attorneys were combined to form the factual base of the *Miranda* decision.

> In each, the defendant . . . was cut off from the outside world. In none of these cases was the defendant given a full and effective warning of his rights at the outset of the interrogation process. In all the cases, the questioning elicited oral admissions . . . which were admitted at their trials.[28]

The cases, for Warren, thus "share[d] salient features—incommunicado interrogation of individuals in a police-dominated atmosphere, resulting in self-incriminating statements without full warnings of constitutional rights."[29]

Warren's opinion for the Court in *Miranda* was divided into three parts. The first part developed the ethical basis of the decision, the second part attempted to ground the decision in orthodox constitutional doctrine, and the last part promulgated a code of police conduct. A striking feature of the opinion was its disproportionate attention to the first and third parts rather than to the second.

Warren began *Miranda* with a thirteen-page discussion of "the nature and setting of . . . in-custody interrogation."[30] Relying on police manuals and comments by law enforcement officials, Warren detailed current techniques of incommunicado interrogation, which included beatings, intimidations, the denial of food and other necessities, psychological pressures, false statements, and bogus overtures of good will. Officers were instructed how to harass suspects, "persuade, trick, or cajole [them] out of exercising [their] constitutional rights,"[31] make them feel dependent, isolated, and abandoned, and generally "put the subject in a psychological state where his story is but an elaboration of what the police purport to know already—that he is guilty."[32] The "interrogation environment" that Warren described was "created for no purpose," he felt, "other than to subjugate the individual to the will of his examiner."[33]

In contrast to Warren's lengthy compilation of police excesses under the existing rules governing custodial interrogation, the portion of his *Miranda* opinion grounding the decision in precedent was short and cryptic. The section consisted of eight and one-half pages, four and one-half of which were devoted to a brief historical review of the privilege against self-incrimination. Warren saw the privilege as an embodiment of "the respect a government . . . must accord to the dignity and integrity of its citizens."[34] In the remaining four pages, Warren cited as precedent three cases[35] incorporating Fifth Amendment privileges against the states, none of which involved custodial interrogation. He also cited *Escobedo*, a case that had been decided wholly on Sixth Amendment grounds. Justice Harlan, in dissent, called Warren's approach a "trompe l'oeil," stating that it failed "to show that the [*Miranda* decision was] well supported, let alone compelled, by Fifth Amendment precedents,"[36] and noted that *Escobedo* "contain[ed] no reasoning or even general conclusions addressed to the Fifth Amendment."[37] Warren's use of *Escobedo* seemed "surprising" to Harlan.[38]

The last thirty-two pages of Warren's *Miranda* opinion concerned themselves with a feature of judicial decisionmaking that Warren

particularly liked, the formulation of practical rules and their application to concrete fact situations. First, Warren announced the "*Miranda* warnings": that a suspect "must first be informed in clear and
unequivocal terms that he has the right to remain silent"; [39] that the
warning "must be accompanied by the explanation that anything said
can and will be used against the individual in court"; [40] that counsel
may be "present during any questioning if the defendant so desires"; [41] and that "if [the suspect] is indigent a lawyer will be appointed to represent him." [42]

These privileges could only be "knowingly and intelligently
waived": [43] the fact of lengthy interrogation or "incommunicado incarceration before a statement" would be "strong evidence that the
accused did not validly waive his rights." [44] Any evidence that a suspect was "threatened, tricked, or cajoled into a waiver" [45] would ne-
gate voluntariness. The privileges took effect "when the individual is
first subjected to police interrogation while in custody at the station
or otherwise deprived of his freedom of action in any significant
way." [46]

Warren then applied the new warnings to the four cases before
the Court in *Miranda*. He invalidated Ernesto Miranda's conviction;
Miranda had not been advised that he had a right to have an attorney
present during his interrogation, [47] and therefore his confession was
inadmissible. He invalidated the conviction of Michael Vignera in
*Vignera v. New York*, because Vignera was not given any warnings
before being questioned by a detective and an assistant district attorney. [48] He reversed the conviction of Carl Westover in *Westover v.
United States*, because Westover had been given no warnings prior to
questioning by local police, even though subsequently he had been
given warnings by FBI agents and had written a confession after
receiving those warnings. [49] Finally, he affirmed the invalidation of
the conviction of Roy Stewart in *California v. Stewart*, because nothing in the record indicated that Stewart had been advised of his
rights. [50]

Warren's *Miranda* opinion almost bristled with his sense of the
injustice of coercive police practices and his conviction that an omnipresent government could rob its citizens of their dignity and humanity. The energy of the opinion was directed at building up evidence of the psychological imbalance between interrogating officers—
comfortable in their setting, able to come and go freely, possessing
access to external information, imbued with the authority of their

profession—and the interrogated defendant, cut off from all forms of aid. Warren declared after *Miranda* that "the prosecutor under our system is not paid to convict people [but to] protect the rights of people in our community and to see that when there is a violation of the law, it is vindicated by trial and prosecution under fair judicial standards."[51] It seemed as if in *Miranda* he was simply assuring that prosecutors and criminal defendants would be on an equal footing. His ethical posture presumed that police investigation of a suspect was inherently coercive; that "fairness" required some advantages to the suspect so that this coercion could be lessened; and that the combination of a lawyer and self-imposed silence would redress the balance.

The problem with Warren's posture, of course, is that there is no compelling reason why law enforcement officers and suspects should be on an equal footing. Although interrogated persons are technically presumed to be innocent of potential charges, as a practical matter a system of law enforcement has to presume potential guilt in detained suspects to justify their detention. Prosecuting authorities and persons charged with crimes may be "equal" in court, but by definition they are not equal in the station house, because law enforcement authorities are assumed to concentrate their efforts on persons whom they think have some connection with a law enforcement violation. Warren's application of the fairness principle in *Miranda* seemed to ignore the practical necessity of giving law enforcement personnel some advantages in doing what society has asked them to do.

*Miranda*'s code of police conduct was widely denounced by law enforcement officials and commentators, was made an issue in the 1968 election, and was apparently undermined in practice. Unlike *Brown v. Board of Education* and the reapportionment cases, in *Miranda* there was no moral void for the Court to fill that legislatures had created by their own inactivity. At the time of the *Miranda* decision the American Law Institute had produced a draft of a Model Code of Pre-Arraignment Procedure, the President's Commission on Law Enforcement and the Administration of Justice had been established, with the reform of criminal procedures one of its goals, and some state legislatures had liberalized procedures in custodial interrogation.[52] The *Miranda* decision brushed aside those developments and established a uniform code of police procedure.

In recalling his years with Warren in the Alameda County District Attorney's Office, J. Frank Coakley suggested that the genesis of the *Miranda* decision was Warren's own understanding of the decisive imbalance between a competent, prepared, indefatigable prosecutorial force and an isolated suspect. The very vigor with which Warren prosecuted criminals and the considerable skills he had in interrogation, Coakley felt, made him aware of the opportunities for coercion in the custodial setting.[53] Warren, we have seen, looked at different criminals differently: While organized crime and "moral" crimes, such as gambling or prostitution, outraged his sensibilities, he did not seem unsympathetic to criminals who had a history of social and economic disadvantages. It was as if he projected the single criminal suspect, isolated in a custodial setting, into the position of the "debased" voter in *Reynolds v. Sims* or the expatriated private in *Trop v. Dulles:* the lone individual struggling against organized, powerful interests. Such an image regularly galvanized Warren's sympathies and produced judicial rhetoric with a heavy emphasis on the unfairness of the predicament.

Warren regularly became annoyed when colleagues or commentators suggested that the impact of Warren Court innovations in criminal procedure was to turn criminals out of jails to threaten the public. In *Miranda*, Justice White said in dissent that "[i]n some unknown number of cases the Court's rule will return a killer, a rapist or other criminal to the streets . . . to repeat his crime whenever it pleases him."[54] Warren believed, in contrast, that the new criminal procedure rules were largely directed at the police. They were rules of more enlightened law enforcement, not rules designed to allow guilty criminals to go free. Warren felt that thorough preparation and the vigorous pursuit of a case, not coercive or deceptive interrogations, should be the prerequisites for effective law enforcement. Rule changes such as those made in *Miranda* were designed to focus the attention of law enforcement officials on hard work and fair play.

Nonetheless Warren was aware, throughout the years of his Court's reform of criminal procedure, that a side effect of the rule changes would be fewer convictions and, consequently, more criminals released to society. An issue that squarely raised this feature of rule changes was their retroactive effect. Historically, new interpretations of constitutional law had been made retroactive once initiated, the rationale being that previous interpretations were somehow jurisprudentially defective and could no longer be treated as "law." But with the acceptance of the jurisprudential theory that new interpretations of law were in some sense "made" by judges, the possibility that different generations of judges could differ in their interpretations of the Constitution was recognized, and older interpretations were seen as perhaps unsound but nonetheless having stature as laws.

It seemed, especially in the field of criminal procedure, unjust to penalize law enforcement officials who had relied on an existing interpretation of the Constitution only to find, years later, that they had acted in an unconstitutional manner. The issue of retroactive versus prospective application of new constitutional interpretations thus surfaced most clearly in Warren Court criminal procedure cases. The issue also had its greatest practical impact there, since many criminals had been jailed under a set of procedures that were now found to be constitutionally defective. Should these persons have the benefit of retroactive application?

Four years after *Mapp v. Ohio*,[55] a case that invalidated state convictions secured through illegal seizures of evidence, the Warren Court declared in *Linkletter v. Walker*[56] that the *Mapp* decision would be prospectively applied. The date selected in *Linkletter* was the date the *Mapp* decision was announced, but only convictions that had become final by that date were treated under the new rules. Cases on direct appeal received the benefit of the *Mapp* rule. The *Linkletter* majority, which included Warren, was candid in acknowledging the potential wholesale release of prisoners should *Mapp* be applied retroactively. Among the criteria the Court devised for ascertaining whether a rule change should be given retroactive effect was "the effect on the administration of justice."[57]

The reform of criminal procedure instituted by the Warren Court had largely been the result of the efforts of four justices, Black, Douglas, Brennan, and Warren, who picked up a changing fifth member—sometimes Tom Clark, sometimes Arthur Goldberg, sometimes Abe Fortas—for their majority decisions. The positions of the four reformers on retroactivity diverged sharply. Black and Douglas, beginning with *Linkletter*, announced that all rule changes should have retroactive effect and maintained that position in subsequent cases. Brennan and Warren, however, were early and firm supporters of prospective application, joining the majority in *Linkletter* and voting for prospective application in all cases between 1965 and 1969 where the Court was seriously divided on the issue. In their support for prospective application, Brennan and Warren had some curious partners: Goldberg and Fortas, supporters of the new rules; Clark, White, and Stewart, who wanted to limit their effect; and, for a time, Harlan, who eventually revised his view and endorsed retroactivity even though he conceded that his stance had the consequence of widening the impact of rule changes he deplored.[58]

Retroactivity issues invited the application of an approach toward constitutional adjudication with which Warren was comfortable—the open balancing of conflicting social interests. In *Johnson v. New Jersey*,[59] a 1966 decision declaring that *Escobedo* and *Miranda* would be given prospective application, Warren read *Linkletter* as "establish[ing] the principle that in criminal litigation concerning constitutional claims, 'the Court may in the interest of justice make the rule prospective . . . where the exigencies of the situation require such an application.' "[60] "Exigencies" in the *Johnson* case were the reliance of law enforcement agencies on decisions, prior to *Escobedo* and *Miranda*, that permitted incommunicado interrogation; the potential "retrial or release of numerous prisoners found guilty by trustworthy evidence in conformity with previously announced constitutional standards";[61] and the unjustifiable burden[s] on the administration of justice."[62] Indeed, Warren concluded, the *Escobedo* and *Miranda* rules should not even be applied to cases on direct appeal at the times those decisions were announced. The effect of *Johnson* was thus to make *Escobedo* and *Miranda* purely prospective rule changes.

After some uncertainty the Court announced a more liberal cut-off date for retroactive application, the date the newly prohibited practices had occurred.[63] Warren defended the more stringent standard of *Johnson* as consistent with "society's legitimate concern that convictions already validly obtained not be needlessly aborted."[64] In *Jenkins v. Delaware*,[65] a 1969 decision, Warren limited the retroactive impact of *Miranda* even further. The *Jenkins* case raised the question whether the *Miranda* rule changes would apply to cases where a defendant had been brought to trial before *Miranda* but had not had a conclusive determination of his or her fate and had been retried after the *Miranda* decision. Such defendants, Warren held for the Court, did not get the benefit of the *Miranda* rules.

In *Jenkins*, Warren acknowledged "fortuities" in the process of prospective decisionmaking. One defendant might get the benefit of new rule changes and another not, merely because of "the progress

of their cases from initial investigation and arrest to final judgment."
These "incongruities," however, had to be "balanced against the impetus the technique provides for the implementation of long overdue
reforms, which otherwise could not be practicably effected."[66] The
implementation of *Miranda*, Warren reasoned, was made easier by
limiting its application. He recognized "society's interest in the effective prosecution of criminals."[67] He felt that applying *Miranda* to
retrials would place an "increased evidentiary burden . . . upon law
enforcement officials."[68]

Warren's approach to the retroactivity cases was reminiscent of
his technique of lawmaking as a California public official. Once he
conceived of legislation as embodying a principle—such as that of
protection for the ordinary citizen against debilitating illness, as embodied in compulsory health insurance—he was willing to compromise on details in order to further the principle's implementation. In
retroactivity cases, the principle was the restraints on unethical police conduct signified by the *Escobedo* and *Miranda* rules. Just who
would get the benefit of the principle and when the principle would
take effect were matters Warren was willing to negotiate; he was not
willing to negotiate the principle itself. Legislatures and executive
commissions were free to develop even more stringent safeguards
from those laid down in *Miranda;* states were free to extend *Miranda*
protections to retrial defendants whose earlier trials had predated *Miranda.* But there was to be no backsliding on the "long overdue reforms" themselves.

✑

It is possible to look at Warren's law enforcement cases as he perceived them—as efforts not to cripple law enforcement but to purge
it of its coercive and unethical features. If the purpose of a prosecutor's office was not to convict criminals but to see that the law was
"vindicated by trial and prosecution under fair judicial standards,"[69]
then the Warren Court reforms were principally directed toward
making law enforcement practices more closely approximate ideals
of justice and fairness. The Court was not hampering law enforcement, it was ennobling it. Late in his life, Warren remembered (conveniently, some thought)[70] that the office he ran as district attorney
had emphasized hard work, thorough preparation, and no unethical
shortcuts. As a district attorney he did not see himself as a hard-

nosed convicter of criminals, but as a representative of the public interest in decency, honesty, and respect for law against organized criminal forces.

When Warren perceived, as a judge, that there were prosecutors and police who did not share his vision of law enforcement, but rather used their office to trespass on the dignity and humanity of disadvantaged persons, he sought to correct that misguided view of the purpose of a system of laws. Law enforcement was supposed to vindicate the equality of persons before the law and to reinforce the commitment made by American civilization to the values of humanity, dignity, and fairness. It was not supposed to be a means whereby the strong oppressed the weak. Warren believed that his concern as a judge was not that criminals go free, but that they retain their humanity and dignity even when "put in the toils of the law." Confessions were the classic example of a solitary citizen confronted with the coercive power of law enforcement. Warren wanted that power exercised in a way that vindicated his ideal of the American system of criminal justice.

As evidence of the seriousness with which he endorsed law enforcement values, Warren might have pointed to the case of *Terry v. Ohio*,[71] a 1968 decision in which he held, for the Court, that a policeman's "stop and frisk" of a suspect on a public street, while concededly a "seizure" and "search" of the suspect's person under the Fourth Amendment, was not constitutionally "unreasonable" if the officer had reason to fear for his own safety or the safety of others. Although the facts of *Terry* suggested that the officer had sufficient reason to detain and question the suspect—Terry had walked past the same storefront window twenty-four times and conferred regularly and earnestly with his companions after doing so—the rules laid down in *Terry* seemingly ran counter to Warren's attitudes about the proper role of government vis-à-vis the private lives of its citizens. Under the *Terry* rules an aggressive officer could simply single out a suspect, detain him, fling him up against a wall and make a thorough search of his person, release him, and claim that the stop and frisk had been undertaken out of fear for himself or the safety of others. The rules seemed to give police officials license to perform random assaults on citizens.

Warren spoke in *Terry* of "the wholesale harassment by certain elements of the police community, of which minority groups, particularly Negroes, frequently complain."[72] He cited arguments that acceptance of the *Terry* rules would "constitute . . . an encouragement of . . . substantial interference with liberty and personal security by police officers."[73] He called the stop and frisk a "severe . . . intrusion upon cherished personal security" and an "annoying, frightening, and perhaps humiliating experience."[74] He nonetheless held that a stop and frisk based on a reasonable fear for one's own or others' safety was constitutionally permissible when conducted by a police officer.[75]

Warren was fully aware that police coercion could manifest itself as easily in "street-strutting" as in incommunicado interrogation. He believed strongly in protecting citizens from assaults upon their person: An involuntary blood test,[76] a bugged bedroom, or unduly long confinement were unethical and consequently unconstitutional police practices. To tolerate the stop and frisk Warren must have assumed that it was a necessary part of effective police work: He did not want unduly to deter police officials from vigorously investigating suspicious circumstances. Justice Douglas, dissenting in *Terry*, said that to give the police stop-and-frisk power was "to take a long step down the totalitarian path" and suggested that if "the police can pick [a person] up whenever they do not like the cut of his jib, . . . we enter a new regime."[77] Warren took those sentiments seriously; his support for the result in *Terry* can be read as a considerable bow in the direction of law enforcement.

As Chief Justice, Warren saw law enforcement from the other side. Instead of pressing vigorously for prosecutions, he scrutinized the conduct of prosecutors. Instead of singling out issues where he could assume the role of protector of the public against corrupt or criminalized special interests, he put himself in the shoes of people who were harrassed or debased by the criminal justice system. As a district attorney and an attorney general, he could ignore the fact that "access to justice" was "unequal in parts of our country." He knew then that "[s]uspects [were] sometimes arrested, tried, and convicted without being adequately informed of their right to counsel." He knew that "many a citizen" was denied full legal rights because he "[could] not afford to exercise [them]."[78] It was not his job to reform the system of criminal justice in those days; his role had merely been to ensure that his office stayed within its legal and ethical bounds.

Seeing law enforcement from the perspective of Chief Justice of the United States meant, for Warren, seeing the gap between an unequal, potentially coercive existing system and the ideal of a truly enlightened law enforcement apparatus, where officials pursued and prosecuted criminals with vigor but with honor and equanimity. This perspective led Warren simultaneously to idealize his years as an agent of law enforcement—thereby hazing over the times when he may have bent laws or ethics for his own purposes—and to recall the power of the police and the prosecutor with a vivid clarity. Looking at the law enforcement apparatus from the vantage point of a solitary, unsupported defendant revealed the fundamentally unequal bargaining position of the incarcerated criminal and his incarcerators. The idealized vision and the harsh memories of his law enforcement years fused to create in Warren a powerful impetus to restrain unethical police conduct.

Nevertheless Warren did not see his stance on the Court as undermining support for law enforcement. It angered him to find critics of his decisions contrasting his prosecutorial stance in California with his allegedly permissive attitude toward criminals as Chief Justice. Warren belived that his decisions were not weakening the system of law enforcement, but rather strengthening it. The new restraints imposed on police by the Warren Court curbed sloppiness, brutality, and intimidation. As such they oriented law enforcement activity to its proper tasks—hard work, persistence, thoroughness, preparation, vigilance. When Warren said in 1968 that he was stronger for law enforcement than he had ever been, he also could have said that he believed his decisions had made law enforcement in America a stronger profession. That many others disagreed with that assessment, and that one of his old enemies campaigned for the presidency on the theme that the Warren Court was "soft on crime" infuriated Warren. He saw himself as perfecting law enforcement, not undermining it.

# 13

# Warren the Economic Theorist: Labor and Antitrust

The theoretical dimension of Warren's opinions, we have seen, was largely ethical in nature. He tended to approach cases not from the perspective of political or social theory, but rather with an interest in achieving a fair and humane outcome in a given case. While there were some areas of law, such as obscenity, in which Warren's perspective did not require an examination of the equities of a given case—Warren was not much interested in the contents of allegedly obscene works once he had determined that pandering was not at issue—for the most part Warren's jurisprudence was designed to give him flexibility to reach "just" results without being overburdened by theoretical baggage.

Cases concerning the economy were a striking example. The labor and antitrust areas are ones in which, for the most part, the Court is asked to focus on statutory interpretation rather than the language of the Constitution. In such a context a judge's attitudes toward issues of economic policy—whether labor should receive favored treatment, to what degree economic concentration is desirable, whether economic policies should be uniform throughout the nation—are apt to control his votes. Warren did not come to the Court with fully developed attitudes toward such issues; his stance as a California public official was not a clear harbinger of the stance he would take as a judge. But in working through cases affecting the economy, Warren did develop a predictable posture. Once again his perspective reflected the surfacing of instincts that he had acquired early in life and that had been suppressed or modified during his years as a California public official.

Warren's early life was dominated by economic issues. His father worked for a corporation that was an omnipresent economic force in Southern California; Matt Warren's livelihood was dependent on the Southern Pacific Railroad. Matt belonged to a union, and when the union struck or its members were laid off or discharged, the Warren family was affected. The Warrens' relocation in Sumner had been a result of a labor dispute; Methias' new employment was also with the Southern Pacific; one of Earl's early employers was the same enterprise. The dominance of the railroad, the dangerousness of its work, its apparent indifference to its employees, and its resistance to unionization were themes of Warren's youth. When he emerged as a force in California public life he did not forget the thralldom of the Southern Pacific, and he remembered his own membership in a musician's union. Organized labor was a symbol of resistance to the economic hardships of his early life.

But Warren had come to be identified in his public career with forces—the Republican party, law enforcement, anticommunism— that were generally hostile to the labor movement in California. As a district attorney, an attorney general, and a candidate for governor, Warren had kept a watchful eye on political militancy in labor unions, making capital of that issue in the *Point Lobos* case, had opposed the New Deal and supported the free enterprise system, and had been perceived as no friend to organized labor, whose endorsements went to Culbert Olson in the gubernatorial campaign of 1942. No one anticipated that Warren would be supportive of labor as governor, and while his administrations were not identified with the business community as much as opponents had predicted, they were also not as receptive to labor organization in California as Olson's administration had been.

With respect to issues of economic concentration, Warren had no particular record as a California public official. His concerns as attorney general had been with the prosecution of organized crime and with preparedness and civil defense: He had done no "trust-busting" at all. He had shown no inclination to repeal taxes on corporations imposed by previous administrations, but his attention as governor had been on issues, such as highway construction, prison reform, and health insurance, that did not, at least at the point of initial formulation, involve questions of economic concentration. One might have noted, however, Warren's consistent concern as governor with the hegemony of "special interests" in the California legislature. While his attacks on special interests were primarily ethical in tone—he claimed that special interests unfairly interjected their particular concerns into issues affecting the general public—the interest groups Warren attacked represented concentrations of economic power. In addition, the original basis of his attraction to Progressive efforts to curb special interests had been his sense of the economic power of the Southern Pacific.

So if one reflected on Warren's instincts about matters of economic policy, the fact that he would come to support the principle of unionization and view with hostility economic concentration did not seem surprising. The ingredient that came to be added to Warren's perspective in his years on the Court was support for federal regulation of the economy. Perhaps because of his cynicism toward state legislatures, perhaps because as a federal judge he had become more aware of the provincialism or self-interest of state governments, Warren came to see the federal government as a beneficent and necessary force in carrying out economic policy.

This perception was reflected in a striking record of support for the National Labor Relations Board in labor cases and for the Justice Department and the Federal Trade Commission in antitrust cases. It made Warren's performance on the Court capable of being characterized as strongly anticorporate, both with respect to labor issues and with respect to monolithic economic practices. The very issues that the Southern Pacific's economic dominance symbolized—employee thralldom and monopolistic power—became rallying points for Warren's jurisprudence. He emerged on his Court as a good friend of organized labor and a staunch ally of trustbusters.

К

When Warren came to the Court, the principal formative issues of a national labor policy had been resolved. The National Labor Relations Board's jurisdiction over employer-employee bargaining had been established: Issues on the Warren Court involved the scope and extent of the board's power.[1] An exemption for labor unions from the antitrust laws had been carved out, although the Court continued to attempt to define the limits of that exemption.[2] Various statutes designed to supplant the common law of employer-employee relations,[3] to regulate certain industries,[4] or to provide employers with remedies against unfair union practices[5] had appeared, and the Court sought to determine the scope of their coverage or their interaction with one another. In the years of the Warren Court organized labor moved from a politically controversial interest group to a fixture in American life. Labor organization was regarded neither as socialistic heresy nor as another form of excessive economic concentration: The Warren years were a halcyon point in its history. Indeed the principal thrust of Warren Court decisions was to insure that the regulatory powers of the federal government would be used to maintain the competitive position of organized labor.

Warren's contribution to this trend was made in two doctrinal areas: preemption and unfair labor practices. In the years of the Warren Court a finding that the National Labor Relations Board had exclusive jurisdiction over a matter, or that state courts were preempted from acquiring jurisdiction, had the effect of being a victory for labor, since the NLRB was considerably more sympathetic to the interests of unions than were state courts.

Warren's early labor decisions, while supportive of the interests of unions or workers, were relatively inconsequential. He construed, for example, the activities of shower-taking and knife-sharpening as being "principal" within the meaning of the Portal to Portal Act, and hence compensable under ordinary wage scales.[6] In the 1956 term, however, he emerged as a strong supporter of federal regulation in industries engaged in interstate commerce. Hints of this position had appeared in *Amalgamated Clothing Workers of America v. Richman Brothers, Co.*,[7] a 1955 decision, where Warren dissented from a Frankfurter opinion holding that an employer could secure a temporary injunction in a state court against union picketing instead of filing an unfair labor practice charge before the NLRB, thereby preventing the board from adjudicating the dispute. Warren said that the majority decision "seriously frustrates a comprehensive regulatory scheme established by Congress for the resolution of the kind of labor dispute involved here."[8] He would have allowed either the union or the NLRB to sue in federal court to obtain exclusive jurisdiction by the board over the controversy.

In five cases decided in 1957 and 1958 Warren revealed himself as a vigorous supporter of exclusive primary jurisdiction of the NLRB in cases affecting interstate commerce. Three of the cases involved the board's power to decline jurisdiction over a dispute and yet prevent state labor boards or state courts from hearing it. In *Guss v. Utah Labor Board*[9] Warren held that the only instance in which the NLRB did not have exclusive jurisdiction over a labor dispute "affecting" interstate commerce was when it had ceded jurisdiction to a state board under a section of the Taft-Hartley Act. That cession was possible only where states "have brought their labor laws into conformity with federal policy."[10]

The same theory precluded state courts from offering relief to employers in cases where the NLRB had declined to exercise jurisdiction over a controversy. In *Amalgamated Meat Cutters v. Butcher Workmen of North America*[11] Warren found that an Ohio meat market that received approximately one-ninth of its revenue from outside the state "affected interstate commerce" and that therefore the market could not secure an injunction in state court against union picketing of its facilities. Warren maintained that "[t]he conduct here restrained—an effort by a union not representing a majority of its employees to compel an employer to agree to a union shop contract—is conduct of which the National Act has taken hold."[12] He noted that although the union had charged the employer with labor practices that the NLRB might well have adjudged to be "unfair," the Ohio court took no account of such practices in its granting of the injunction. This discrepancy provided Warren with "an excellent example of one of the reasons why . . . Congress has expressed its judgment in favor of uniformity"[13] in interstate labor disputes.

The third preemption case of the 1956 term was *San Diego Building Trades Council v. Garmon*,[14] a forerunner to a more famous Warren Court preemption decision two years later. In *Garmon I*, a California company operating two lumber yards refused to sign a "union shop" contract, claiming that a majority of its employees had not selected any union as their collective bargaining agent. The company's business was concededly interstate in character. The union picketed the company upon its refusal, and the company filed suit in a California state court, asking that the picketing be enjoined and that the company be awarded damages. The court awarded the injunction and one thousand dollars in damages; meanwhile the NLRB declined to adjudicate the question of union representation. The California Supreme Court found that the state courts could act on the matter in light of the NLRB's declination and affirmed the decision against the union. Warren reversed the decision, citing *Guss* and *Amalgamated Meat Cutters.* He declined to reach the question of whether the damage award was permissible since damages were not among the remedies provided by the NLRB.

The 1957 decision left the field of interstate labor activities in an odd state. Since state agencies and courts were precluded from entertaining jurisdiction over labor disputes "affecting" interstate commerce, in those instances where the NLRB declined to adjudicate a controversy, the parties were without a remedy. Justice Burton, dissenting in *Guss,* said that the decision and its companion cases created "an extensive no man's land within which no federal or state agency or court is empowered to deal with labor controversies." [15] Warren himself had recognized this in *Guss* and invited Congress "to change the situation at will." [16] In 1959 Congress did this by providing in the Landrum-Griffin Act that the board could not "decline jurisdiction under the standards prevailing upon August 1, 1959" and that where it did decline jurisdiction, the states could entertain controversies. [17]

Prior to congressional action, however, the Court exhibited a concern, in cases like *Garmon I,* that the absence of remedial relief from the board or the board's refusal to adjudicate a matter could leave employers or employees without a remedy. In two preemption cases in the 1957 term, a majority of the Court allowed state courts in California and Alabama to award contract and tort damages to employees for being illegally expelled from a union and for being harassed in connection with employment. In the first case, *International Association of Machinists v. Gonzales,* [18] the Court found that neither the National Labor Relations Act nor the Taft-Hartley Act protected union members from "arbitrary conduct by unions and union officers." [19] Since the NLRB had no power to award damages for mental or physical suffering and state courts were permitted to reinstate wrongfully discharged members of unions, the damage remedies could be taken as supplementary to, not competitive with, the federal statutes.

In the second case, *Automobile Workers v. Russell,* [20] a nonunion employee in an Alabama copper company crossed a picket line and was threatened and barred from entering the premises. He sued in tort for interference with his occupation, and recovered compensatory and punitive damages. The NLRB, as in the *Gonzales* case, had power only to award back pay for unfair labor practices; it could not award damages. The *Russell* majority found the Alabama state court's remedies not competitive because they were different from those of the NLRB and because they were not directed specifically at labor disputes.

Warren dissented in both cases. In *Gonzales* he referred to state court damage awards in labor disputes as "at war with the policies of the Federal Act,"[21] having potential for "disturbing the delicate balance of rights and remedies"[22] envisaged by national labor policy, and encouraging "local procedures and attitudes toward labor controversies."[23] In *Russell* he claimed that state court damage awards "upset the balance"[24] of employee, union, and management interests envisaged by federal legislation, had "an unfavorable effect upon the uniformity [federal policies] sought to achieve" by encouraging "differing attitudes toward labor organization . . . by juries in various localities,"[25] and discouraged "resort to the curative features of . . . federal labor law." Warren could "conceive of nothing more disruptive of congenial labor relations than arming employee, union and management with the potential for 'smarting' one another with exemplary damages."[26]

The diverse results reached by the early Warren Court in preemption cases were superficially reconciled by the decision in the second *Garmon* case,[27] in which Warren joined the majority. *Garmon II* recast the Court's test for whether the NLRB's jurisdiction over a labor dispute was exclusive. Instead of focusing on the adequacy of the relief sought by an injured party, the Court after *Garmon II* focused on the nature of the activity that gave rise to the dispute. If the activity was "arguably" prohibited or "arguably" protected by a section of the National Labor Relations Act, the NLRB had exclusive jurisdiction. Under this test, state-court relief was limited to cases where violence or the threat of violence occurred or where the activity in question was of "merely peripheral concern" to the act's policies.[28] *Garmon II* thus read the *Gonzales* and *Russell* decisions as being based on the potential for violence. After *Garmon II* the opportunities for supplementary relief in state courts were significantly reduced.

Warren's stance in preemption cases was thus consistently one of support for federal regulation of labor disputes. His opinions invoked rationales such as uniformity, efficiency (avoidance of conflicting remedies), and predictability (avoidance of confusion among disputing parties about their rights and responsibilities). They also took note of the hostility of some state courts toward organized labor. In general, Warren believed that the more state courts were prevented from involving themselves in labor disputes, the more likely those disputes were to be settled in a satisfactory manner.

The theme of support for the interest of organized labor also surfaced in Warren's decisions on unfair labor practices. In *Fibreboard Paper Products Co. v. Board*[29] Warren faced the question of whether or not a decision on the part of an employer to "contract out" part of its operations was a "term and condition of employment" under Sections 8 and 9 of the National Labor Relations Act. Under those sections employers had a duty to bargain collectively with unions that had become the exclusive representative of their employees. Fibreboard Paper Products, after unsuccessful efforts to reduce the costs of maintenance work performed by its own employees, hired an independent contractor to perform the work, discharged its own maintenance workers, and refused to negotiate with the designated union over its actions. The union filed unfair labor-practice charges against the company. The charges were upheld by the NLRB, and the company was ordered to reinstate the employees as maintenance workers, to give them back pay, and to bargain with the union about a new maintenance contract.

The facts of the case showed that Fibreboard had not really given the union an opportunity to respond to its concerns about the high costs of maintenance and that its decision to subcontract was based on these high labor costs. Instead of discussing the issues, Fibreboard had announced that it was "contracting out" its maintenance work, hired another company to perform the work, and then stated that "negotiation of a new or renewed agreement [with the union] would appear to us to be pointless."[30] Justice Stewart, concurring, said that these actions on the part of Fibreboard "frustrated collective bargaining . . . by [a] unilateral act of subcontracting the work."[31] The case was not far, Stewart suggested, from one where the employer "had merely discharged all of its employees and replaced them with other workers willing to work on the same job . . . without . . . fringe benefits."[32]

Nonetheless Warren's opinion for the Court seemed to suggest that all employer decisions to subcontract work were a proper subject for collective bargaining and failure to bargain about such decisions might constitute an unfair labor practice. Warren confined his holding to "the facts of this case"[33] and said that it "need not and does not encompass other forms of 'contracting out or subcontract-ing.' "[34] But he called the "primary issue" in *Fibreboard* "whether the 'contracting out' of work being performed by employees in the bar-gaining unit is a statutory subject of collective bargaining under [the National Labor Relations Act],"[35] and he concluded that "the re-placement of an independent contractor" was a statutory subject of collective bargaining.[36] Stewart said that the opinion "radiate[d] im-plications of . . . disturbing breadth"[37] and could be taken as an effort to bring "such larger entrepreneurial questions as what shall be produced, how capital shall be invested in fixed assets, or what the basic scope of the enterprise shall be"[38] within the statutory framework.

On balance, Warren's *Fibreboard* opinion was probably no more than another example of his penchant for accompanying broad gen-eral dictums with sharply qualified resolutions of the point at issue. While a decision not to continue production of a given make of au-tomobile, if it resulted in a reduction of the number of workers staff-ing an automobile manufacturing plant, would surely affect the "terms and conditions" of ordinary employment in the company, it is hard to imagine that *Fibreboard* was intended to embrace that situation. In the *Fibreboard* case, however, the decision to contract out mainte-nance had as one of its primary purposes the discharge of existing workers. The company claimed in its argument that the maintenance union had not "joined hands with management in an effort to bring about an economical and efficient operation";[39] accordingly, it re-solved to discharge all the maintenance workers. Warren could have been more circumspect with his language, but that was not Warren's usual tendency in opinion writing.

Warren's next major unfair labor-practice decision, *NLRB v. Great Dane Trailers*,[40] was somewhat more transparently prounion. Great Dane Trailers, Inc., and a local ironworkers union had entered into a collective bargaining agreement that provided, among other things, for vacation benefits. Negotiations to extend the agreement broke down, and the union struck the company for a period of about seven months, beginning in May, 1963 and ending in December of that year. During the strike the company operated with nonstrikers, re-placements for strikers, and employees who abandoned the strike.

The signed contract provided that vacation benefits would be paid around July 1 every year, and in July, 1963, after a number of strik-ers had demanded vacation pay, the company announced that it

would grant vacation benefits to all those employees who had reported to work on July 1, 1963. The union filed an unfair labor-
practice charge, claiming that the company had discriminated against
strikers for the purpose of "discouraging union membership." The
NLRB found that an unfair labor practice had been committed, although it found no evidence that the company's purpose was to discourage union membership.

Warren, for the Court, found that while the company's decision
was "discrimination in its simplest form," it was not "inherently destructive of employee interests" and was thus justifiable, if there was
no proof of antiunion motivation, if it served a "substantial and legitimate business end."[40a] This last requirement was something of a
gloss on previous decisions: It required "discriminating" employers
to provide affirmative justifications for their conduct. The Court of
Appeals for the Fifth Circuit, finding for the company, had speculated about some legitimate business purposes possibly being behind
the company's decision, such as discouraging any leaves immediately
before vacation periods, but Warren brushed these aside, stating that
"[t]he company simply did not meet the burden of proof, and the
Court of Appeals misconstrued the function of judicial review when
it proceeded nonetheless to speculate upon what *might have* motivated the company."[41] Justice Harlan, in dissent, called the "legitimate and substantial business justifications" requirement "an important shift in the manner of deciding employer unfair labor practice
cases" that he found "substantially inaccurate" and "unwise."[42]

*Great Dane* is a good example of Warren's impatience with legal
technicalities in the face of a perceived injustice. He thought that the
court of appeals' attributions of "legitimate" business purposes to the
company were nonsense: The purpose of the vacation benefits decision, he believed, was plainly to reward nonstriking employees and
to discourage strikers from continuing their efforts. The problem was,
however, that it was not clear that the National Labor Relations Act
or the Court's prior decisions prohibited this kind of employer tactics. The tactics were arguably not "inherently destructive," since
they did not penalize strikers but only sought to encourage them to
stop striking. Moreover, Great Dane successfully broke the strike by
hiring replacements and inducing striking employees to return to
work; thus, it was hard to claim that their decision about vacation
pay was antiunion or punitive.

Since no proof of an antiunion purpose had been made out, Warren was faced with a situation where established construction of the statute would not yield an unfair labor-practices claim. Accordingly, he created another category of employer discriminations—ones where impact was "comparatively slight"—and asserted that in those cases the employer had to show affirmative evidence of a "legitimate business" justification. But Great Dane had no notice of any such requirement. Justice was done, for Warren, by reading the National Labor Relations Act to proscribe employer discriminations he believed it obviously intended to reach, although it had not said so.

*Fibreboard* and *Great Dane* paled, however, in comparison with Warren's unanimous decision for the Court in *NLRB v. Gissel Packing Co.*[43] *Gissel* raised three significant issues: the use of employee authorization cards as an indication of the status of a union as a bargaining agent, the consequences of an employer's refusal to accept the validity of authorization cards, and the First Amendment rights of employers in controversies about union representation. The first two issues combined to raise the significant question of whether or not union representation status could ever be secured without an NLRB representation election. The third issue was significant because of the Warren Court's extension of First Amendment rights to forms, such as defamation and privacy, that were not linked to the political process.

The NLRB had developed a practice with respect to authorization cards that went as follows. When a majority of the employees of a company signed cards authorizing a particular union to serve as their bargaining agent, the employer was required to commence bargaining with the union unless he had a good-faith doubt that the union in fact had such majority support. The NLRB conceded that some cards were ambiguously worded so that their endorsement only constituted an employee's agreement to vote on the issue of unionization, and that the potential for union coercion of employees with respect to card signing existed since employee preferences were not made in secret. Despite these concerns the NLRB allowed the use of cards, but some circuit courts of appeal, such as the Fourth Circuit, found cards "inherently unreliable" and ruled that employers could ignore them and insist on an election to determine the status of a union.

The *Gissel* decision consolidated four cases in which unions had claimed majoritarian status on the basis of authorization cards and employers had refused to recognize them. The employers then insti-

tuted "vigorous antiunion campaigns that gave rise to numerous unfair labor practice charges."[44] One company interrogated employees
about their union activities and planted an agent at a union meeting;
another offered union supporters new jobs at higher pay if they would
use their influence to "break up the union";[45] a third told employees
that if a union came in "a nigger would be the head of it;"[46] and the
fourth warned that unionization and a strike would jeopardize the
continued operation of the plant.[47] In each case the NLRB found
that the companies had engaged in unfair labor practices and ordered
the companies to bargain with the union. In three of the cases the
Fourth Circuit reversed the board, holding that authorization cards
were inherently unreliable and that an employer could insist on an
election for union certification. In the fourth case the First Circuit
upheld the board's bargaining order.

Warren held that authorization cards were a permissible basis for
determining the representative status of a union, but that an employer could insist on a secret ballot election unless he engaged in
"contemporaneous unfair labeling practices likely to destroy the
union's majority and seriously impede the election."[48] Since all the
employers in *Gissel* had engaged in unfair labor practices on being
informed about the authorization cards, they were not relieved of
any duty to bargain. He also found that the use of authorization
cards was a permissible means of recognizing a union, despite conceding "that there have been abuses" by union organizers in the process of securing employee authorization. Having sanctioned the use
of cards, Warren then held that the NLRB could rely on a combination of cards and an employer's unfair labor practices to issue a
bargaining order that compelled the employer to negotiate with a
union even though there had been no election of union representatives. He remanded three of the cases to determine whether the
bargaining-order issues had been based on a combination of a union
majority in authorization cards and coercive and antiunion employer
tactics.[49]

In the fourth case the First Amendment issue surfaced. The president of the Sinclair Company, a producer of mill rolls and wire,
had initially responded to a union's organizing campaign by speaking
to his employees and warning them against joining the union; had
distributed letters emphasizing that the company might leave Massachusetts if unionization occurred; had sent employees a pamphlet

two weeks prior to the scheduled union election that called the union a "strike happy outfit" under "hoodlum control"; and had told his employees the day before the election that "the Company's financial condition was precarious" and that in the event of a strike the continued operation of the plant would be jeopardized and reemployment would be difficult.[50] He claimed his comments were constitutionally protected.

Warren argued that employer expression in the labor relations area needed to be analyzed "in the context of its . . . setting."[51] That setting revealed "the economic dependence of the employees on their employers" and "the necessary tendency of the former . . . to pick up intended implications of the latter." Employer speech thus had far greater opportunities for coercion than did, say, speech about political issues or speech criticizing public officials. The latter type of speech could be listened to "objectively"; employer speech could not be. While the "general views" of an employer about unionism were protected, those views could not be accompanied by "threat[s] of reprisal or force or promise[s] of benefit." The latter category of statements were "without the protection of the First Amendment."[52] In *Gissel* the observations of Sinclair's president about the financial condition of the company, the proclivities of the union to strike, and the difficulties that a shutdown of the plant would pose for employees amounted, in the context of a labor dispute, to "threat[s] of retaliatory action." The speech was "coercive" in nature. It was intended to "mislead . . . employees."[53]

*Gissel* was a decision wholly favorable to the interests of organized labor. It allowed union organizers to claim representative status, under certain circumstances, by a technique—authorization cards—that emphasized the public disclosure of an employee's sentiments, and thus the employees were open to coercion. It forced employers either to accept union representation once majoritarian status was claimed on the basis of cards or to call an election, and it narrowed the range of permissible employer comments about elections. More generally, it apparently restricted employer comment about a prospective union or the possibility of unionization to those views that were susceptible to proof. And it distinguished between the rights of newspapermen or politicians to comment on matters of public concern and the rights of employers to comment about matters peculiarly affecting the operation of their businesses. It suggested, in fact, that union organi-

zers could make a variety of unsubstantiated promises about the beneficial effects of unionization but that contrary statements by
management would constitute unfair labor practices.

One must remember, however, that the theory of labor legislation in American history has not been that employees should be made
"equal" to employers, but rather that they should be given some
comparative advantages in the process of bargaining in order to offset
the starting advantages of employers. The National Labor Relations
Act, with its creation of an overseeing board and its various remedies
against unfair labor practices, was not an effort, as were the civil
rights acts of the 1960s, to eliminate inequities in an area of American life, but rather to compensate for inequities that were natural
features of the employer-employee relationship. American labor legislation has not sought to eradicate self-interest, prejudice, and coercion in labor-management relations, but rather to give employees opportunities to vindicate their own self-interests and engage in their
own coercive tactics.

Only by this recognition of the purposes of labor legislation can
the one-sided character of Warren Court labor decisions be understood. When one negotiating party's adverse comments on another
are regarded as simply part of the bargaining process but another's
constitute unfair labor practices, an assumption has been made that
the restricted party has far greater opportunities to be coercive. A
theory of employer dominance has marked the Supreme Court's
labor-law decisions: The widening scope of employee remedies can be
seen as a continuing recognition of the myriad ways in which employer dominance can be expressed. In *Gissel* the Court pressed that
theory to the point where it was wisest for an employer confronted
with a purported union majority based on authorization cards to say,
"no comment." The assumption Warren and his Court made in *Gissel*
was that the potential for abuses among union organizers produced
by the *Gissel* rules was much less serious than the potential for employer coercion if the rules did not exist.

In assuming that the principal purpose of labor law was to shore
up the bargaining position of employees vis-à-vis their employers,
Warren and a majority of his fellow justices demonstrated their acceptance of the prevailing conception of labor relations from the New
Deal through the 1960s. That conception identified the employer-
employee relationship as the pivotal focus of labor relations, thereby
minimizing the significance of other relationships, such as the rela-

tionship between union leaders and union membership. Unionization was assumed to be a beneficial and necessary part of economic
life, and unions were assumed to reflect the needs of their members.
The capacity of union leadership to develop interests independent of
the rank and file was minimized, as was the capacity of employees
to define their interests as coexistent with those of their employers
but not coextensive with those of a union. Warren thought that the
opportunity to join a union ought to be part of the birthright of
every American; he equated union membership with full participation in American economic life. He equated vigorous union involvement in labor relations with greater economic freedom for workers; he did not suspect that such involvement could restrict that freedom.

Thus if Warren's labor decisions are flawed, they are not for the
reason conventionally brought forth in criticism of his judicial performance: "result-orientation." Warren's opinions were clearly designed to reach results that he thought fair and just, but they were
also, for the most part, results that the institutions charged with developing national labor policy—Congress and the NLRB—would
have endorsed. To be sure, Warren engaged in some glosses of statutory language, as in *Great Dane*, changed the rules on employers in
midstream, as in *Gissel*, and attributed to Congress a dazzlingly strong
interest in occupying the field of labor relations, as in his dissents in
*Gonzales* and *Russell*. But his results were designed to afford employees and their unions special protection against the unequal bargaining power of employers. Such results were consistent with post-New
Deal labor policy. Warren did not substitute his intuitive preferences
for legislative mandates in the labor relations area; those mandates
embodied his preferences.

֍

While labor legislation and antitrust legislation were products of the
same era and instituted by persons with common assumptions about
government regulation of the economy, the thrust of the two regulatory schemes was different. Labor legislation had as its primary
purpose the protection of a hitherto disadvantaged class of persons;
antitrust legislation had as its primary purpose the restoration of "free
competition." The ultimate goal of labor policy may have been to
create an environment in which representatives of management and

labor could bargain "freely," but its first concern was to neutralize the assumed advantages of employers by advantaging their employees. Antitrust policy, in contrast, did not, for the most part, seek to create a class of advantaged persons to compete with trusts and conglomerates. The only example of "advantaging" was the remedy of triple damages for adversely affected private persons, designed to give them an incentive to bring suit against antitrust violations. In general, antitrust policy responded to economic imbalances by attempting to dissolve "unfair" market advantages and restore equilibrium.

The different thrusts of the two regulatory schemes revealed themselves most clearly in the areas where their policies competed. Advantaging labor meant, among other things, allowing labor unions an exemption from the antitrust laws.[54] Such an exemption assumed that the excessive size of a labor union or its disproportionate control of employment in a particular industry were not areas of national concern. Market dominance in an enterprise was an evil that national antitrust policy was designed to combat; market dominance in a union was simply one of the ways by which labor was advantaged.

Justices on the Warren Court who were friendly to both labor unions and antitrust enforcers eventually had to confront the clash of policies in labor antitrust, and they resolved the issue in a fragmented and disorderly fashion. Earl Warren's resolution was to try to strike a balance between advantaging labor and enforcing the antitrust laws: He was sympathetic to both policies.[55] Indeed Warren's antitrust decisions suggest that he, like many Progressive sponsors of labor and antitrust legislation, preferred to see no conflict between the goals of those two efforts. Warren believed that the average working man would not get fair treatment without unionization and union advantages. He also believed that market dominance by an industrial enterprise was presumptively suspect. In his ideal society, economic privilege was suspect whatever form it took.

In antitrust cases the personification of Warren's ideals was the federal government in its role as dissolver of excess corporate power. Whether represented by the Federal Trade Commission or the Justice Department, the prosecuting government body was doing what Warren wanted done, combating economic royalism by restoring competition in a market. Warren was not particularly interested in the technical economic aspects of antitrust cases, such as efficiency analysis or market definition. He was more interested in protecting small producers and consumers from the allegedly deleterious consequences of economic size. The principal theoretical perspective Warren brought to antitrust cases was his conviction that corporate size and power were to be distrusted and eradicated.

Raw statistics rarely illuminate a judge's performance, but Warren's voting record in antitrust cases may be an exception. In 101 antitrust cases decided between 1953 and 1968 he voted against liability only nine times. He supported the Justice Department's position in forty-three out of forty-five cases and the Federal Trade Commission's position in twenty-three out of twenty-five cases. His percentage of support for liability was the highest of all the justices on his Court.[56] Of Warren's fourteen opinions in antitrust cases only two were dissents, and both of those upheld a position favorable to enforcement of the antitrust laws. The draftsmen of the Sherman and Clayton Acts might well have marveled at how fully and completely their fears of economic privilege had been absorbed.

Occasionally Warren seemed to be doing more than merely articulating congressional fears of concentrated economic power. In *United States v. McKesson & Robbins, Inc.*,[57] one of his early decisions, the impact of two subsequent statutory "fair-trade" provisions of the Sherman Act was considered. The Miller-Tydings Act[58] and the McGuire Act[59] had exempted certain resale price maintenance agreements, forbidden by the Sherman Act, provided the parties were not "wholesalers" and not "in competition with each other." *McKesson* raised the question whether a manufacturer of drugs could contract with wholesalers to fix the prices of its merchandise if it was itself a wholesaler. Warren read the language of the acts literally and found that McKesson was a "wholesaler"; that the resale price maintenance agreements were between persons "in competition with each other" since McKesson as a wholesaler competed with other wholesalers; and therefore that the contracts were not exempted from the Sherman Act.

Warren's opinion did little more than state that any form of price fixing violated the Sherman Act and that "wholesaler" meant all wholesalers, even if they were also manufacturers. The opinion ignored language in the Senate debates over the McGuire Act that "for a manufacturer to say that a certain product will sell at a certain price from the manufacturer down to the retailer is legal."[60] Justice Harlan, dissenting, said that Warren's opinion contravened the purpose of the Miller-Tydings and McGuire Acts, which was "to permit a manufacturer to set the resale price for his own products" to "protect the goodwill attached to a brand name."[61] He suggested that "[l]ack of sympathy with [the Miller-Tydings and McGuire Acts]" was the basis for Warren's decision.[62]

If Warren's opinion for the Court in *McKesson* was perhaps unresponsive to the business context of the case, his dissent that same term in *United States v. du Pont & Co.*,[63] the famous *Cellophane* case, was vitally concerned with that context. In a free-swinging opinion, Warren ridiculed the majority's finding that du Pont had not "monopolized" the market for cellophane by manufacturing seventy-five percent of the cellophane sold in the United States because the "relevant market" for cellophane was "flexible packaging materials."[64] This finding made cellophane a competitor of greaseproof and vegetable parchment papers, waxed papers, sulphite papers, and aluminum foil. Warren emphasized that two of these products were opaque, whereas cellophane was transparent; that two had no resistance to grease and oils; that three had a low bursting strength; and that none approximated cellophane's unique combination of attributes.

In addition Warren demonstrated that buyers and even manufacturers of other flexible packaging materials were relatively indifferent to substantial price differences between cellophane and its purported "competitors." Between 1923 and 1947 the production of cellophane grew substantially and its price sharply dropped, but the prices of other flexible packaging materials increased. Neither the sales of cellophane nor the sales of other flexible packaging materials declined. Du Pont itself regarded other flexible packaging materials as not competitive with cellophane[65] and took steps to preserve the secrecy of its formula. The sole American competitor for cellophane, Sylvania, entered the market only when two du Pont officials absconded with the secret process, and the Sylvania Corporation emerged as the consequence. Sylvania, which controlled twenty-five percent of the market for cellophane, "absolutely and immediately followed every du Pont price change." Producers of other flexible packaging materials, in contrast, "displayed apparent indifference to du Pont's repeated and substantial price cuts."[66]

Between 1924 and 1951, Warren noted, there were only two sellers in the cellophane market. In 1951 between seventy-five to eighty percent of all the cigarettes manufactured in the United States were wrapped in cellophane. The only two sellers in the market, one a licensee of the other, charged virtually identical prices. By ignoring the unique character of the cellophane market, Warren maintained, the majority had held "in effect that, because cellophane meets competition for many end uses, those buyers for other uses who need or want only cellophane are not entitled to the benefits of competition."[67] The conduct of du Pont and Sylvania between 1924 and 1951 illustrated for Warren "that a few sellers tend to act like one and that an industry which does not have a competitive structure will not have competitive behavior."[68]

Warren seems to have been correct in his view that there was a unique "inner market" for cellophane within the larger market for flexible packaging materials. Since du Pont not only controlled seventy-five percent of that market outright but also had licensing arrangements that enabled it to profit from the sales of the other manufacturer, it was a cellophane monopolist. Whether the public was worse off for du Pont's position is speculative: Cellophane prices generally went down between 1924 and 1951, and the quality of the product did not deteriorate. But the economic benefits of monopoly are not given any credence by the Sherman Act: Monopolies are made illegal. Thus Warren's dissent in the *Cellophane* case was faithful to the purposes of national antitrust policy. It was also an example of how his instinct for grasping the essence of situations could combine with his interest in the factual details of a case to produce an opinion, the empirical logic of which had considerable power. The *Cellophane* dissent was among Warren's best.

While Warren's principal contribution to antitrust law was to come in the area of monopolistic practices under the Sherman and Clayton Acts, some of his other opinions reveal his general stance toward antitrust issues. In three cases involving the Federal Trade Commission's power to police discriminatory and deceptive commercial practices, Warren defined the commission's role as that of a protector of small businesses and the individual consumer. He found, for a unanimous Court, that a brewery could not sell premium beer at one price outside of St. Louis and at a lower price within the St. Louis market.[69] The Clayton and Robinson-Patman Acts, he noted, were "born of a desire . . . to curb the use by financially powerful corporations of localized price-cutting tactics" that would "impair the competitive position" of less-advantaged sellers.[70] He also found that the FTC could prevent a manufacturer of shaving cream from demonstrating the product's ability to "soften sandpaper" by means of an undisclosed television "mock-up" in which sandpaper was not used.[71] Warren took the opportunity presented by that case to support the FTC's complete ban on the use of undisclosed mock-ups in television advertising.

Finally, Warren held that the FTC could prevent suppliers of supermarket goods and their retailers from making promotional discount agreements unless wholesalers who sold to competing retailers were also given the benefit of promotional discounts.[72] Warren again found the Robinson-Patman Act's purpose that of "prohibit[ing] all devices by which large buyers gained discriminating preferences over smaller ones by virtue of their greater purchasing power."[73] Warren felt that "smaller retailers whose only access to suppliers is through independent wholesalers" would be disadvantaged by the discount practices.[74] He thus held small retailers who bought through wholesalers to be "customers" of a supermarket supplier even though they did not buy their products directly from the supplier. Harlan, in dissent, said that this was the equivalent of reading "customers" in the Robinson-Patman Act to mean "non-customers who the Court thinks need protection." Warren, he charged, had "read into the Act [his] own notions of how best to protect 'little people' from 'big people'."[75]

Warren's conception of the FTC as a watchdog for small business was consistent with his general view of the purposes of the antitrust laws. He saw antitrust policy as designed to "protect 'little people' from 'big people' " by dismantling bastions of economic privilege and scrutinizing discriminatory conduct by large enterprises. In cases where Congress or administrative agencies had not sufficiently facilitated those goals, Warren was prepared to take on the role of antitrust enforcer himself. In a 1967 dissent from a denial of certiorari, Warren chided the Court for not reviewing the antitrust implications of fixed minimum rates for commissions on the New York Stock Exchange. He asked whether the rate-fixing practice was compatible with "this Nation's commitment . . . to competitive pricing" and invited "the Solicitor General to participate in argument so that the public interest may be fully explored."[76]

Similarly in *Utah Public Service Commission v. El Paso Natural Gas Co.*,[77] Warren went out of his way to order complete divestiture by El Paso Natural Gas of a subsidiary company it had acquired in violation of the Clayton Act. El Paso, after having been found to have violated the act, had secured an agreement with the Justice Department whereby the illegally acquired assets were transferred to a new company. This agreement was set aside by the Supreme Court.[78] A federal district court then attempted to redistribute assets without complete divestiture. Warren held that this also failed to comply with the original Clayton Act decision. He required the severance of all managerial and financial connections between El Paso and any new company through a cash sale.

Warren announced this without full oral argument, without the submission of complete briefs, and in the face of a decision by the Justice Department and all the interested parties to terminate the case. Harlan, dissenting, said the decision made the Court "an investigating body with a roving commission to travel the length and breadth of this land policing its mandates."[79] Warren, in response, stated that "our mandate directed complete divestiture"[80] and complete divestiture had not yet been achieved. The restoration of competition was not to be thwarted by compromises among powerful corporations and the government.

Warren's most significant antitrust opinion involved guidelines for anticompetitive mergers under the Clayton Act. In 1955 the Justice Department filed suit to block a proposed merger between the Brown and Kinney Shoe Companies. Brown at that time produced approximately five percent of the nation's shoes and Kinney about one-half percent. The case was the first brought under the amended Section 7 of the Clayton Act (1950), which provided that corporations could not acquire other corporations where "the effect of such acquisition may be substantially to lessen competition or to tend to create a monopoly."[81] A federal district court found that the proposed merger of the shoe companies violated section 7. The district court based this holding on a "definite trend" toward economic concentration in the shoe industry and on a finding that the merger might tend to lessen competition substantially in the retailing of men's, women's, and children's shoes. In *Brown Shoe Co. v. United States*[82] Warren upheld the district court.

The *Brown Shoe* case was in some respects like other controversial and innovative decisions of the Warren Court—school prayers and racial segregation, for example—in that it seemed conscious of couching a dramatic step in moderate and ambiguous terms. Abolition of prayers in the classroom was accompanied by espousals of the established American belief in religion and a supreme being; abolition of segregation was accompanied by the "deliberate speed" for-

mula. *Brown Shoe*, as well, "seemingly by design contain[ed] something for everyone."[83]

Warren suggested, for example, that mergers between healthy
and failing companies were not prohibited, nor were ones "between
two small companies to enable the combination to compete more
effectively with larger corporations";[84] that congressional concern was
with "the protection of *competition* not competitors," and thus the fact
that mergers adversely affected other companies did not make them
presumptively suspect;[85] that mergers needed to be examined on a
case-by-case basis; and that mergers had to be viewed "in the context
of [the] particular industry."[86] These comments suggested that a
merger in a nonconcentrated industry that did not seem hostile to a
new competitor might well not violate section 7.

On the other hand Warren's reading of the antitrust laws as a
charter of protection for "viable, small, locally owned business"[87]
showed up visibly in *Brown Shoe*. In response to the argument that a
large integrated manufacturer-retailer chain could eliminate wholesalers and thereby lower prices, Warren stated that Congress in passing the antitrust laws "appreciated that occasional higher costs and
prices might result from the maintenance of fragmented industries
and markets" and "resolved these competing considerations in favor
of decentralization."[88] While the shoe industry was fragmented, there
was a "tendency toward concentration": Congress's mandate was that
"tendencies toward concentration in industry are to be curbed in
their incipiency."[89] Thus while mergers would be considered on a
case-by-case basis under section 7 and the mere fact of a merger
would not necessarily give rise to an antitrust violation, the fact that
a merger increased efficiency or lowered prices in the industry would
not in itself be a justification.

The significance of *Brown Shoe* was increased by the Warren
Court's subsequent merger decisions, which Warren joined and which
all tightened the impact of *Brown Shoe*. *United States v. Philadelphia
Nat'l Bank*[90] invalidated a consolidation of two banks even though
Section 7 of the Clayton Act had exempted bank assets acquisitions
from its coverage. *United States v. Aluminum Co. of America*[91] prevented a merger in a concentrated industry even though the merger
would have increased concentration only slightly. In *United States v.
Pabst Brewing Co.*[92] the Court read section 7 to "outlaw mergers which
threatened competition in any or all parts of the country."[93] And in
*United States v. Von's Grocery*[94] the Court stated that the principal

purpose of section 7 was "to prevent economic concentration in the American economy by keeping a large number of small competitors in business."[95] As in the race relations area, a moderate and vague Warren opinion had been followed by aggressive, decisive, and cryptic Court enforcement. Section 7 of the Clayton Act had become the small businessman's emancipation proclamation.

✍

In piecing together Warren's decisions on the economy, one finds the presuppositions of Progressivism lending a superficial unity to results that seem to proceed on divergent theoretical grounds. The protectionist rationale of labor law was reconciled with the free trade rationale of antitrust through the conviction that the individual member of the labor force—whether worker or small businessman—needs help against the forces of economic privilege. The tendencies of organized labor to combine, to restrict competition, and to monopolize the market for skilled industrial work were not regarded as inconsistent with antitrust policy but as a special case deserving favored treatment. Similar tendencies in members holding seats on the New York Stock Exchange were deplored.

Depersonalization of individual workers by unions was not noticed, while the coercive effects of management's economic privileges were stressed. The substantial foreclosure of retailers' access to a competitor's products was deplored as economic coercion; the fact that such foreclosure could yield lower prices to consumers was minimized. Economic privilege and gross economic power remained the evils; freedom for the lone economic man remained the goal. Once again Earl Warren was combating the special interests and striking a blow for the economically disadvantaged.

If all these glosses on national economic policy were merely Warren's idiosyncratic theories, one might find his decisions on issues affecting the economy just Progressive cant. But the two legislative areas where early twentieth-century economic reformers were most successful in putting their social assumptions into practice were labor and antitrust. An assumption of unequal economic power between employer and employee was built into labor legislation from its origination. A fear of economic concentration was likewise built into the antitrust laws. If Warren sought to protect the average worker through unionization and the NLRB, so did Congress; if Warren saw the antitrust laws as a charter for small business, so did several of their sponsors.

One cannot accuse Warren the economic theorist of substituting his impassioned ethical judgments for the more considered deliberations of a legislature; he simply pressed the legislators' economic theories to the limits of their logic. The fundamental question Warren's decisions on the economy raises is not the currently familiar one of the value of judicial usurpation; it is rather that of the inherent soundness of the assumptions that Warren was determined to make the controlling forces in matters of national economic policy. Early twentieth-century reformers were determined to protect the working man and the small corporation in a market they saw dominated by "combinations of capital." Warren had seen such domination first-hand in his Bakersfield days; in supporting organized labor and small business on the Court, he was voicing his satisfaction that Congress had finally caught up with the Southern Pacific.

# FOUR

*THE LEGACY*

# 14

# The Last Years

As the 1967 term began, Warren and the Warren Court stood at a high-water mark in their jurisprudential histories. With Frankfurter's retirement in 1962 and the succession of Arthur Goldberg and then Abe Fortas to Frankfurter's seat, the "hard core liberal wing of the Court" had acquired a fifth vote, and Warren found himself with the majority in most controversial decisions. The five years after 1962 had also witnessed some of the Court's most visible commitments to a philosophy of equal justice: *Loving v. Virginia*,[1] striking down statutes forbidding interracial marriage; *Harper v. Virginia Board of Elections*,[2] invalidating economic restrictions on the right to vote; *Jones v. Mayer*,[3] resurrecting an 1866 civil rights law to prohibit racial discrimination in housing; and the *Escobedo-Miranda*[4] sequence in criminal confessions. Warren had been sympathetic to all those efforts: He had written the opinions in *Loving* and *Miranda*.

In March, 1966 the *New York Times Magazine* had used the occasion of Warren's seventy-fifth birthday to publish an article about his achievements on the Court. In the article Fred Rodell, a Yale law professor who was an admirer of Warren, suggested that Warren "seems more mellow, more gently sure of himself, more immune to side-line criticism than ever before." Warren had "spectacularly succeeded" as Chief Justice, Rodell felt; he "exud[ed] . . . quiet and confident power." In Rodell's judgment Warren was "the greatest

Chief Justice in the nation's history."[5] While some other commentators were less enthusiastic in their praise of Warren and less sanguine about trends on the Warren Court,[6] there was no question that Earl Warren had taken his place among the leading judges in American history. A 1972 poll of academics was to rank him among the twelve "great" justices to sit on the Supreme Court; a subsequent study was to rank him fourth in "greatness" among all the justices who had sat on the Court.[7]

For Earl Warren the praise was not, perhaps, without its own, even more personal, satisfactions. Once again he had proved himself a far better judge of his own abilities than had his critics, whose limited expectations for him he had far exceeded. By 1967, he had proved that even as Chief Justice of the highest Court he had held his own among a group of independent and talented men. Now he was approaching his seventy-seventh birthday; despite old age, his sense of timing and his political instincts were as sharp as ever. Within a year, these instincts would tell him that the time was right for one last political decision.

Warren's posture as Chief Justice had decisively cut him off from partisan Republican thought early in his tenure as Chief Justice. His repeated endorsement of decisions that supported beneficiaries of Lyndon Johnson's Great Society, especially the economically and socially disadvantaged, caused others to identify him with the liberal Democratic politics of the Johnson years. In fact, Earl Warren and Lyndon Johnson had become, during their coterminous periods of office in Washington, D.C., good friends. Johnson had presented Warren with a picture inscribed to the "greatest Chief Justice of them all" on Warren's seventy-fifth birthday,[8] and the Warrens, who did not normally attend Washington social functions, made themselves available for official evenings at the White House. When Johnson, under extreme pressure because of growing public dissatisfaction with the conduct of the war in Vietnam, announced on March 31, 1968 that he would not run for reelection, Warren wrote him that "I had hoped that you would decide differently, as you have earned another term of more tranquil years." "In the past six years," Warren continued, "you have done more to make our citizenship conform to the American Dream than any of your predecessors."[9]

Then, within three months of Johnson's speech, two shocking assassinations took place—that of Martin Luther King on April 4 and then that of Robert F. Kennedy on June 5. Those events and the war issue suggested that the Democratic party might find itself at a distinct disadvantage in the forthcoming November elections. Meanwhile Richard Nixon, in an astonishing political comeback, had transformed himself from a defeated California gubernatorial candidate in 1962 to the front-runner for the Republican presidential nomination in 1968. On June 11 Justice Fortas called the White House to say that Warren had requested an interview with Johnson.

Warren met with Johnson at the White House on June 13. An aide's memorandum of the conversation reportedly stated that the meeting took fifteen minutes, that Warren told Johnson that he was planning to retire from the Court, and that he wanted Johnson "to appoint as his successor someone who felt as Justice Warren did." [10] Warren then wrote Johnson two letters, dated that same day. The first stated, "I hereby advise you of my intention to resign as Chief Justice of the United States effective at your pleasure." The second indicated that Warren was retiring "not because of reasons of health or on account of any personal or associational problems, but solely because of age." [11] Johnson did not make the letters public until June 26, but the *New York Daily News* reported on June 22 that Warren was planning to retire. [12]

When Johnson released Warren's letters of resignation, he also released his letter of acceptance. In that letter he stated that he would "accept [Warren's] decision to retire effective at such time as a successor is qualified." [13] At a July 5 press conference, the only one he held as Chief Justice, Warren explained that he had left the date of his resignation uncertain because "there is a lot of administrative work [as Chief Justice] . . . and if I selected a particular day and the vacancy was not filled it would be a vacuum." Warren conceded, however, that "if the first of October rolls around and there is no successor, I suppose I would be obliged to act as Chief Justice." [14]

Despite Warren's disclaimers, commentators took the circumstances of his resignation as an effort to influence the choice of his successor. Four days after Johnson announced Warren's resignation the *New York Times* called the language in Warren's and Johnson's letters "calculated vagueness," and the *New Republic* said that "life tenure as a Chief Justice is one thing. Life tenure with a right to influence confirmation of a successor is rather another." [15] The day after Warren's visit to the White House, the *Wall Street Journal* had reported that Warren was considering leaving the Court, and that "he hoped to have a voice in naming his successor." [16] This story outraged Senator Robert Griffin of Michigan, who said two years later that at the time he had thought "what business does Warren have, having anything to say about his successor. It smacked of collusion." On June 22 Griffin announced that he would oppose any nominee of "a 'lame duck' President" for the chief justiceship.[17]

Johnson's nominee was announced on June 26, when the exchange of letters between Warren and Johnson was made public. The nominee, as Warren expected, was Abe Fortas. Warren and Fortas had developed a friendship during Fortas' three terms as an associate justice. Fortas found Warren "a rare and wonderful man" and admired his "straight, uncluttered, unsubtle" way of making decisions; Warren was said to have been impressed with Fortas' "astuteness and self-assurance."[18] At the July 5 press conference Warren denied knowing that Johnson was planning to appoint Fortas and said that "I left politics fifteen years ago and would hate to have political matters injected into stories about my retirement."[19] But the press conference itself marked the beginning of an aggressive campaign to secure Fortas the nomination.

Before nominating Fortas, Johnson, in private conversations, had apparently secured the support of James Eastland, chairman of the Senate Judiciary Committee, Richard Russell, the most senior Democrat in the Senate, and Republican Everett Dirksen, the Senate's minority leader. When Johnson announced the Fortas nomination, he also nominated Judge Homer Thornberry of the United States Court of Appeals for the Fifth Circuit to fill Fortas' position as associate justice. Thornberry, an intimate of Johnson who had served as a congressman for fourteen years prior to being named to a federal district judgeship in 1963, had twice had his nominations for judgeships easily confirmed by the Senate.

Despite Johnson's preparations and his experience with internal Senate politics, Fortas' nomination was imperiled from the outset. There was, first, Johnson's "lame-duck" status: Only seven out of sixteen Supreme Court nominations made by presidents in the last year of their tenure had been confirmed by the Senate. There was, in addition, the charge of "cronyism." Fortas admitted to having consulted with Johnson on a variety of policy matters while serving as an associate justice, and Thornberry had been picked by Johnson to represent his Texas congressional district when Johnson first ran for the Senate in 1949. Other presidents, most recently Truman, had nominated close acquaintances to the Court, but Johnson, in

1968, was particularly susceptible to that charge because of his loss of credibility in connection with the escalation of the war in Vietnam.

Finally, there was the timing and the manner of Warren's retirement. The exchange of letters between Johnson and Warren was made public in late June, less than six weeks before the Republican convention. The Republican nominee, who was almost certain to be Nixon, was likely to make a campaign issue out of the alleged permissiveness of a majority of the Warren Court justices, which included Warren and Fortas, toward criminal defendants. Warren was known to be an old political enemy of Nixon, and Johnson was known to be a supporter of the "liberal wing" of the Warren Court. The language of the resignation letters seemed to give the Senate the option of confirming Fortas or having Warren remain as Chief Justice. On the first day of Senate hearings on Fortas' nomination, Senator Sam Ervin of North Carolina devoted a full day to debating whether a "vacancy," within the meaning of the Constitution, actually existed on the Court because Warren had not yet retired and had not fixed a date for his retirement.

All these themes and Congress's impending recess for the 1968 nominating conventions combined to undermine the Fortas nomination. When Congress adjourned for the summer, a Republican filibuster against the nomination, led by Griffin, was brewing, and Russell and Eastland had given indications that they would not oppose such a filibuster. Fortas supporters reportedly had the votes to confirm his nomination should it come to a vote, but the two-thirds majority required to invoke Senate cloture against a filibuster was uncertain. There the situation stood when Congress recessed in late July.

When Congress reassembled after Labor Day the situation had worsened for Fortas. The Democratic convention had been marred by clashes between demonstrators and the Chicago police; the general unrest in the country and the unsatisfactory progress of the war made Democratic nominee Hubert Humphrey the distinct underdog in the November presidential election. Nixon, in his speech accepting the Republican nomination, had suggested that "some of our courts" had "gone too far in weakening the peace forces, as against the criminal forces."[20] Fortas was thus associated with two institutions, the Johnson Administration and the Warren Court, the political stock of which was low. And in addition Griffin's office learned, through a tip from an employee of American University, that Fortas had accepted $15,000 for a series of law school seminars over the summer.[21]

The revelation of Fortas' seminar fee dashed any hopes on the part of the Johnson Administration that cloture could be invoked. While the seminars had been publicized, while Supreme Court justices had taught in law schools in the past, and while the subject matter of the seminars was not concerned with cases pending before the Court, the amount of Fortas' fee and the manner in which the funding had taken place were embarrassing. Fortas' salary as an associate justice was then $39,500; the $15,000 was to be paid for a total of nine seminar sessions. The money for Fortas' fees had been raised by Paul Porter, Fortas' former law partner at the firm of Arnold and Porter, from contributions by five individuals. One was the chairman of the New York Stock Exchange, one the chairman of Braniff Airways, one a senior partner in Carl M. Loeb, a stockbrokerage firm, one the vice-chairman of Federated Department Stores, and one the vice-president of the Philip Morris tobacco company. Three of the individuals were clients of Arnold and Porter. Porter raised a total of $30,000 from these five men, of which $15,000 went to Fortas and $15,000 was retained by American University to finance future seminars.[22]

The seminar-fee incident seemed to draw into sharp focus a series of interrelated themes that had troubled Fortas' candidacy. The "cronyism" theme had suggested that his relationship with the president was too close to satisfy the spirit of separation of powers. The "lame-duck" theme had suggested that Fortas would be a Chief Justice beholden to a discarded political point of view. And the "no vacancy" theme had suggested that Fortas, Warren, and Johnson were not above playing politics with the constitutional process under which justices to the Supreme Court were selected. None of the themes seemed very damaging in itself, at least in the sense that Supreme Court justices had repeatedly been involved with politics. But the seminar fee incident seemed yet another example of Fortas' indifference to potential conflict-of-interest issues.[23] While he reportedly did not know the sources of his fee, he accepted it, knowing it to be a strikingly high figure from a moderately endowed institution. While he reportedly did not discuss Court business at the White House, he was repeatedly present at Johnson Administration policy meetings.

After the fee incident became public, Warren's attempted retirement was quickly turned into a summer vacation. On October 1 the Senate failed to invoke cloture; on October 2 Fortas wrote Johnson to ask that his nomination be withdrawn; on October 10 Johnson formally asked Warren to remain on the Court "until emotion subsides."[24] The Court began its 1968 term with Warren as Chief Justice.

There were still two moments of drama left in the story of Warren's retirement. In the presidential election in November, Nixon defeated Humphrey. Nixon's apparently insurmountable early lead had rapidly dwindled in the fall, and some political observers felt at the time that had the election been held a week later the outcome might have been reversed. Warren had, of course, hoped for a Humphrey victory, not only because of his profound dislike and distrust of Nixon but also because there was no chance that Nixon, who had campaigned against the Warren Court, would name a successor sympathetic to Warren's views.

Nixon's election presented Warren with an acute dilemma. He could withdraw his resignation and stay on as Chief Justice, hoping to deny Nixon a chance to replace him. Actuarial tables were against him on this issue, since he would be eighty-one when Nixon ended his first term, and Nixon was eligible to run for a second. In addition, Warren's letters to Johnson had identified age as his sole reason for leaving the Court, and he could not remain on the Court indefinitely without that statement being brought up by someone seeking to embarrass him. Moreover, Warren had tried to disassociate his retirement from politics and to claim that he had "left politics" behind on coming to the Court; to offer his resignation to a Democrat and to withdraw it from a Republican would have undermined his credibility. Pulling against these strong reasons for retiring were Warren's entrenched opposition to Nixon and the humiliation of having to give Nixon the opportunity to name his successor.

Eventually in late November, 1968, Warren sent a message via his son-in-law, television personality John Daly, to William Rogers, whom Nixon was to name as Secretary of State. The message, apparently transmitted during a round of golf in which Daly and Rogers participated, was that Warren was planning to repeat his offer of resignation to Nixon. Warren, however, did not want to resign in the middle of the Court's term: This would undermine the credibility of his statement made during the Fortas nomination that the administrative duties of the chief justiceship required that its occupants be continuous. Nixon accepted Warren's terms, and the arrangement was announced on December 4.[25] Warren was to await the ironic task of swearing in a new president who was his old political enemy and then waiting to see whom that president would choose to replace him as Chief Justice.

Warren's response to his prospective retirement demonstrated that despite the remarkable achievements of his public life, he had never become comfortable with criticism, and despite his calculated non-participation in politics while on the Court, he had remained acutely aware of the political implications of his actions. Independence, control of events, and planning for the future had been central values for Warren the California public official: They remained important for Warren the Chief Justice. He resented the public criticism of the Warren Court and expected that Nixon would seek to add to that criticism as a presidential candidate. In contrast, he responded warmly to Johnson's admiration for what the Court was trying to accomplish. Warren feared being dependent on age, health, or an old enemy, and chose to retire so as to give the choice of his successor to a new friend. He looked forward to a retirement in which his legacy of activism on behalf of social justice would be carried on, even in the face of a potentially hostile executive branch. He sought, as he had frequently sought in his career in California, to influence events by anticipating the future. He was no more indifferent to the timing of his retirement than he had been to the timing of the Warren Commission Report.

There is a theme of hubris that runs through the narrative of Warren's retirement. All the principal participants can be said to have assumed that they were especially capable of controlling their destinies. Johnson had become used to having his own way as president through manipulation, bargaining, and a certain amount of bullying. Fortas' shrewdness and intellectual acumen had so repeatedly rewarded him in the legal profession that he did not expect that going on the Supreme Court would seriously diminish his opportunities for continued success in two features of his profession that he especially liked—making money and exerting backstairs political influence. Warren had grown accustomed, during a lengthy career as the head of an office, to controlling events and maximizing his own opportunities. He had also repeatedly shown an ability to anticipate political change and to profit from it.

The plan whereby Johnson would name Fortas to succeed Warren as Chief Justice thus appeared to be another example of the ability of three powerful and resourceful men to control the future. The plan failed because each participant made a serious misjudgment about his ability to exercise that control. Johnson believed that his "lame-duck" status and the controversial image of his presidency would not override his mastery of Senate politics: He did not sense how much stature he had lost. Fortas believed he could serve on the Court and continue as Johnson's adviser or as an extrajudicial consultant: He did not think that the conventional restraints on judges' conduct applied to a person as broad-gauged and talented as he. Warren believed that he could thwart Nixon and perpetuate his vision of the Supreme Court by an artfully timed retirement: He did not expect others to see his motives so clearly.

The consequence of this collective hubris was that for the first time in Warren's public life he was helpless to prevent an office with which he had been intimately associated from becoming staffed by a person whom he counted as one of his opponents. He had won every office he stood for in California, and in moving on to another office he had either helped choose his successor in his past office or thwarted opponents from replacing him in his present one. He had outflanked Goodwin Knight, he had struck back at Culbert Olson, he had kept Republican partisans from coopting him or sabotaging him. Even his unsuccessful 1952 campaign for the presidency had been turned to his advantage: Eisenhower won the nomination, but Warren made himself indispensable to the ticket. There was a bitter irony in the fact that Richard Nixon, whose vulnerability in 1952 had helped provide Warren with the leverage he needed to surface as a candidate for the Court, would have Warren's resignation in his hand for most of the 1968 term. Nixon, who had worked to keep Warren from the presidency only to have Warren wind up on the Court, was now in a position to appoint someone who would side with the "peace forces" against the "criminal forces."

&

Warren thus retired in June, 1969, his massive opinion in *Powell v. McCormack*,[26] which chastened Congress for refusing to seat Congressman Adam Clayton Powell, serving as his last lesson in civics. Powell had been accused of misusing travel expenses and making illegal salary payments to his wife; the voters of his district reelected him in the face of those accusations. The House of Representatives then voted, 307 to 116, to exclude Powell from its membership. Article I, Section 5 of the Constitution provides that "each House [of Congress] shall be the judge of the . . . Qualifications of its own members" and shall "expel a member" with the concurrence of a two-thirds majority. Warren held that the House vote was not a vote to "expel" Powell, but merely to exclude him, and that the decision was not definitively committed to the legislative branch of government and could be reviewed by the Court. Since Powell had met the constitutional requirements for being seated in the House, Warren argued, the House had a duty to seat him, and since the House had refrained from doing so, the question was whether the breach of that duty was constitutionally justifiable. Resolution of that question required an analysis of the House's constitutional power to exclude duly elected members; such an analysis was "the responsibility of this Court" in its role "as the ultimate interpreter of the Constitution."[27]

Warren's interpretation of the House's power found that "in judging the qualifications of its members Congress is limited to the standing qualifications prescribed in the Constitution."[28] Powell had met the constitutional qualifications for office, yet the House sought to exclude him. It was without power to do so. The *Powell* holding so narrowed the "political questions" doctrine, which *Baker v. Carr* had undermined, as to invite speculation as to whether there were any decisions that Congress or the executive could make that were insulated from judicial review.[29] Among such decisions was a decision on the impeachment of a president, which the Constitution's text had committed to the Senate. *Powell* was in this sense a representative Warren opinion in its assumption that the Court's "responsibility" to interpret the Constitution gave it an almost limitless review power. Warren's final civics lesson was that on matters of civics, the Supreme Court knew best.

Nixon chose as Warren's replacement Warren Burger, a judge who during his tenure on the United States Court of Appeals for the District of Columbia Circuit had given evidence of being considerably more solicitous of law enforcement authorities than had the Warren Court in its last years. Beyond that issue Burger's views were less clear: He could not be said to be the antithesis of Warren in every respect. Burger's appointment was nonetheless a symbolic rebuff to Warren, given Nixon's 1968 campaign rhetoric. Warren, who had reportedly predicted Burger's appointment in advance,[30] resolved to disassociate himself from the Burger Court.

For the most part, Warren's disassociation was not made public. He continued his practice of not granting interviews to the press, except in prearranged and comfortable circumstances, and he avoided public commentary on the Court. He nonetheless took steps to convey his sense that the Court was now in enemy hands. He suggested to clerks that the Court, on which four Nixon appointees sat by the 1971 term, would move quickly to scuttle Warren Court precedents. He saw Burger as "a rather doctrinaire law-and-order judge who would try to find any conceivable way to uphold convictions and narrow the impact of Constitutional protectors for defendants."[31] And he began a pattern of retirement activities that deemphasized his status as a former Chief Justice.

It has been customary for retired justices to be allotted chambers in the Supreme Court building, to continue to select law clerks, and to perform some judicial functions. Retired justices are eligible to sit on federal district courts and courts of appeal; during several years of Warren's retirement, Justice Tom Clark, who had retired in 1967, sat regularly. Other justices do not engage in judicial functions and rarely make use of their chambers; some do not request chambers. In such cases the justices' clerks, if they select any, are routinely assigned to the sitting Chief Justice, who then sometimes assigns them to other justices and sometimes retains them for himself.

Warren's retirement arrangements differed from this general pattern. He retained chambers in the Court building and continued to employ Margaret McHugh as his secretary. He also continued to select law clerks. But his law clerks were hired with the express understanding that Warren would not be sitting on any law cases and that they would not be assigned to Burger's office. Warren's clerks were to perform functions similar to those performed by some members of Warren's staff in California: speech writing, research on his memoirs and other projects, and other functions of an aide. They were to be beholden to Warren and not to members of the Burger Court.

The principal project that Warren embarked upon on his retirement was the writing of his memoirs. No justice in the history of the Court had ever published a memoir, and Warren was eager, while his health remained intact, to set out an account of his long career in public life. The mood in which he approached his memoirs was partially that of a person looking back with satisfaction over a life-time of accomplishments and partially that of a person determined to settle some scores, "set the record straight," and establish his ac-complishments in history.

Warren's ambivalence toward his memoirs resulted in his ap-proaching them with reluctance and even a certain amount of pique. He postponed work on them to do a short book on citizenship for the *New York Times* book division,[32] to accept speaking engagements from academic institutions and from organizations such as the Amer-ican Civil Liberties Union and the NAACP's Legal Defense Fund, to attend World Peace Through Law conferences, to write an occa-sional lecture or tribute to one of the Warren Court justices, and to return periodically to California in order to see family members, meet with old friends, and give interviews to the Earl Warren Oral His-tory Project at Berkeley. The Warrens retained their residence at the Sheraton Park Hotel after his retirement, and Warren maintained his work habits, reading at home in the early morning, arriving at the Court at about nine-thirty, working in his chambers, where he was served a light lunch, remaining until after the evening traffic had thinned out, and sometimes, when pressed to meet a deadline for a lecture or a speech, working on weekends.

The Warrens regularly went to the West Coast for the Christmas holidays and for intervals during the summer, often staying as long as six weeks at a time. These holidays, the uneven pattern of War-ren's work habits, and his changing projects made life somewhat un-predictable for his clerks. When a deadline for some form of written presentation approached, clerks could expect Warren to act in a man-ner consonant with the experience of California staffers. After mull-ing over a proposed topic for a speech or a lecture, Warren would—often perilously close to the deadline—summon in a clerk and give an impressionistic summary of what he wanted to say. The clerk would then repair, sometimes in a state of desperation, to produce a draft. When the draft was submitted, Warren would again summon the clerk, and the draft would be "reworked," which normally meant that Warren would rewrite it nearly from scratch. The clerk's role in this rewriting process was essentially to reinforce Warren's judg-ment that changes improved the presentation and to hear the final draft read out loud. It was a role that fostered an unconscious, un-anticipated bond between clerks and Warren's earlier speechwriters.

Speeches, ceremonies, visitors to the chambers, lectures, conferences, and pleasure trips occupied much of Warren's retirement. But all the while two problems nagged at him—the problem of writing his memoirs and the problem of his antipathy to the Nixon Administration and its purported efforts, through the appointive process and other means, to modify the jurisprudence of the Warren Court. The memoirs reminded him of some of the triumphs and pleasures of his life, but they also provided memories, made all the more vivid by his powers of recall and his capacity to bear grudges, of the persons who had sought to thwart him in his efforts to use the powers of his offices to "make progress." The Nixon Administration and the Burger Court reminded him of his pledge to disassociate himself from politics and to keep silent on public issues, but they also confirmed for him the treacherous nature of Richard Nixon and the "doctrinaire" character of the new Chief Justice.

The frustrations of Warren's otherwise placid retirement emerged in two episodes—the structuring of his memoirs and his attack on the 1972 proposal for a National Court of Appeals. Warren had signed a contract for his memoirs in July, 1970, and the manuscript was expected to be completed in eighteen months. By the fall of 1971 Warren had finished most of the first six chapters, amounting to about half of the eventual manuscript, and was in the process of writing a seventh, dealing with his years as governor. The tone of these chapters was for the most part upbeat, factual, and bland. Warren was circumspect in his treatment of opponents, relatively modest in detailing his accomplishments, and eager to relive experiences in considerable detail. One reviewer of the eventual book noted that "these were good years for Warren and he remember[ed] them as such."[33]

The governorship chapter, as well, was upbeat: Its title was "Problems to Overcome," and its emphasis was on the "spate of . . . concerns" that Warren confronted as governor and the "extensive projects" that he initiated in response to them. Warren staff members and appointees to state administrative positions, such as Wilton Halverson, James T. Dean, Warren's director of finance, and Charles H. Purcell, the director of public works, were singled out for praise, and only a few enemies were sniped at, such as Goodwin Knight, who was "hoist by his own petard" on a clemency case, and Mario Giannini of the University of California Board of Regents, who resigned "in a fury" to "form an organization of 'vigilantes' " during the loyalty oath controversy. The general tone of the first seven chapters was captured in a sentence late in the governorship chapter.

"I derived a great satisfaction from my work and soon realized I was a politician," Warren wrote. "I was never embarrassed by being classified as a politician, in fact I was proud of it." [34]

Beginning with chapter eight, however, a different tone began to creep into Warren's memoirs. In that chapter he described his difficulties with the Dewey campaign and the efforts of the Hearst papers, Knight, and "my determined enemy," the "vitriolic multimillionaire" Bill Keck, an oil company president, to "proclaim my political demise." He recounted the activities of Nixon on the 1952 campaign train and the "bitter speeches" at the 1952 convention of Senator Joseph McCarthy and Everett Dirksen, who "treated Tom Dewey . . . as though he were a pickpocket." He surmised that "the despotic manner" in which the Taft forces attempted to run the convention "lost Senator Taft a sizable number of votes." [35] Remarks such as these increasingly appeared in the last four chapters of the memoirs.

At the same time Warren demonstrated a disinclination in the memoirs to give detailed accounts of events that were pivotal in his career. His account of the deliberations of *Brown v. Board,* and for that matter his entire treatment of his tenure on the Supreme Court, revealed very little information on the Court's deliberations. This was consistent with Warren's strong belief in the confidentiality of Court proceedings and with his expressed reluctance to dwell on personalities in his account. He had expressed outrage that Chief Justice Stone had made his private Court papers available to a biographer, Alpheus Mason, and made it clear that his memoirs would reveal no secrets. This attitude, when coupled with the somewhat adversarial tone of his later chapters, lent a curious cast to the memoirs.

At first glance the memoirs created an impression that Warren's career was notable for the remarkable harmony and tranquility with which affairs were conducted. "We were all impressed [in the segregation cases]," he said at one point, "with their importance and the desirability of achieving unanimity if possible." This impression, however, was made incongruous by an undertone of bitterness, evidenced in such comments as "the Court was under attack by powerful interests nearly all the time I was there." On the one hand there was only one incident that "greatly disturbed" the justices during the entire sixteen years of Warren's tenure; on the other hand the Court "was made the target for widespread abuse" simply because it "sought to make our criminal procedures conform to the relevant provisions of the Constitution and be a reality for the poor as well as the rich." [36] The juxtaposition of these themes in Warren's memoirs baffled reviewers, especially those who admired Warren. L.A. Powe, Jr., who had clerked for Justice Douglas the first year of Warren's retirement, found the memoirs "surprisingly defensive and self-serving" and wondered why "the man whose place in history seems so secure" had written "a ponderous, extended apology." [37]

Warren's memoirs illustrated the complex interaction between his persona and the emotions that lay beneath its surface. In his reluctance to reveal information in the memoirs, even, as in the *Brown* episode, when the revelations would have served him well, his persona as the cautious, genial, inoffensive politician predominated. This persona was most evident in two kinds of events that Warren remembered from his past: events that he found discomforting or inconsistent with his self-image while he was writing his memoirs and events that were drawn from memories of the inner life of the Supreme Court. He did not mention the Max Radin affair or the 1950 loyalty oath for California state employees; he trod carefully over the Japanese relocation decision; he downplayed the charged internal deliberations over the *Brown* decision; he did not mention the Burger Court. In the retelling of such events Warren adopted a calculated posture of nondisclosure. He apparently did not want to hurt others, and he did not want to expose himself.

Beneath this level of presentation ran the prideful, combative reactions of Warren, who remembered slights, was sensitive to criticism, and tended, when his sympathies were engaged, to see issues as stuggles between antagonists. In remembering his life Warren recalled many events from this perspective. His tendencies to see the world as a struggle between hostile interests were perhaps reinforced, at the time he was writing his memoirs, by a conviction that an antagonistic administration and a potentially antagonistic Court were threatening values he had fought to solidify in American jurisprudence.

Barred by his sense of propriety and his political instincts from making direct attacks on those antagonists in his memoirs, Warren let loose his spleen on pettier foes. The American Bar Association and "southern racists" did "much to discredit" the Court. Persons "with ghoulish minds" criticized him for not including the pictures of John Kennedy's head in the Warren Commission Report. "Ex-

tremists" tried to make ideological capital of the *Point Lobos* decision.[38] The result, for one reviewer, was that Warren's memoirs conveyed an "image . . . of a man who was insecure, self-serving,
unwilling to deal . . . with the tough decisions he had faced or with
the living people who might fight his interpretations."[39]

The later chapters of the memoirs, written between 1971 and
1974 and never fully completed, conveyed a sense that Earl Warren
was not going to be completely fulfilled in a life where he did not
hold power to strike back at his antagonists. As a retired Chief Justice, and one dedicated to preserving the confidentiality and integrity
of the Supreme Court, there was not very much Warren could do
about the Nixon Administration and the Burger Court.

In the fall of 1971 Chief Justice Burger, in his capacity as chairman of the Federal Judicial Center, appointed a study group to examine the Supreme Court's ostensibly heavy workload and to make
possible recommendations to alleviate it. The study group produced
a report, released publicly on December 19, 1972, that included
among its recommendations the creation, by federal statute, of a National Court of Appeals, which would serve as an intermediate court
between the United States courts of appeals for the eleven judicial
circuits and the Supreme Court. The National Court of Appeals was
to be staffed from judges on the existing circuits, who would serve
on a rotating basis. The new court would have two main functions—
the decision of cases on which the circuits had shown themselves to
be in conflict, and the selection of a pool of 450 "certworthy" cases
for consideration by the Supreme Court in the exercise of its certiorari power. Of these cases it was expected that the Supreme Court
would select about 100 to 150 cases on which to hear argument each
term.

The study group, which consisted of four law professors and three
practicing lawyers, one of whom, Peter D. Ehrenhaft, was a former
Warren clerk, had finalized its deliberations on the proposed National Court of Appeals earlier in the fall, and in the first week of
November a story appeared in the *National Observer* detailing the proposal.[40] Early critical reaction surfaced. That same month Henry J.
Friendly, Chief Judge of the United States Court of Appeals for the
Second Circuit, said that he was "profoundly unconvinced" by the
National Court of Appeals proposal, and Justice Potter Stewart was
reported by the Harvard Law School student newspaper as believing
that the Supreme Court's current caseload was "neither intolerable

nor impossible to handle."[41] The study group responded to this early criticism by holding a televised press conference in which two of its members, Harvard law professor Paul Freund and former American Bar Association President Bernard Segal, defended the proposal. Freund then wrote an article in the March, 1973 *American Bar Association Journal* in which he argued for implementation of the proposal.[42]

Between the time of the study group's formation and the issuance of its report, its work was public knowledge within the Supreme Court. The study group interviewed law clerks and justices about the Court's workload, and Warren was well aware of its activities. From the outset he was suspicious of the project. The study group had been commissioned by Burger and contained two law professors, Alexander Bickel and Charles Alan Wright, who had been publicly critical of the Warren Court, and a past president of the American Bar Association, from which Warren had resigned in 1957 after concluding that the ABA had become a forum for criticism of the Supreme Court.[43] While the report was in preparation Warren indicated that if it contained proposals designed to "curb the Court" he would speak out publicly against it, despite his decision not to comment on matters affecting the Court during his retirement. "I'm going to have to speak against it," he said at one point. "I can't take this lying down."

When word of the study group's recommendations was first leaked to the press in November, 1972, Warren responded by writing an outraged letter to his former clerk Ehrenhaft. His letter expressed dismay that Ehrenhaft would have served on the study group in the first place and that he would join a group whose recommendations were so obviously designed to limit public access to the Court and to curb the Court's discretionary powers. The text of Warren's letter and Ehrenhaft's response were circulated to all Warren's former clerks. A subsequent *National Observer* article alluded to Warren's actions and the Warren-Ehrenhaft exchange, calling Warren's circulation to his clerks "a clear attempt to mobilize opposition to the proposal."[44]

Warren then accepted an invitation to address the Association of the Bar of the City of New York on May 1, 1973.[45] He opened the address by charging that the National Court of Appeals proposal "would cause irreparable harm to the prestige, the power, and the function of the Court." He then ticked off a list of objections to the

322

proposal and the process by which it had come into being. The study group had not included any "past or present member of any court." Its "very existence as well as [its] *modus operandi* . . . was kept highly secret." Its meetings were "sporadic" and in "closed sessions." It did not interview any of "the six living former justices of the Court" nor any lower court judges. It was "an *ad hoc* proposal, released through a televised press conference, by seven individuals speaking only for themselves." Changes in the Court's jurisdiction, Warren maintained, "should not be premised solely on the comments and impressions of any group of outsiders."[46]

Warren then turned to the "factual and statistical premises" on which the proposal rested. He called the study group's calculations "simplistic" and its use of numbers "facile and unevaluated." The "false impression" the calculations created was "reminiscent of the McCarthy days." "Bare numbers," he suggested, "tell us little about the actual workload." Much of the "increased workload" figures cited by the study group involved *in forma pauperis* applications for certiorari, the "overwhelming majority" of which were "totally and obviously without merit" and could be quickly disposed of. Moreover, Warren had never seen "the slightest evidence that any member of the Court was distracted from full devotion to the decisional processes" by having to consider petitions for certiorari. The study group had "simply misunderstood and misdiagnosed the capacity of the Supreme Court to manage its decisional processes." The report could "only be characterized as naive."[47]

Naivete was not all that was involved in the National Court of Appeals proposal, Warren felt. Decisions on certiorari were a matter of "feel"; they were intimately related to "the concerns and interests and philosophies of the Supreme Court justices." The process by which a justice acquired "feel" was part of the larger process by which he became acclimated to serving on the Supreme Court and came "to have complete confidence in his position." The "outlook and approach" that Warren equated with "effective" judging on the Supreme Court was an "innovative" one. Such an approach was "acquired only by those who serve on the Supreme Court." Lower court judges had a "more limited and conservative . . . judicial experience." They would tend "to deny review of those decisions that . . . seem correctly decided in terms of precedent and settled law."[48]

In these last remarks lay the gravamen of Warren's complaint. He saw the National Court of Appeals proposal as an attack on "innovative" judging, an exercise in substituting "honorable and dedicated" but "limited and conservative" court of appeals judges for justices with a "broad overlook and an innovative approach to the law and the Constitution." He saw the proposal as an effort on the part of a group commissioned by Chief Justice Burger to "pack" the Supreme Court with judges who followed the "limited and conservative" approach to judging that Warren associated with his successor. "Even the justices themselves do not come to the Court fully equipped to understand and to execute the awesome functions of the Court," Warren noted.[49] He meant by that statement that he had come to realize the "awesome" opportunities to "innovate" that befell a Supreme Court justice, beholden to none but his conscience and his conception of the Constitution. The Warren Court's tendency to innovate, not its workload, was in Warren's eyes the central concern of the study group.

Warren's attack on the National Court of Appeals was a vintage performance. In it, some of the major "lessons" of his public life were restated. He was reminded of the tendency of sinister interest groups to operate in secret and to exclude affected parties from their deliberations; he reaffirmed his sense of the importance of human "concerns and interests and philosophies" in Supreme Court decisionmaking; he emphasized again the obligation of public officials to reach past their own technical concerns to hear the claims of the people. The choler of his last years was channeled into a characteristic moral exhortation. The Supreme Court, he maintained, was properly seen as an institution that was "open to all people and to all claims of injustice," an institution "always there to right the major wrongs . . . and to advance our precious constitutional liberties and privileges." In the place of such an institution of the National Court of Appeals proponents wanted to establish "the final judgment of a chance group of unknown and temporary subordinate judges."[50]

In his attack on the National Court of Appeals Warren conveyed, far more effectively than in the latter chapters of his memoirs, the tone of his contribution as Chief Justice of the United States. In the halcyon years of the Warren Court he was reaching out to do justice, "innovating," righting wrongs, protecting liberties and privileges, "testing every case . . . to see if justice [had] truly been done."[51] He had engaged in an "awesome" but exhilarating exercise of power, and he had become convinced that he had exercised that power for noble purposes. In his last years on the Court and during his retirement, "castigators," as he called them in his memoirs,[52] had sought to attack his innovations; now they were seeking to change the Court's opportunities to exercise power. The "special interests" were threatening once again to place their selfish concerns ahead of the public's yearning for decency, fairness, and justice. Warren the crusader was once more prepared to fight them.

𝒦

Warren's address was published in the American Bar Association Journal in July, 1973, along with a tepid defense of the study group's proposals by Warren Burger, who said that the recommendation of a National Court of Appeals was "in the spirit of provoking discussion" and that "I for one will defer my own conclusions until all the arguments are in and all alternatives have been explored."[53] The National Court of Appeals proposal, to this date, has not advanced beyond the stage of a recommendation.

As the National Court of Appeals proposal was being debated, the Watergate crisis came to a head, and Warren began increasing his speaking engagements. After the fall of 1973, when the existence of concealed tape recordings of conversations at the White House became known, Warren was aware that impeachment of President Nixon was a serious possibility. He refused to comment on impeachment issues and declined to offer private encouragement to advocates of impeachment.

Warren's emphasis, in his public addresses, was on the self-correcting features of the American political system, the positive qualities of public service, and the need for "a new commitment to the rights of man through a modern Magna Charta for governmental conduct." He saw his role in the speaking engagements as "plead[ing] with [students] not to become involved in the current cynicism resulting from the Watergate scandals" and reminding them that "adherence to the Constitution of the United States" was what had "brought us to our present great stature." "We are only in great national trouble," Warren suggested, "when people violate or circumvent the Constitution."[54]

Although Warren was increasing his public appearances in 1973 and early 1974, his health began to show signs of deterioration. Since his operation for stomach cancer in 1951 Warren's health had remained generally sound. He battled a tendency to gain weight, attending a health spa in January, 1972, where he lost eighteen pounds. That visit had been prompted by the discovery that he had some angina pectoris and some occlusion of his coronary arteries. These conditions persisted throughout his retirement, although they were not particularly acute or discomforting.

In January, 1974, Warren was on one of his periodic holiday visits to the West Coast when he developed chest pains and was hospitalized for a week. Angina and coronary artery disease were diagnosed, and he was instructed to cut back on his activities. He declined to do so, speaking in Washington on April 17, New Orleans on April 27, Palo Alto on April 30, and Santa Clara on May 11. By the seventeenth of May he was too exhausted and incapacitated to fulfill a speaking engagement in New York. On May 21, however, he delivered a commencement address at Morehouse College in Atlanta, repeating his criticism of the National Court of Appeals and reminding students that under the American governmental system "when [the people] have made a mistake, they can rectify it."[55]

On May 23, after he had returned to Washington, Warren's chest pains recurred, and he needed to be hospitalized. His first choice was Bethesda Naval Hospital, which was available to retired Supreme Court justices on permission from the White House. Nixon, now in the last stages of his presidency, allegedly refused to sign the executive order admitting Warren, and consequently Warren was admitted to Georgetown University Hospital. While he was there Nixon then instructed his personal physician to offer to have Warren transferred to Bethesda, but Warren refused. He remained hospitalized until June 2, when he returned to his rooms at the Sheraton Park Hotel.

Warren's condition continued to deteriorate for the next month, and on July 2 he was again admitted to Georgetown, this time with congestive heart failure. In the last week of his life his conversations with visitors centered on the Nixon Administration's claim of executive privilege in the Watergate tapes case,[56] on which oral arguments would be heard before the Supreme Court on July 8. Former Justice Arthur Goldberg visited him on that day, and Justices Brennan and Douglas on the following day. In both conversations Warren declared his belief that the Court would not and should not sustain the executive privilege claim. Two hours after the visit with Brennan and Douglas ended, at about 7:30 P.M. on July 9, Warren's heart condition significantly worsened, and by 8:10 P.M. he was dead of cardiac arrest. Nina and Honeybear were with him at the time.

For two days after his death Warren's body lay in state in the foyer of the Supreme Court building, the first time such a ceremony had taken place on the death of a Supreme Court justice. The foyer was opened so that the public could file past the casket; over nine thousand persons paid their respects. On Friday, July 12, the casket was transported from the Court to the National Cathedral, where a memorial service was held. Nixon and all the justices of the Court attended, with Nixon escorting Nina Warren out of the cathedral at the conclusion of the service. Warren was then buried with military honors at Arlington Cemetery.

Twelve days after Warren's interment Chief Justice Burger opened a session of the Supreme Court by recognizing Warren's accomplishments as Chief Justice. Burger then announced the decision in the Watergate tapes case. The decision, with Burger writing the opinion, was unanimously against the Nixon Administration's claim of executive privilege: The Watergate tapes were to be surrendered. Nixon surrendered the tapes on August 5 and resigned the presidency on August 9. An ironic linkage between the careers of Earl Warren and Richard Nixon had persisted to the very end of Warren's life.

The public viewing at the Supreme Court and the interment at Arlington Cemetery bore witness to the fact that on Earl Warren's death he was a national figure. He had gone from Bakersfield to Berkeley to Oakland to Sacramento and from there to Washington. On that last journey he became Chief Justice of the United States, and it was as Chief Justice that his public life took on its symbolic dimension. He had become the personification of his Court, both as the figure on John Birch Society billboards and the man past whose casket thousands filed, and in personifying his Court he personified the values that lay at the foundation of his public career. His death, coming when it did, seemed the death of a certain spirit about public officeholding in America: the idea, attributed to Warren and expressed by William Sweigert, that public office was to be treated as a public trust. Warren seemed to be the last major American public servant who continued to believe that the future would be better than the past, that public morality and private morality were the same, and that under the American system of government, justice would eventually prevail.

# 15

# *From Progressive to Liberal*

We have seen that Warren came to maturity at the height of Progressivism's political influence in California and that his first public hero was Hiram Johnson, the most visible Progressive official in the state. We have also seen that Warren saw firsthand the evils Progressives sought to combat: the presence of omnipotent special interests, the corruptibility of public servants, the oppressed condition of laborers, the growth of urban political machines. In addition, we have noted that Warren's own career in public life was to benefit from the triumph of Progressivism in California politics. The departisanization of the California electoral process, through the cross-filing system, had made it possible for Warren, who could not have financed campaigns himself and would not be beholden to lobbyists, to become a viable candidate. The deep suspicion with which Progressives viewed local legislatures allowed strong executives and administrators to flourish; Warren entered elective politics with a background of executive and administrative experience. The Progressives' distaste for corruption and venality in all walks of life had given public credence to the roles of "crusading" law enforcement officer and of "incorruptible" politician, two roles Warren was temperamentally suited to perform. Finally, the Progressives' impatience with the political status quo had made activism by governmental officials respectable, allowing Warren to make political capital of his version of selective activism.

Progressivism in California and elsewhere was more than a political ideology and more than a reform movement. It represented a reorientation of educated American thought, a new synthesis of established and novel social assumptions. In this broader sense, Progressivism becomes a catchword for a distinctive set of assumptions and values that influenced educated American thought in the first three decades of the twentieth century. Many persons who held that set of assumptions and values did not see themselves as political supporters of the Progressive party or even as social reformers. Moreover, their shared value premises were often concealed by specific disagreements on social issues; their ideological coherence appears stronger to us than it did to them. Warren was a Progressive in this expanded sense. His perspective on social issues was that of the Progressive generation.

The attitudes and values that distinguished Progressivism can be subdivided into two types: attitudes of belief and attitudes of policy. Attitudes of belief were insights that Progressives extrapolated from their experience—what Warren called "lessons." The process by which such attitudes emerge is one that has fascinated and baffled historians of ideas for some time: It is far easier to note the existence of such attitudes than to explain how they came into being. Since this book is not an intellectual history of Progressivism, my treatment of Progressive attitudes of belief and attitudes of policy, which were attitudes about how governmental institutions should function so as to implement attitudes of belief, will be generalized and attenuated; further documentation is provided in sources cited in the notes.[1] After sketching Progressive attitudes of belief and attitudes of policy, I shall consider Warren's relationship to both types of attitudes.

Progressivism sought a restoration of the nation's moral tone. In a book entitled *The Old Order Changeth*, written in 1910, William Allen White claimed that "for ten years there has been a distinct movement among the American people [that stems from] . . . a conviction of their past unrighteousness." The movement "is called variously: Reform, The Moral Awakening . . . the Uplift."[2] All around them, Progressives saw moral issues: the issues of corruption in government, the issue of licentiousness or drunkenness among the general public, the issue of excess size or power in economic life.

They believed that moral codes of conduct were discernible and persisted over time, that citizens could judge the morality of other citizens, that the moral climate of the nation had deteriorated, and that public officials should help restore it.

A belief in morality was linked by Progressives to a belief in patriotism. The concept of patriotism in Progressive thought combined two assumptions—the assumption that America was the pinnacle of world civilization and the assumption that America's greatness was associated with the dominance of "native" rather than "foreign" stock. The two assumptions were sometimes contradictory: The image of American civilization as a "melting pot" stirred the popular imagination at the same time that legislation was being proposed and passed on the state and national levels excluding "undesirable" aliens and popular literature fostering Anglo-Saxon supremacist attitudes was being published. In one sense patriotism meant what Warren in 1972 called "love of country . . . something fine and noble to which all of us can subscribe regardless of minor differences among us";[3] in another sense it meant nativism, racism, and xenophobia.

The third distinctively "Progressive" attitude of belief was a belief in progress itself. By "progress," Progressives meant the interaction of the inherent perfectability of mankind with the permanently dynamic quality of American society. Having experienced the rapid industrialization of late nineteenth- and early twentieth-century American society, Progressives expected continued technological change to take place, but they were not fearful of the prospect. They believed that as technological advances occurred, advances in the human condition would also occur. They thus equated change not with decay but with evolution: Life in America was getting better and better.

The values of morality, patriotism, and progress can be seen as stabilizing forces in early twentieth-century American culture. The first twenty-five years of the twentieth century were years in which the American population grew rapidly, became increasingly mobile, and became ethnically heterogeneous. They were years in which unprecedented numbers of aliens sought to enter the country. They were years of massive technological change and relative prosperity. And they were years in which an increasing number of Americans became literate and therefore more aware of one another. In this context morality, patriotism, and progress can be said to have functioned as buffers against rapid change. To believe in those values was to believe that massive transformations in American culture were positive developments since they had not altered the essential character of the nation.

That politicians sensitive to changing values paid lip service to the themes of morality, patriotism, and progress was not nearly enough for those already committed to this cause. They wanted to see evidence that these values were actually adhered to by those in positions of public trust. Hence Progressivism gained mass public attention as a reform movement: Its adherents were interested not only in articulating their attitudes of belief but in implementing them. Their attitudes of belief fostered attitudes of policy, and in their policymaking, Progressives drew upon a critique of their nineteenth-century experience.

The deficiencies in American society that Progressives identified—corruption in government, squalor in urban life, unjustifiable economic inequalities—could each be linked, they felt, to a single late nineteenth-century development: the emergence of "special interests." The relative passivity of government and the rapidly changing character of the economy in the late nineteenth century had resulted in an unholy alliance of giant corporations, interested solely in their private gain, and political bosses, interested solely in their personal power. Immigrants had become pawns in this alliance, and most of America's ills could be traced to it.

The Progressives sought to break up this alliance, restore morality in American political and economic life, and channel technological change so that it became progress. To achieve these goals they sought to transform the character of American government. Progressive policymaking envisaged an ideal twentieth-century form of government that possessed three features not present in the nineteenth-century version. Government was to be nonpartisan, it was to be a champion of the "public" interest as distinguished from special interests, and it was to be an activist force in the economy and in society at large. Each of these policy goals followed from the Progressives' belief in morality, patriotism, and progress.

Nonpartisanship was a technique designed to rid government of boss rule. The political parties had become corrupted by boss-dominated immigrant machines, Progressives felt, with the result that large segments of the public simply voted on the basis of party affiliation. This prevented the average citizen from making independent assessments of the qualifications of public servants. The remedy, for

Progressives, was to eliminate party affiliation as a significant feature of political candidacies. A candidate should appeal directly to the voters and speak directly to the issues. Candidates who were selected by such a process were more likely to be honest, articulate, and mindful of public opinion. Nonpartisanship was thus a moral policy, since it sought to eliminate influence and corruption; a patriotic policy, since it relied upon the traditional American democratic process to produce enlightened public servants; and a policy that facilitated progress, since it was intended to produce more responsive and abler public officials.

The persons who were to staff the government envisaged by Progressives were expected to be no more affiliated with special interests than with corrupt partisans. A familiar Progressive vision of politics contrasted the "public interest," which was innately enlightened, with special interests, which were inevitably selfish and shortsighted. Part of the difficulty with nineteenth-century politics had been the ease with which special interests had manipulated government: Politicians curried favor with and were supported by groups who sought to further their short-run economic advantage. Government, as Progressives conceived it, was to function as a watchdog on special interests and a promoter of the public interest. Its officials were to be open and responsive to public opinion while being suspicious of lobbyists and officials in their control. The more public servants trusted the instincts of the public and resisted pressures from organized interest groups, the Progressives reasoned, the more enlightened their policymaking was likely to be.

A governmental preference for the public interest over special interests was moral, patriotic, and progressive for reasons similar to those that linked morality, patriotism, and progress to nonpartisan government. Since special interests were innately corrupt, and the public interest instinctively wise, trust in the latter and contempt for the former were morally sound positions. They were also patriotic positions, since America had been a civilization where pressure from a virtuous populace had periodically exposed those who sought to use power to further their own selfish pursuits. And, of course, they were positions that facilitated progress, because special interests, unlike the public at large, were dedicated to preserving a status quo that advantaged them. As their influence diminished under the scrutiny of public officials, more opportunities for enlightened policymaking would be created.

But what kind of enlightened policymaking was to take place?

Here Progressives did not speak with one voice. Denouncing corruption, calling for an end to partisanship, and championing the public interest were activities that most Progressives could endorse with impunity: Indeed these attitudes helped define the character of Progressive reform. But Progressives found it harder to agree upon the precise nature of the programs that a new class of public servants was to enact after the rotted brushwood of nineteenth-century government had been cleared away. Some Progressives proposed a regulatory role for government, epitomized by agencies such as the Federal Trade Commission and its state equivalents. Others wanted to dismantle large-scale industrial enterprises to restore more perfect competition and sought to use antitrust legislation to achieve these goals.[4]

Some Progressives felt that morality could be restored through direct legislation prohibiting "immoral" conduct, such as excessive drinking. Others felt that patriotism demanded closer attention to immigration patterns and proposed restrictionist immigration legislation. Some Progressives believed that government should attack the problem of economic inequality by modest redistributions of wealth, such as the creation of a federal income tax that would provide revenues that could be used to provide relief to economically disadvantaged persons. Others felt that such redistributions would diminish incentives for persons to escape their disadvantaged condition and proposed legislation placing industrial workers in a more competitive or less exploited condition.

These disparate policies had in common the idea that government could not trust the marketplace to produce an ideal society. Progressive government was intended to be active government, whether regulatory or punitive, paternalistic or redistributive. Government, in the Progressive polity, was intended to intervene in order to further morality, secure the nation's heritage, and promote progress. Its officials were to be activists. An older approach to government had emphasized inactivity; the results had been corruption, decadence, and economic disequilibrium. A new class of public officials was to take action to set things right again. Active government was to be a moral, patriotic, and Progressive force.

ᛕ

We have seen that a striking concordance existed between these Progressive attitudes of belief and attitudes of policy and the views expressed by Warren in his public career. The following pages briefly summarize that concordance. The restoration of morality in American life was a fundamental theme of Warren's career. As a law enforcement official he was a bitter foe of gambling, an opponent of prostitution, and an enemy of chicanery, conflicts of interests, and corruption among public officials. He prosecuted gambling ships, confidence men, and corrupt sheriffs not only because he was performing his prosecutorial duties but also because their presence was an affront to civilized society. As governor he rehabilitated prison and mental health facilities because he thought conditions in them violated standards of common decency.

On the Court, Warren struck out against purveyors of obscene literature and gamblers, finding them less entitled to Bill of Rights protections than other persons. His lack of sympathy for the alleged First Amendment rights of pornographers or the alleged Fifth Amendment rights of gamblers came from a judgment that they were people who had not acted in a decent fashion. In 1972, aroused by the "deteriorat[ion] in moral and cultural standards," he called for the "reestablishment of a moral tone for our nation."[5]

In another passage from the book just quoted, Warren advocated a renewed attention to "love of country." "America is our home," he wrote, "and we want it to be wholesome, peaceful, dignified and satisfying." The term "love of country," Warren said, could be equated with "patriotism,"[6] a value that he had held throughout his public life, from his brief service in the First World War, through his interest in preparedness, civil defense, and the relocation of the Japanese during World War II, to his "putting on his uniform again" for Lyndon Johnson through service on the Warren Commission and his choice of Arlington Cemetery as a place to be buried. In his later life Warren ceased to equate patriotism with militarism, opposing the war in Vietnam and becoming engrossed in the concept of world peace through law. But he never lost his conviction that he lived in the freest and fairest and most promising of nations.

Patriotism had sometimes motivated Warren, especially in his law enforcement days, to be suspicious of radical organizations, aliens, and other potential subversives. His sponsorship of the Japanese relocation program grew out of his concern for civil defense and his conviction that Japanese Americans, unlike Americans of German or Italian descent, were not "assimilable" and hence potential saboteurs. On the Court, Warren had shown some initial inclinations to favor the government in cases where national security was at issue,[7] but this tendency ceased after his clashes with the House Un-American Activities Committee in *Watkins v. United States.*

After *Watkins,* Warren's recurrent support for the constitutional rights of "Communist sympathizers," his notoriety with the John Birch Society and other "patriotic" groups, and his break with the American Bar Association tempered his public reputation as a patriot. But Warren retained his patriotism throughout these episodes, equating it more with an American belief in fairness and justice for all than with xenophobic impulses. "Our country," he wrote in 1972, "has shortcomings as well as virtues; it has had failures as well as successes; but it does contain all of our hopes, our ambitions, our cherished values, our treasures."[8]

The concept of progress in Progressive thought had, as noted, combined an assumption that human nature was perfectible with an assumption that life was in a constant state of change. These two assumptions were central to the adoption by Progressives of environmental theories of behavior. According to such theories, which were intended as alternatives to the individualistic theories of the late nineteenth century, the conditions of poverty, degradation, and even criminality were deemed to be a consequence of social conditions rather than a failure of individual character. Progressives believed that if disadvantageous social conditions were eliminated (a state of affairs they saw as capable of being realized, given the potential of modern technology), American citizens could develop their innate potential.

In a 1969 interview with Anthony Lewis, Warren identified himself as subscribing to each of the assumptions on which a belief in progress rested. He said that he was "optimistic . . . about the United States"; that he felt that America was "a young nation, that we haven't yet reached our potential, . . . that we are still going through the growing process." He believed that in the future "we will emerge a better nation and a stronger nation." He identified crime with the wartime atmosphere of international politics and the lack of employment in urban ghettos. The eradication of crime, he argued, could be achieved by "improv[ing] the condition of our cities. We must get rid of the ghettos, we must see that every youngster . . . is afforded a decent education and is given some skill through which he can compete in the market." Such persons must also have "the opportunity to get a job, [so] [they] must be able to join a union." Crime, he believed, was a social condition that could be reduced by the elimination of the environmental factors that encouraged its growth. With these factors removed, the natural talents of American citizens could be used in productive ways.

Warren's faith in human perfectibility stayed with him throughout his tenure as a public official. He sponsored compulsory medical insurance and prison reform because he wanted to eliminate conditions that prevented humans from realizing their potential. He supported the University of California and the California system of public education because he believed that education fostered individual talent and developed individual self-esteem. He thought racially segregated schools inconsistent with fundamental principles of American society because, among other things, they fostered a negative self-image in black children. He took the rights of criminal defendants seriously because he refused to believe that most criminals were other than victims of an unfortunate environment. Only in the case of "professional" criminals, such as gamblers, dealers in smut, or members of organized crime syndicates, did he forswear rehabilitation.

Part of the basis for Warren's belief that his programs as governor and his decisions in the Court were inherently just was that in countless instances he was seeking to give people opportunities to better their lives. He looked beyond the criticism of his "self-interested" opponents to the accolades of the public, who, he believed, would understand his motives. "He thinks," Lewis said in the 1969 interview, that "judges and other men are obligated only to do their best—and if in conscience they do, they need not lie awake at night."[9]

Warren's attitudes toward policymaking were also characteristically Progressive. He adopted, from the outset of his public career, the belief that partisanship in politics was corruptive and to be discouraged. He identified bosses and legislatures as seedbeds of illicit influence and chicanery, noting that partisan politics in California had contributed to the tendency of the legislature to reflect special interests and to the rise of political machines and corrupt electioneering practices. He supported and profited from the reforms originated by Hiram Johnson, especially the statewide primary and cross-filing. He played the role of nonpartisan politician to the hilt, entering himself in the primaries of both parties each time he ran for governor, using the slogan "leadership, not politics" in his 1942 and 1946 campaigns, making appointments to administrative posts on a bipartisan basis, declining campaign contributions from partisan Re-

publican groups, and refusing to endorse other Republican candidates, including candidates for lieutenant governor and Richard M.
Nixon in his 1946 and 1950 campaigns.

When President Truman visited California after having defeated
Dewey and Warren in 1948, Warren posed for photographs with
him; Truman told the press that Warren was a Democrat and didn't
know it. Warren's nonpartisanship, we have seen, prevented him from
being a more effective force in national Republican politics. In the
1948 presidential campaign he complained bitterly about Dewey's
canned speeches and refused on occasion to mouth conventional Republican doctrine. Warren had, after all, become a Republican, as
had Hiram Johnson, primarily because California was then "an overwhelmingly Republican state"; [10] his nonpartisanship was a matter of
genuine conviction.

Warren's vision of political life had at its core the Progressive
juxtaposition of special interests against the public interest. In the
1920s he saw himself as a law enforcement official whose task was to
attack corruption and influence peddling in the name of common
honesty and decency. Fifty years later he generalized this insight.
"The people," he wrote in 1972, "are often in advance of their leaders. This is consistent with the realities of the present and with the
American tradition." Persons entrenched in public positions were
susceptible of being "complacent and satisfied with conditions contrary to the public welfare so long as there is no public outcry" and
of being "subservient to powerful vested interests whose only aim is
to strengthen their dominant positions." The remedy was for public
servants to bypass special interests and appeal directly to the people:
It was "in the spirit of our institutions," Warren felt, "for the people
to lead in governmental affairs." [11]

Warren's career had been a series of appeals to the "forward-
looking" public. He had cultivated public outrage about lax or corrupt law enforcement practices, increasing the power of his offices
in the process. As governor he had attempted to shame or to embarrass lobbyists by refusing to make compromises with them and by
revealing their presence to the public. And as Chief Justice he had
attempted, buttressed by his reading of the ethical imperatives of the
Constitution, to dismantle discriminatory legislation drafted by racists and others to implement their shortsighted prejudices. He had
not only sought to invalidate such practices, but to eliminate the
political institution from which they originated, the malapportioned

legislature. *Reynolds v. Sims* was his most significant opinion, he thought, because it insured that henceforth elections would reflect the collective public interest—embodied in the "one-man, one-vote" standard—rather than the machinations of special interests.

Like many Progressives, Warren was more confident about the potential of an active government to foster progress than he was clear about the precise character or direction of affirmative governmental action. While he had repeatedly been an activist in his own public offices, expanding his own power as he extended his office's reach, he was late to convert that activism into a philosophy of affirmative government. As a gubernatorial candidate and a Republican party official he had criticized the New Deal and Culbert Olson's efforts to increase the scope of California state government. Once elected governor, however, Warren began to institute affirmative programs on a selective basis: prison reform, mental health care, and compulsory health insurance were examples. Characteristically, these programs were in areas where human initiative was being thwarted by environmental factors such as poverty, disease, or criminality; Warren used the arm of the state to make a given environment more conducive to human growth.

Warren did not endorse affirmative government without retaining a sharp concern for the intrusive effects of government on human dignity. In this sense he was not as sanguine as Progressive supporters of social welfare programs about the inherent benevolence of government administrators. He continued to use language extolling the virtues of independence, self-reliance, and autonomy as a California public official, and on the Court we have seen that his opposition to the information-gathering techniques of investigative agencies was grounded on a belief in the privacy and dignity of the individual. When the possibility that Lee Harvey Oswald was an FBI agent surfaced during the Warren Commission's deliberations, Warren's fears about investigative excesses surfaced; he alone among the commissioners advocated an independent investigation of the FBI's sources. One can also trace the theme of governmental intrusiveness in Warren's criminal procedure decisions; there he came to identify with the solitary, friendless individual overwhelmed by governmental investigatory power.

Thus Warren preferred affirmative government only when humans could not be expected to help themselves or one another, and he retained some of the suspicion that he directed at powerful private interests for powerful organs of government. When he felt, as he did in the reapportionment cases, the segregation cases, and to some extent in the school prayer cases, that government was in league with special interests, he would protest on behalf of the individual citizen; when, however, he felt that government was combating special privilege, as in the antitrust area, he welcomed affirmative action.

The antitrust cases and the labor cases demonstrated that Warren was prepared to combat special privilege by support for redistributive and paternalistic economic policies. Redistributive and paternalistic economics seemed to Warren, as Chief Justice, to be common-sense solutions to the problems of poverty, dependency, and degradation that he had associated with the Southern Pacific's empire. Too much wealth and power in a corporation had resulted in its being indifferent to those who were dependent on it for their livelihood. The absence of economic bargaining power and of job marketability had caused men to work in occupations that endangered their health for low wages and no job security. Injured employees were simply discarded from the labor force; disgruntled or oppressed employees were told to look elsewhere when they had no alternatives; loyal workers were laid off at the railroad's pleasure. This long memory of grievances gave Warren an emotional attachment to policies that sought to reduce the power of vast industrial enterprises, and he did not make distinctions among the policies, supporting regulation, divestment, or the creation of countervailing bargaining units. On the Court he came to support the positions of unions far more vigorously than he had as a California public official, and more vigorously, for that matter, than had most California Progressives.[12]

Warren's antitrust policies typified the position of one influential segment of Progressive thought, which sought to use the antitrust laws as a weapon by which economic privilege could be countered and excessive conglomerations fragmented. Warren's decisions demonstrated an emotional attachment to small business and the lone entrepreneur and an inherent suspicion of mergers. Such views reflected an attitude that associated economic bigness with monopoly and monopoly with predatory conduct. In the antitrust area Warren was prepared to use government as a weapon for attacks on special interests. His rationale was that such attacks were designed to give otherwise disadvantaged competitors greater opportunities. The economic arena, like the political arena, inevitably engendered selfishness and corruption if it were not scrutinized and regulated. The intrusiveness of a paternalistic and regulatory government was not a matter of concern to Warren in antitrust cases because governmental intervention was furthering economic opportunities rather than restricting them.

ɤ

One of the striking features of Warren's Progressivism was the degree to which he was able to retain Progressive attitudes of belief and policy past the time when they were dominant in national politics. Nationally Progressivism had reached its apex in the 1912 election, when Theodore Roosevelt, William Howard Taft, and Woodrow Wilson each defined themselves as Progressives, and declined as early as 1920, although Robert LaFollette mounted a visible, if minor, campaign for the presidency in 1924. The New Deal signified another "reform" movement, but one sufficiently different from Progressivism to arouse the opposition of numerous former Progressives.[13] The New Deal also marked the triumph of liberalism as a consensual reform ideology, one that borrowed from but also departed from Progressivism. By the 1940s national politicians were known as "liberals" or "conservatives": Aspiring presidential candidate Warren was asked in 1948 to define liberalism rather than Progressivism.

Yet when Warren defined liberalism in 1948, we have seen, he equated it with Progressivism. The term "progressive," he maintained, represented "true liberalism." He then went on to repeat some of the central tenets of Progressivism, such as the idea that "democracy is a growing institution," and the "subordinat[ion] of . . . private interest to the common good."[14] Warren was able to make this facile equation of Progressivism with liberalism, which ignored basic differences between the two ideologies, because California political culture had been receptive to his Progressive style of leadership and because prior to the 1948 campaign he had not exposed his thinking to a national audience.

Morality, patriotism, and progress had been central values of Warren's administrations in California; nonpartisanship, attacks on special interests, and activism had been characteristic of his style of governing. He had functioned as a Progressive public official in the 1940s and fifties; in few other states could he have thus functioned. Mid-twentieth-century politics in California was dominated by the

twin legacies of Hiram Johnson's administration—a political process
that deemphasized partisan affiliation and a provincial nativism that
sought to exclude "aliens" from participation in the island paradise
that was California. Warren was heir to both of these traditions, and
signaled, in behavior that ranged from advocacy of compulsory health
insurance to support for the internment of Japanese, his acceptance
of them. So complete had been the triumph of Progressivism in California that its politics remained "Progressive" in style well after the
demise of the Progressive movement nationally.

The extent to which Warren retained Progressive attitudes and
values well after they had declined in influence was to form one of
the crucial factors in his metamorphosis from an alleged "middle-of-
the-road" Republican politician to a militantly liberal Chief Justice.
Warren was neither a 1950s "liberal" nor a 1950s "conservative" on
being appointed to the Court; he was an early twentieth-century
Progressive who had survived the passing of the movement from
which he drew his insights on public life.

The uniqueness of California politics and Warren's ability not to
expose himself on issues had enabled him, in his pre-Court career,
to avoid fully coming to terms with two issues that were to distinguish the world of the Progressives from the world of mid-twentieth-
century liberals. One issue was the place of civil rights, especially
the rights of ethnic and racial minorities, in a society that prided
itself on being a democracy but also took pride in its Anglo-Saxon
heritage. The other issue was the legitimacy and scope of affirmative
governmental action in a pluralistic society whose members were dependent upon yet independent of state control.

Progressives, including Warren in his pre-Court career, had not
seriously confronted either of those issues. They had equated civil
rights with the rights of native Americans, thereby implicitly excluding aliens and subversives. They had also assumed that government
could enforce majoritarian moral, political, or economic policies as
part of the process of bettering the lives of individual citizens. But
both issues were to pose troublesome dilemmas for liberals, and in
the course of addressing them as Chief Justice, Warren was to modify his own thinking.

In Warren's description of liberalism in 1948, he asserted that
"civil rights, representative government, and equality of opportunity
were all part and parcel of the liberal tradition" and that most
Americans were "progressive and liberal in this sense." [15] In framing

this definition, Warren, an aspirant to national office, was principally attempting to associate himself with uncontroversial, "popular" points of view. But his statement also revealed that a fusion of two ideas had taken place in his mind, a fusion that can be said to be the chief characteristic of twentieth-century liberalism in America. Warren had fused the idea of rights against the state, chiefly personified by the claims of dissident minorities, with the idea of affirmative action by government to protect disadvantaged persons.

No comparable fusion had marked Progressivism. In the idealized modern state of Progressive social theorists, a beneficent government, staffed by nonpartisan experts, fought poverty or illness or malefactors of great wealth, but it did not protect rights against the state. Rights against the state was an outmoded nineteenth-century idea, a rhetorical justification for the power exerted by entrenched special privilege. Legal scholars influenced by Progressivism, for example, deplored judicial protection for entrenched interests based on the "liberty of contract" doctrine. The doctrine itself was premised on a belief in unrestricted economic freedom; that belief, Progressives felt, nurtured special privilege.[16] Nor were many Progressives sensitive to discriminations against racial or ethnic minorities: In early twentieth-century California, we have seen, political reform went hand in hand with Orientalist racism.

The fusion of a belief in rights against the state with a commitment to affirmative government created a sensibility that shaped the political perspective of liberalism from the Second World War through the 1960s. The liberal sensibility was noticeably responsive to inequalities of condition and chose to attempt to redress them by creating equal opportunities. Rights against the state became translated into a presumptive claim on the part of all Americans to be treated equally. Recognition of this abstract claim, in standard parlance, came to be labeled protection for civil rights and civil liberties. In practice, such protection meant the opportunity to seek governmental redress if one were discriminated against. The idea of rights against the state and the idea of affirmative governmental action were fused in governmental policies sanctioning discrimination.

The first significant expression of a liberal sensibility on the Supreme Court of the United States came with Warren's opinion for the Court in *Brown v. Board of Education*. *Brown* identified an inequality of condition (black schools and white schools, if separate, could never be equal) and sought to alleviate it by restoring equality of opportunity to both blacks and whites (integrated schools). *Brown* also implicitly treated the right to be educated free from stigmatizing conditions as a right that the state could not infringe, and used the constitutional review power of the judicial branch of government to declare those conditions constitutionally invalid and to compel their eradication. *Brown* was an extraordinarily bold decision not only because of the social changes that it compelled but also because it announced that the Supreme Court, rather than a popularly elected branch of government, could take the lead in identifying and alleviating inequalities of condition in American society. Previous twentieth-century Courts had primarily used their constitutional review power to invalidate affirmative governmental action; the Warren Court was using its power to force government to act.

*Brown*, as noted, marked the beginning of Warren's emergence as a liberal judge, dedicated to identifying and alleviating inequalities in American society. Warren became a liberal judge, where he had not been a uniformly liberal governor, for two reasons, one having to do with the special obligations liberalism imposed on twentieth-century judges and the other having to do with his belated acceptance of the liberal fusion of civil rights and affirmative government.

Judicial activism became a necessary feature of liberal theory as an incompatibility between thoroughgoing protection for rights against the state and deference to the judgments of majoritarian democracy became apparent. Democratic majorities often repressed minorities, as the Japanese relocation policy, the loyalty oath controversy, and the practice of segregating public schools illustrated. The judiciary had been the branch of government historically charged with protecting minority rights, but in the early twentieth century, as legislatures had had their reformist programs invalidated by recalcitrant activist judges, Progressive theorists had designed a new role for judges, the role personified by Felix Frankfurter. The role envisaged judicial deference to the reformist actions of legislatures, the governmental representative of democratic majorities, in order to facilitate social progress.

A growing majority of Warren Court justices, including Warren, eventually rejected this stance of "judicial restraint" for one that made the judiciary a liberal watchdog of other branches of government, scrutinizing legislative and executive actions to see that they were not unduly infringing upon minority rights. In Warren's case, rejection of a Frankfurterian stance for this activist, liberal posture was a combination of his growing disaffection with Frankfurter as a symbol, his long and largely successful history of activism in office, and his growing belief in the goals of mid-twentieth-century liberalism. Whereas Warren had compromised or ignored the claims of minorities in the Japanese relocation program, the apportionment of the California legislature, and the loyalty oath controversy, he was to vote against racism, malapportionment, and loyalty oaths as Chief Justice. Whereas Warren had treated suspected Communists and criminals as less deserving of legal protection than other persons during his years as a California public official, he was to find, as Chief Justice, that even those classes of persons had presumptive rights against the state.

The stages of Warren's evolution to a judicial advocate of liberal theory have already been traced. In summary, he underwent a baptism by fire in *Brown*, realizing that the Supreme Court could counter a combination of deplorable social practices and legislative inactivity by a bold, although politically crafted, declaration of a constitutional principle. Warren then confronted the implications of *Brown* for his stance as a judge. Did the *Brown* experience lead to activism in defense of an ethical reading of the Constitution, or did it lead to a more deferential, politically safer stance, which envisaged the Court as framing compromises with Congress and state legislatures? In a few short years, after his experiences in *Cooper v. Aaron*, *Watkins v. United States* and other pivotal cases, Warren had made his choice: He was to be an activist judge, trusting his ethical impulses even in the face of political controversy.

The last stage in Warren's emergence as a liberal theorist was the development of a consonance between his ethical impulses and liberal policies. While such a consonance did not fully develop—Warren refused to protect the civil liberties of persons, such as gamblers or panderers, whose activities he deplored—for the most part Warren's impulses told him to reach results that seemed to implement the liberal fusion. His much greater attachment to the preservation of civil rights and civil liberties as a judge than as a governor can be attributed, principally, to an enhanced sense after *Brown* that he had no need as a federal judge to submerge his intuitive empathy for the claims of disadvantaged people.

As an attorney general and a governor Warren had believed that "the American concept of civil rights should include . . . the absence of arbitrary action by government in every field" and the prevention of "government from using ever present opportunities to abuse power through harassment of the individual."[17] But he had had to temper that belief to the exigencies of California and national politics and to what he regarded as the imperatives of the Second World War and the Cold War. Now he was seemingly alone with his conscience and the Constitution, encouraged by other liberal activists around him, offended by the cerebrated agonies of a judicial compromiser, Frankfurter. In politics, he was to say, "progress could be made and most often was made by compromising"; on the Supreme Court "the basic ingredient of decision is principle, and it should not be compromised and parcelled out."[18] Constitutional principles, ethical intuitions, and liberal policies began to be lumped together in his consciousness.

In short, Warren as Chief Justice retained the Progressives' belief in affirmative government, shifting the locus of affirmative governmental action to the federal judiciary, and extended his commitment to civil rights and civil liberties well past the point Progressives had been prepared to go. In so doing, he completed the last transition of his public life—from Progressive governor to liberal judge. And in so doing, Warren engaged in the most influential work of his long career. His Court ended a variety of customs and practices in America—segregated public facilities, compulsory prayers in the public schools, rural domination of state legislatures, and coerced confessions represent only some—and made the judiciary an affirmative participant of social policy decisions to a degree unparalleled in American history.

✒

Yet the Warren Court can be seen not only as a dramatic new chapter in the history of American judging, but as the culmination, and perhaps the end, of a twentieth-century liberal sensibility. The assumption by the Warren Court of an activist policymaking role was made necessary, as noted, by the dual goals of liberalism. Only through judicial intervention could the claims of minorities against the state be fully protected. The necessity for judicial intervention brought the internally contradictory nature of the liberal fusion into sharp focus. The idea of affirmative government, pressed to the limits of its logic, meant that rights against the state were held at the sufferance of those who formulated and implemented governmental policies. One could not insure protection through governmental intervention for some classes of persons without circumscribing the rights or liberties of other classes. Thus the idea of affirmative government presumed a wide discretion in a class of allegedly enlightened policymakers to choose to advantage one class of persons at another's expense. Inevitably, this meant less than full protection for the civil rights of somebody.

For a time liberal theory seemed to have confined this contradiction by making intermediate sets of distinctions. Property rights were less deserving of protection than rights of personhood; the civil rights of victimized persons, such as blacks, were more deserving of protection than the rights of those who had done the victimizing. Some kinds of affirmative governmental actions, such as the welfare system or the poverty program, were justifiable, even though they disadvantaged the employed and the affluent, because they sought to eradicate inequalities of condition (joblessness, poverty) and create equal opportunities, and "right-thinking" Americans would agree to disadvantage themselves in order to promote equality of opportunity.

It has become increasingly apparent, however, that not only has the traditional liberal political "coalition" of disparate classes of persons broken down, but that a distinctive liberal sensibility, with its fusion of potentially opposing ideas, has also fragmented. Protection for rights against the state has been shown, in areas as removed from one another as gay rights and taxpayer revolts, to be an idea that is persistently hostile to affirmative governmental action based on a majoritarian consensus and administered through the discretion of experts. Affirmative governmental action, for its part, has been shown to be an idea that coexists poorly with extensive freedom of individual choice.

As the liberal sensibility has fragmented, the sets of intermediate distinctions made by liberal theory to help effectuate the fusion of its dual impulses have come under attack. Affirmative action programs in higher education furnish an example. The practice of responding to inequalities of condition (dramatic differences in the early educational experiences of blacks and whites) by attempting to create equalities of opportunity through racially sensitive admissions standards has been said to violate the equality principle itself. Critics of affirmative action have argued that creating enhanced opportunities for a handful of persons through preferential admissions policies does nothing to "equalize" American higher education; it simply widens the educational opportunities of some disadvantaged persons by narrowing the opportunities of other advantaged ones. The Supreme Court has held, notwithstanding this argument, that the Equal Protection Clause of the Constitution permits such a discretionary narrowing and widening.[19] But affirmative action programs do not reinforce the liberal fusion of protection for civil rights with enlightened governmental activism; they rather subject that fusion to stress.

The major Warren Court decisions, decisions in which Earl Warren strongly believed, can be seen to be based on the assumptions of a liberal sensibility. The reapportionment decisions assumed that the proper way to protect voting rights was through a system of representation in which each voter's vote counted the same, and then insisted that legislatures adhere to that system, even if the constituents of legislators had voted a preference for some less "equal" system. The school prayer decisions assumed that a majoritarian policy of compulsory prayers unfairly discriminated against the civil rights of nonbelievers, and took action to invalidate the policy. The segregation decisions assumed that the practice of racial segregation would be constitutionally impermissible even if racial groups chose to educate themselves separately and a state sanctioned that choice, and insisted that public schools be integrated.

In each of these examples the policy judgment made by the Court rested on a device intended to reconcile prospective contradictions in liberal theory, the device of an intermediate distinction. The distinction came at the point of analysis where claims based on rights against the state needed to be reconciled with claims based on affirmative governmental action if the analysis was to yield a policy solution. In the reapportionment cases the distinction was between representation based on population and other forms of proportional representation: The former was preferred as more democratic. In the school prayer cases the distinction was between the choice of a majority to ritualize the recognition of a deity and the choice of a minority to deny that recognition: The latter was preferred as more libertarian. In the area of racial segregation the distinction was between the desires of some whites and some blacks to have a racially integrated educational experience and the desires of some whites and some blacks to limit their educational experiences to persons of their own race. The former desire was preferred as being based on a theory of the educational process that minimized the relevance of race and was therefore deemed more enlightened.

The use of these intermediate distinctions preserved a fusion of protection for rights against the state and affirmative governmental action by implicitly declaring that some rights against the state were worth more than others. The right to have one's vote counted equally was worth more than the right to choose another system of representation. The right to choose not to pray to God as part of one's exercises in school was worth more than the right to so choose. The rights of some citizens to insist that race be made a condition of attendance at public schools was worth far less than the rights of other citizens to insist that race not be made a condition.

In each of the examples, once the intermediate distinction is made and a claim that one set of rights is worth more than another is accepted, affirmative governmental action to preserve that set of rights is faithful to the twin goals of liberalism. One may notice that some of the cases may currently seem to be easier exercises in balancing the worth of competing rights against the state than others, and in those cases the intermediate distinction posited by liberal theory seems less strained. The example of race in the public schools currently seems the case in which the intermediate distinction necessitated by liberal theory appears least vulnerable. In 1954, however, that distinction was perceived as far less obviously justifiable, and judicial intervention on behalf of integration was therefore perceived as a delicate political problem. That Americans currently take the right to condition educational attendance on skin color much less seriously than in 1954 illustrates that in this instance an intermediate distinction necessitated by liberal theory has retained its power to persuade. That distinction seems threatened, however, by affirmative action programs.

While some of the intermediate distinctions made by Warren Court majorities in their efforts to implement liberal theory seem to have retained their legitimacy, the tendency of liberal theory to embrace internally contradictory goals seems more apparent now than in the years of Warren Court activism. The problem with a theory of government based on simultaneous support for affirmative government and for rights against the state is that the more seriously one takes the latter, the greater the number of individual choices seem infringed upon by the former; whereas the more sanguine one is about the former, the less seriously one seems to be taking the latter.

Earl Warren, once he became confident in the role of a liberal activist judge, had no difficulty with imposing his policy judgments on others. He had never minded being paternalistic or authoritarian as a public official if he thought his cause was just; and while he had originally disliked extensive governmental intervention at the federal level, he came to feel that intervention by the federal judiciary was not only necessary but desirable. He did not feel much tension between the twin goals of liberalism because he invariably persuaded himself that the course of action to which he was committing the Court was necessary to protect constitutional rights to their fullest extent. He reached that position of self-assurance by ranking some "constitutional" rights higher than others in his mind, the test being his ethical sensibilities. He assumed the role of liberal activist judge without much strain, with enthusiasm, and with obvious pleasure.

The relative lack of anguish with which Warren made the intermediate distinctions that enabled him to choose one set of rights over another and the untroubled spirit in which he approached the boldly activist judicial role he had designed for himself mark Warren as a prototypical mid-twentieth-century liberal judge. No group of judges before the Warren Court had sought simultaneously to encourage affirmative government and to protect rights against the state. No group of judges had been willing to depart so markedly from canons of judicial restraint and at the same time to champion so vigorously the rights of disadvantaged and dissident persons. Warren and his fellow members of the liberal majorities on the Warren Court took those positions, in the face of severe professional and lay criticism, because they believed in the rightness and justice of their undertaking.

Warren, in this sense, may have been the only, as well as the last, liberal Chief Justice of the twentieth century. The intermediate distinctions he so easily made do not seem to be likely to have a long life. The confidence in affirmative government he held has been evaporating since the Johnson Administration, and his instinctive sense of which rights should prevail over others has become muddled as legal issues have evolved from the level of abstract principle to the level of practical implementation. We are, in the 1980s, beyond the bold pronouncements in *Brown* that race is an invalid basis on which to deny persons access to educational institutions and have moved on to the cautious suppositions in *University of California Regents v. Bakke* [20] that race is not always an illegitimate basis on which to grant persons that access. Contemporary America seems to be a culture receptive to starkly communitarian or starkly libertarian social theories, but not particularly receptive to a theory, such as liberalism, that seeks to fuse a communitarian governmental apparatus with a dedication to libertarian ideals. As the internal contradictions in liberalism come to light, the choices made by Warren and other advocates of judicial liberalism appear less as ordained principles of justice and more as vulnerable policy judgments.

# 16

## Ethics and Activism

One of the arguments of this study of Earl Warren's life is that a deeper consistency can be discerned beneath the apparently contradictory positions he espoused. One way to approach the theme of consistency in Warren's public career, previous chapters have suggested, is to see his political orientation as that of an early twentieth-century Progressive; another is to see Warren as a representative figure in the history of twentieth-century reformist thought, perhaps the last of a line of "liberal" public officials who conceived the period between World War I and the Vietnamese War as one of a continuing vindication of libertarian and egalitarian principles. This chapter discusses a third means of encapsulating Warren: his consistent dedication to the active promotion of ethical principles by whatever office of government he happened to represent. Activism on behalf of ethics was a feature of Warren's conduct throughout his career, but it took on greatest significance, this chapter suggests, during his tenure as Chief Justice.

We have seen that on the Court, Warren cast legal controversies in ethical terms, identified instances of injustice, and sought to use the powers of his office to provide a remedy. He functioned on the Court much as he had as governor, identifying needed reforms and seeking to undermine the position of those opposing such reforms by emphasizing that their opposition perpetuated injustices. He was no more put off as a judge by characterizations of his decisions as excessively activist than he had been deterred by criticism of his gubernatorial programs as excessively socialistic.

There was, however, a dimension in the criticism of Warren's judicial performance that had not existed in the comments on his service in California. Nowhere in the assessments of Warren's governorship had there been a suggestion that his programs improperly extended the powers of his office. Various groups—the state medical association, oil companies, and other apostles of free enterprise—had opposed the substance of Warren's policies without suggesting that he lacked power to formulate them. "Activism" and "self-restraint" were not terms used in debates about gubernatorial policies.

On the Court, however, the principal criticism of Warren as a judge was directed at his interpretation of his office rather than at the substance of his results. Various critics, both from within the ranks of the judiciary and elsewhere, have intimated that the long-term consequences of Warren's activism may be to make the public fearful of tyrannical judges, expose the subjective dimensions of judicial decision making, provoke resistance to Court decisions, and lower the Court's public reputation. The criticism tends to lump together two discrete propositions: first, that judicial activism of any kind is a jurisprudentially indefensible and politically unwise stance; second, that even if some judicial activism is defensible, Warren's variety, being based on ethical judgments rather than on legal principles, is not. This chapter considers those two propositions.

࿐

Activism is a distinctly twentieth-century term of art. John Marshall's decisions, which are now sometimes referred to as "activist," were not given that label at the time. Activism assumes that repose or passivity is the normal state of affairs for the judiciary, an assumption that belies the nineteenth-century experience, where statutory lawmaking by legislatures was relatively uncommon and where major political disputes—slavery, competition in transportation, currency reform, the existence of an income tax—were settled in the courts. Indeed, one strand of nineteenth-century constitutional jurisprudence, which identified the Constitution as the source of fixed principles and the judiciary as their guardian, held that legislative efforts to upset established constitutional principles were impermissibly "activist." The programs of the New Deal, according to this line of reasoning, were invalid not because they were wrongheaded but because Congress had no power to intervene in private affairs. Congressional activism had overreached itself.[1]

The identification of the term "activism" with the judiciary is a product of an early twentieth-century identification of judges as lawmakers. Preceding theories of judicial review had not assumed that judges had the power to change the meaning of constitutional language. Whatever judges did—whether it was to "discover" finite legal truths or to "interpret" the technical meaning of legal words for less learned persons—they did not make law; "the law" was already there. Several judges, from Marshall through David Brewer, said this explicitly: The courts had no lawmaking powers.[2]

By 1905, in a case where a state legislature had been, according to a majority of the Court, impermissibly activist, a new theory of judicial review had appeared. For Holmes, in *Lochner v. New York*, the Constitution was "made for people of fundamentally differing views."[3] Constitutional interpretation, he argued, was not a process in which judges articulated their social and economic theories. Whether a judge liked or disliked a piece of legislation was irrelevant to his making a decision about its constitutionality. What was relevant was the legislature's purpose and the permissibility of majoritarian sentiment as there expressed. The judge was a passive expositor of the legislature's will as long as the basis of the legislation was rational. Any other stance in the judiciary entailed the substitution of its judgments for those of the legislature.

That substitution, of course, was precisely what judges had been doing, as guardians of the Constitution, in the prevailing earlier theories of judicial review. Holmes found those theories defective and their stance for the judiciary illegitimate because he thought they rested on a theory of judging that was either naive or disingenuous. Constitutional language was vague, open-ended, and susceptible of conflicting interpretations. Sometimes one interpretation prevailed, sometimes another. Sometimes judges created a constitutional doctrine out of whole cloth, as with liberty of contract, the immutable "principle" that invalidated the legislation in *Lochner*. In engaging in this exegesis, judges could not fairly be said to be merely discovering finite truths. They were acting as lawmakers; their actions therefore raised the problem of one unrepresentative and nonelected institution in American society doing what another representative and elected institution was supposed to do. Holmes knew his proper role as a judge and a democrat: He was to defer to the appropriate law-making branch of government.

In the years between *Lochner* and Warren's succession to the Court, Holmes's theory became orthodoxy. Students of jurisprudence, including judges, conceded that the judiciary functioned as lawmakers and conceded its antidemocratic character. Influential judges, such as Brandeis, Cardozo, Stone, Learned Hand, and Frankfurter, adopted various versions of judicial self-restraint. The versions were united by Holmes's premise: The natural posture of the judiciary, where a legislature has acted, is deference because legislators are the proper lawmakers. Only when legislatures act irrationally can the judiciary legitimately intervene. So influential was this line of reasoning and so widely was it accepted that one might have expected Warren, conscious of his own judicial inexperience and his fledgling status on the Court, to adopt it. Any such expectations failed to reckon with Warren's dogged independence, his highly developed views on legislative activity, and the crucible of his experience in *Brown v. Board of Education.*

Warren did not share the twin characterizations of legislative activity made by the orthodox twentieth-century theory of judicial review. Warren's Progressivism had led him to believe that legislatures were neither "democratic" nor "representative" of public opinion. While he had supported and profited from a deemphasis on partisanship in the California elective process, he did not believe that the California legislature had consequently been freed from special interests or that legislators abandoned their private concerns for power and influence, even at the expense of their constituencies. He had seen the capacity of legislators to create issues, to block humanitarian legislation on parochial or partisan grounds, to conceal their activities from their constituents, and to profit from their association with lobbyists. He remained, as few of his peers on the Warren Court did, a thoroughgoing skeptic about the representativeness or democratic character of the legislative forum.

Warren had thus been inclined, as a California executive official, to prefer his own solutions to social problems over those of legislators; his experience in *Brown* further contributed to an activist posture. That experience "reminded rather than informed" him of some features of constitutional adjudication; this reminder would prove crucially important to his subsequent posture on the Court. That a judicial doctrine was well settled, he felt, did not make it inviolate, especially if the doctrine was ethically tainted. Reasoning subscribed to by an impressive majority of the Court could lose its stature with time. Moreover, the Court could not trust legislatures to curb social injustice. Many persons in Congress and in southern states deplored the forcible separation of black and white athletic teams or blacks and whites in hospitals, but no legislative body had moved to correct the problem. If the stigma of racial segregation was to be eradicated, it was up to the Court to act.

Warren became a champion of activism after the decision in *Brown* and the short-run reaction to that decision. His investment in the unanimity of *Brown*, its lukewarm support from the Eisenhower Administration, and the resistance to the Court's decrees from Little Rock and other southern communities had convinced him of the need for the Court to serve as a vindicator of ethical imperatives, a conscience of legislatures, and a protector of the public. The summer after *Brown* was decided, Warren wrote to Frankfurter that he was still "endeavoring to orient [him]self"[4] in the area of due process interpretation; by *Cooper v. Aaron*,[5] four years later, he had settled into the role of vindicator, protector, and conscience. He had learned from *Brown* that literal constitutional language or the customs of history were not compelling in constitutional adjudication. He had also learned that deference to "democratic" branches of government might perpetuate injustices. He had resumed the familiar stance of Progressive champion of the public interest.

As a matter of history, and as a theoretical posture, there was nothing indefensible about Warren's stance. He was returning to a scrutinizing role for the courts that was of longer standing in American life than the role that Holmes helped to originate. To be sure, Warren's justification for activism was dramatically different from those made by his predecessors. While he found various duties and imperatives in the Constitution, he did not revert to the argument that judges were other than lawmakers. His activism was based on the premise that justice needed to be done and that one could not expect, given the realities of experience, that ostensibly more "democratic" institutions would do it.

The audacious feature of Warren's activism was that he did not limit his review of majoritarian decisions to those instances where the Constitution gave a minority explicit textual protection. His decisions asserted a twentieth-century version of natural law as a legitimate basis for judicial review. Warren's "natural law"—his idea of imperatives emanating from the ethical structure of the Constitution—was derived, we have seen, from progressive social theory, from the lessons he had learned in public life, and from his own code of ethics, which he believed the Constitution embodied. Each case for Warren contained its own "essence" and its own resolution according to natural law. He saw his task as discerning that resolution and persuading others to support it.

The strength of Warren's version of activism lay first in his remarkable ability to embody in his ethical judgments the ideals of other Americans and second in the tortured state of the orthodox theory of judicial review in the years of the Warren Court. Over and over again, Warren seemed to discern an injustice and to propose its eradication; repeatedly his premises seemed to rest on a compelling, ideal vision of American society. The legitimacy of separate but equal conditions for blacks and whites was an established constitutional doctrine, but in practice it fostered inequalities and was therefore unfair and immoral. Malapportionment in a legislature was a "political question," but it was a question that would only be answered in one way by the political process. The nonpolitical judiciary needed to frame its own answer because malapportionment was favoring some voters at the expense of others. The House Un-American Activities Committee was performing a valuable service in searching for Communist sympathizers in government, but some of its investigations harassed persons who had not the remotest connection with subversive activity. Congress could not be expected to check the power of one of its own; the judiciary had to do the checking.

America was a nation in which the government existed for the pleasure of its citizens; American citizenship embodied the dominant role of the citizen. It was therefore a contradiction in terms for the government to be able to deprive a citizen of citizenship status. That deprivation required the citizen's consent. Wholesomeness and decency were fundamental American values; the First Amendment was not designed to protect purveyors of perversion from subverting those values by exploiting the public's susceptibility to prurient impulses.

Law enforcement in America differed from law enforcement in less enlightened nations because of the obligation in America of those who upheld the law to obey the law. The vigorous pursuit of criminals was to be encouraged, but only in ways that vindicated the essential fairness of the law enforcement system. That system, if imperfectly administered, tended toward coercion; the very power-lessness of those who came in contact with the system justified their need for protection. Some cynical and degraded professional crimi-nals accepted the system's coercion and took their chances, but the average criminal was a person without social advantages who had blundered into illegal conduct. That type of person was the most easily coerced and the most deserving of protection. No one could fairly expect law enforcement authorities to concern themselves with protection for criminal defendants; their job was to preserve and ex-pand their coercive powers. No one could expect legislative majori-ties to espouse greater protection for criminals; natural justice had to be secured by the courts.

Many of Warren's natural-law arguments confronted a barrier erected by the orthodox twentieth-century theory of judicial review. In the segregation cases, it was the doctrinal argument of established precedent and the institutional argument of legislative autonomy. In the reapportionment cases, the barrier was a doctrine of judicial def-erence toward "political questions" that had been fashioned by twentieth-century judges. In the citizenship cases, the barrier was a positivist theory of sovereignty developed in earlier deportation cases that had both doctrinal and institutional consequences. In the congressional investigation cases, it was the institutional power of Congress to gather information pertinent to its enumerated consti-tutional responsibilities. In the criminal procedure cases, it was the institutional theory of legislative experimentation, one of the justifi-cations for judicial deference. In each of these instances Warren found that adherence to the orthodox theory of judicial review simply fos-tered injustices. He believed that, notwithstanding their commit-ment to majority rule, Americans would recoil against majoritarian excesses. If a majority was threatening its own rights in restricting the rights of others, it needed to be curbed.

In making an appeal beyond the majoritarian component of dem-ocratic theory to the "law beyond the law"—the ethical imperatives of a constitutional democracy—Warren was also exposing an unfor-tunate interaction between majoritarianism and passive theories of judicial review. In instances where majorities restrict the rights of minorities, a passive approach to judicial review serves to legitimize the restrictions. Elites in the generation that produced the Constitu-tion were concerned that legislative majorities might be too excessive in their democratic instincts and restrict the rights of advantaged persons, such as property holders; they were less concerned about majoritarian restrictions on the rights of disadvantaged persons. As a concern for the disadvantaged became a feature of twentieth-century reformist thought, the capacity of legislative majorities to restrict the rights of powerless minorities surfaced as a social problem.

Many of the principal Warren Court decisions involved this version of majoritarianism. Restrictions on powerless or disadvantaged persons were embodied, for example, in the segregation cases (blacks), the reapportionment cases (suburban voters), many of the criminal procedure cases (indigent criminals), the school prayer cases (atheists), and the subversive activity cases (Communist sympathizers). To the liberal twentieth-century mind, the toleration of these restrictions by passive theories of judicial review was far more troublesome than restrictions on the property rights of minorities.

The Warren Court's activism was thus distinguishable from other activist courts in the past in the nonelitist character of its beneficiaries. The beneficiaries of the Warren Court's scrutiny of legislative majorities were persons who had neither the prestige, the power, nor the means to vindicate their claims in a majoritarian forum. In reaching out to protect such persons, Warren and other activists on his Court put liberal advocates of a passive role for judges in an awkward position. Their theories originally had the desirable side effect, from a liberal perspective, of legitimizing legislative efforts to help disadvantaged persons. Now those theories were functioning to justify legislative indifference to the disadvantaged.

The paradox of Warren's activism was its use of elite power to enhance the power of nonelites. In cases where he found the Constitution to be a barrier against majoritarian restrictions, Warren declared that the judgments of a nonelected group of nine justices were to be given greater credence, in a democracy, than the policy judgments of elected majorities. Warren would not have put it so starkly: He regarded the restrictions on majoritarianism imposed by his Court as sanctioned by the Constitution rather than by the social judgments of his colleagues. But in none of the leading activist Warren Court decisions was a constitutional mandate obvious. *Brown v. Board of Education* reversed a long-established precedent on the basis of social science evidence. *Baker v. Carr* found a judicial power to review the political judgments of legislators where none had previously existed. *Miranda v. Arizona* instituted a novel role for the Court as a promulgator of police regulations. *Harper v. Virginia Board of Elections*, which declared poll taxes unconstitutional, was based on modern ideas about the illegitimacy of conditioning voting power on solvency that ran directly counter to the ideas of the framers. In some of its decisions, the Warren Court seemed to be distinguishing between the corrosive elitism of legislatures and its own benevolent elitism, and arguing that the Constitution encouraged only the latter.

✒

A long line of respected twentieth-century judges has taken the position that the proper decision-making calculus for an appellate judge should not include unarticulated notions of fairness and justice. From Holmes's alleged rejoinder to Learned Hand's exhortation to "do justice" ("that is not my job"), through Frankfurter's repeated efforts to rest his decisions on grounds other than "my notions of justice,"[6] to Harlan's remarks on Black's retirement ("[h]e rejects the open-ended notion that the Court sits to do good in every circumstance where good is needed"),[7] effective judging has been associated with qualities ("reason," "craftsmanship," "principled adjudication") that are regarded as counterweights to natural justice rather than embodiments of it.

Warren, we have seen, rejected this interpretation of the judicial function. The content of Warren's "natural law" was a set of ethical principles that Warren believed formed the foundation for constitutional adjudication. Principles of fairness, decency, individuality, and dignity were, for Warren, constitutional imperatives that the Court was under an obligation to consider in its decision making. The "ethical norms" to which Warren appealed yielded a set of constitutional principles that Warren sought to apply consistently. Those principles, however, by their very nature, were incapable of precise formulation. Moreover, consistent application of the principles was rendered difficult for Warren by his belief that they needed to be weighed in situations where they conflicted.

Warren acknowledged the above constraints on judging, but he rejected the constraints of doctrinal consistency and institutional deference, often encapsulated in the phrase "principled adjudication," when he felt that those constraints served to divert attention from the ethical imperatives of a case. He believed, he said, in "open-ended" judging; he thought of himself as having been liberated by becoming Chief Justice. His political life had required "half-loaves" and compromises, but his judicial decisions were based on "principle," and ethical truths need not be diluted.[8] His instinctive reactions to issues had often been suppressed in political life; they came to the fore on the Court. It was as if the absence of an elective constituency freed him to vindicate moral principles without concern for the consequences.

Warren thus equated judicial lawmaking with neither the dictates of reason, as embodied in established precedent or doctrine, nor the demands imposed by an institutional theory of the judge's role, nor the alleged "command" of the constitutional text, but rather with his own reconstruction of the ethical structure of the Constitution. No other influential judge on the Warren Court, or for that matter in the twentieth century, adopted such a view. Several leading judges, such as Holmes, Learned Hand, Frankfurter, and Harlan, were institutionalists; Black was a textual determinist; Benjamin Cardozo and Roger Traynor were creative disciples of reason.[9] Warren stood alone in being an ethicist. The only other Supreme Court justice who approached Warren's jurisprudential posture was Frank Murphy, who never achieved Warren's stature as a force on the Court or as a public figure.[10] Indeed, the same posture that invoked ridicule in Murphy was the source of Warren's strength as a judge.

Warren's definition of judging as an exercise in ethical choices raises two questions, each of which has received preliminary attention, and each of which is here sought to be answered. The first involves the juristic consequences of his stance: Did his approach produce a consistent and meaningful jurisprudence? The second concerns the worth of Warren's approach: Is it essential that it be held by a person of genuinely humane instincts? Anthony Lewis once said of Warren that he "was the closest thing the United States has had to a Platonic Guardian, dispensing law from a throne without any sensed limits of power except what was seen as the good of society. Fortunately he was a decent, humane, honorable, democratic Guardian."[11] Is Lewis right in suggesting that the posture of an ethicist is fatally dependent on the ethicist's own character?

Ethics as the foundation of a political stance may be striking, but ethics as the foundation of a jurisprudential perspective is even more so. A politician insistent upon holding to his ethical principles might be perceived as foolish or rigid, but one could not deny the permissibility of his actions. One might even wish that ethical standards were more widely followed in politics. In judging, however, a perspective based on ethics raises the question of the relationship between ethics and law.

Although Warren asserted that legal principles had as their foundation ethical standards, other American legal theorists denied this. In the early nineteenth century, law was regularly equated with "fundamental principles of natural justice" or theological beliefs, but that equation was sharply questioned by late nineteenth-century theorists. Holmes and other post-Civil War scholars advocated a sharp separation between law and morals, claiming that the first step in a true appreciation of the way law functioned was a stripping of theological or moral postulates from its content. Law existed for purposes—social control, the facilitation of private agreements, the implementation of majoritarian prejudices—that could not be called ethical.

While some twentieth-century theorists sought to reemphasize the congruence of law and morality,[12] few regarded ethics as being a sound basis for decision making. The principal fear of twentieth-century jurists was unchecked power in the judiciary: In a world where judges made law, some means of confining judicial bias had to be found. Ethics was an unjustifiable basis for a judicial perspective because one's ethical judgments were individualized. More "neutral" or general principles had to be resorted to: the principle of fidelity to the constitutional text, the principle of deference to the legislature in close cases, the principle of adherence to precedent, the principle of supplying reasoned professional justifications for results. Ethics was too vague, too idiosyncratic, too capable of being plucked out of thin air, too difficult to reduce to general propositions. A jurisprudence based on ethics would amount to an escalation of a judge's personal preferences to the stature of a theoretical perspective.

On the Court, Warren sought to justify his stance by reversing the relative emphases placed on reasoning and results by twentieth-century academic literature. For a line of influential critics in the twentieth century, the significance of a result in a given case paled in comparison to the reasoning used to justify it. The difficulty with the doctrine of substantive due process, for example, was less that it reached "bad" results than that it was flawed as a technique of reasoning. Broad constitutional language like the Due Process and Equal Protection Clauses had no finite content, and judges had no power to fill them with their subjective preferences. The noblest result could become flawed if its reasoning was illegitimate: The Court's decision in *Brown v. Board of Education* was an example. No one wanted to defend "separate but equal" segregation on moral grounds, but nothing in the Fourteenth Amendment's Equal Protection Clause prevented the practice; to invalidate it on the basis of dubious empirical research was the essence of unbridled judicial glossing of the Constitution. Because the reasoning in *Brown* was flawed, so was the result.[13]

Warren directed his energies in a direction opposite from that urged by twentieth-century academic theorists. He sought to determine the ethically required resolution of a given case and to convince others, using ethical arguments, as to the soundness of his determination. He spent much less energy developing conventional legal reasons to justify the result. His approach to judging called for a close examination of the facts of a case, the actors in it, the record below, and the social issues the case raised. It required comparatively little attention to precedent, the exact language of the Constitution, or academic theories of judicial review. In Warren's jurisprudence the results dictated by the ethical structure of the Constitution were more than simply outcomes—they were good outcomes. Such results carried their own ethical justification. If that justification was not sometimes made explicit, or was couched in vague language, that did not mean it was not important.

In *Marchetti v. United States*,[14] for example, Warren did not openly state the basis for his decision that gamblers filing income tax returns should not receive the protection accorded other taxpayers by the Fifth Amendment's Self-incrimination Clause. The basis of Warren's decision was that income from gambling was ethically tainted; that was the position he argued to his colleagues. If his position had prevailed, Warren would probably not have emphasized it in an opinion.[15]

The reason Warren would have not articulated too openly the basis of his vote in *Marchetti* is that there is no evidence that the Fifth Amendment's Self-incrimination Clause was designed to protect only innocent persons, only some guilty persons, or the disclosure of only some "incriminating" activities. Warren believed, in the abstract, that the Bill of Rights extended to all citizens, but he reserved the right, as a judge, not to extend Bill of Rights protections to some "undeserving" citizens in specific instances. The basis of this reserved power was his belief that the Constitution should conform

to certain standards of ethics. In *Marchetti* he seems to have concluded that because he found gambling to be immoral as well as
illegal, it was unethical for gamblers to trade on the constitutional
rights of more deserving citizens, and therefore that the Constitution
should not countenance this attempt by degraded professional criminals to hide behind its protections.

It seems that in instances such as *Marchetti*, Warren's position
rests on intellectually shaky ground. At a point, Warren's concern
with ethical imperatives comes so close to what appears to be an
apology for subjective preferences in judging as to be unsupportable.
Judges, after all, are supposed to reveal the basis of their decisions,
not conceal them.

Orthodox judicial reasoning, as defined by the juristic conventions of his time, would have compelled Warren to articulate fully
the broad and vague principles on which his decisions were grounded.
Warren's ethical imperatives were undeniably vague: Committing
oneself to a principle of decency did not enable one to resolve immediately all issues that could be said to bear a connection to that
principle. Warren knew well that to propose a vague principle as the
basis of a decision—even when that principle was the foundation of
myriad explicit constitutional doctrines—was to invite censure and
even ridicule from numerous members of the academic and judicial
communities. Sometimes he accepted that censure—where too much
that he cared about was at stake to avoid decision. Other times he
preferred not to reveal the bases of his judgment, and on still other
occasions he used orthodox judicial reasoning in a cavalier fashion.

But for Warren the ethical imperatives of the Constitution were
so important, and so obvious, that it did not matter on many occasions that orthodox justifications for a result were problematic or
virtually lacking. Warren believed that he was dealing with important issues—issues that affected and to some extent even defined the
essence of American democracy. He was not about to be bound to a
mode of judging whereby doctrinal consistency was held to be more
important than vindication of the basic ethical imperatives of the
Constitution.

Some justices enter debate with open minds and are genuinely
moved by the arguments of their colleagues; others approach a discussion with fixed views to which they adhere and explore issues
from a preordained framework. For example, Hugo Black would deny
that because his theory of constitutional interpretation was not the

product of reasoned debate, it was therefore "unprincipled"; nor would he adopt a compromise solution to a controversy if he found it did not square with his interpretation. Thus judicial reasoning for Black, at least in the area of constitutional interpretation, was more of an isolated process in which he considered the text of the Constitution and certain social values rather than a canvass of competing views.

Warren's mode of reasoning on the Court was closer to that of Black than to the ideal articulated by such justices as Harlan and Frankfurter. Warren approached issues within a particular framework—the framework of values that he believed to be the genuine foundation of the Constitution. But in a sense Warren's reasoning resembled that of more orthodox jurists. His reasoning focused on the relationship between salient facts in a case and constitutional principles, as did theirs, and he reasoned from his sense of the facts to a broad justification for a result, as did his orthodox counterparts. The premises on which Warren's justifications rested were ethical imperatives, not "neutral principles" of constitutional doctrine, and in this sense he was unconventional for his time and in the sweep of American judicial history. His lack of conventionality, however, centered more on his view of the meaning of the Constitution and the relationship of ethics to law than on the process of his reasoning.

Warren reasoned about the existence and application of ethical principles first; traditional doctrinal reasoning came later. His modifications of judgment, in most cases, were modifications with respect to implementation, to strategy, or to language, not modifications of "principle." He would tone down language or reduce its doctrinal significance to pick up a vote; he would adopt one or another legal argument in response to others' sense of its analytical stature.

In *Bolling v. Sharpe*, we have noted, his instincts told him that segregation cases in the public schools really involved denials of the fundamental right to an education. When others suggested that, at least in terms of currently fashionable constitutional doctrine, there was no such "fundamental" right, Warren turned *Bolling v. Sharpe* into an unconventional equal protection case. The important thing was to reach a result outlawing segregation in the District of Columbia because its persistence would be "unthinkable" after *Brown*. How that result was accomplished was much less significant.

In an area as open-ended and susceptible to change as constitutional interpretation, who can say that Warren's approach, which settled on the ethically dictated result first and derived traditional justifications later, is less sound than an approach conditioning results on the emergence of "principled" reasoning? One need not assume that judicial decisions based on ethical imperatives only serve to vindicate the personal moral judgments of individual judges. The premises on which ethically dictated results are based count for something, because some premises are regarded, by contemporary wisdom, as better than others. Reasoning in constitutional interpretation may be fluid, but it does not fairly include arguments whose premises made no sense, are outmoded, or are currently regarded as wrongheaded. The most eloquent statement of a white supremacist interpretation of the equal protection clauses would not be likely to gain much current acceptance. The logic might be impeccable, but the premise would be considered flawed.

If the content of premises makes a difference, then result-orientation is not merely a process of deciding what one wants as a judge. It is finding an acceptable basis for doing what one thinks one ought to do. One may deny that reason is an impersonal construct and still concede that some reasons are better entertained than others. A judge (e.g., Hugo Black) may eschew appeals to disembodied rationality ("so much natural law/fundamental-fairness talk") and adhere rigidly to his point of view ("The Constitution says Congress shall make no law restricting free speech and 'no law' means no law"), but he is unlikely to reveal his premises in an unacceptable way ("I don't care what the Constitution says, I don't like sit-in demonstrators and won't protect them"). Indeed, he is very likely to cloak his premises in language that sounds as if he is invoking disembodied reason. ("The Constitution protects speech, not conduct, and the act of 'sitting-in' at a lunch counter is not speech.") One cannot avoid searching, even if one believes in a result-oriented approach to judging, for a pattern of reasoning that will legitimize one's results.

If, on analysis, a given judge's pattern of reasoning tends to reduce itself to judicial protection for that which he believes is deserving of protection, the judge faces what might be called an ethical difficulty. He has not been appointed or elected to office with an expectation that he will simply use his power to do what he thinks is fair and just. Such behavior may be common in judges, but it is not advertised in advance. On the contrary, a judge is expected to "follow the law," "uphold the Constitution," and otherwise subordinate his personal preferences to some more legitimized body of wisdom. The fact that judges contribute to the content of law, or even make glosses on the Constitution, does not mean that they are expected to become independent of the corpus of wisdom to which they are contributing. When they prefer their independent judgments to that corpus, one could say that they are betraying a trust.

Thus if judicial reasoning is to function as the servant of ethical premises, the reasoning must itself establish the acceptability of the premises. Warren's performance as a judge seems vulnerable on this issue. Because his reasoning was often technically imperfect, opaque, or assertive, it suggested that he was primarily interested in results. When one probes the unexpressed bases for his results, values in which he believed—fairness, decency, anticorporatism, individuality, dignity—emerge as a hidden rationale. Many of these values are given great respect by American society: Fairness, decency, humanity, and integrity are ideals to which most Americans would aspire. It sometimes makes one feel good to discover that the unexpressed reason for a Warren decision was consideration for the plight of a disadvantaged person, an insistence on securing fair treatment by a person accused of a crime, or a conviction that American society would be better off if the government did not spy on its citizens.

The difficulty with subjecting Warren's opinions to this analysis is that sometimes his ethical judgments were not inspirational. It is hard to see why, if the law should be the same for the rich and the poor, it should not also be the same for gamblers and other taxpayers, or for pornographers and other dissident publishers. More important, it is hard to know what justification exists for Warren's picking and choosing among beneficiaries of First or Fifth Amendment protection. If one applauds Warren's choice and thinks most Americans would as well, one might say that he could have articulated his premises more clearly; but if one deplores Warren's choice, one might question what empowers him to pick and choose.

The principal difficulty, then, with basing one's approach to judging on ethical principles is not that such an approach is an unrealistic way of going about making decisions as a judge. Indeed, it may more closely approximate the reality of judging than competing approaches. The difficulty is that an approach based on ethics seems inevitably to lead one to justifications of one's ethical stance; the success of those justifications tends to turn on whether one's ethical stance in a given case is perceived as "good" or "bad." The successful judge will be that judge whose ethics seem "right" most of the time.

The stature of a judge who purports to employ ethical imperatives is peculiarly dependent on the ability of those imperatives to embody "correct" moral thinking, which in the end is measured by conventional standards.

Greatness in a judge who adopts Warren's approach thus seems inextricably linked to public acceptance of the rightness of that judge's ethical stance. It is as if the public will forgive unconventional, unorthodox, or even opaque judicial reasoning if the judge reaches the right result, but he had better get the result right. Since the "rightness" of a result tends to turn on intuitive perceptions of morality and justice rather than on "technical or recondite learning,"[16] those judges perceived by the public as having a considerable capacity for reaching "right" results tend to be those who are able to communicate their ethical convictions impressively. Warren was such a judge. While unpopular and controversial among segments of the public, he was generally regarded as a Chief Justice whose decisions sought to do good, achieve justice, and make America a more decent and honorable society.

This perception might have been enough for Warren, and perhaps for most judges. But what of the judge who seeks to adopt Warren's approach and whose ethical judgments turn out to be at odds with the general public? What if a judge currently decided, using constitutional language as the servant of his ethical premises, that a court could, as a means of reducing and equalizing health costs, permit hospitals to decline to treat terminally ill persons in order to treat persons who had a better chance of recovery? Conceivably the text of the Constitution could be interpreted in a way consistent with that result: States have broad public health powers. The problem comes in securing consistency, on such an issue, between the ethical premises of the interpreter and those of the public. Warren's approach, in exchange for a broad definition of "analytical" competence in opinion writing, creates a heavy burden of ethical concordance.

Thus when one compares Warren's approach with that of one, such as that of Hugo Black, which reached comparable results in countless cases, one finds a vulnerability in Warren's perspective that Black's avoids. Black may choose to read the Constitution "literally" in some areas, "liberally" in others, and pretend that no contradiction exists in his choices. He may claim to be "bound" as a judge by imperatives such as the constitutional text, but decline to be equally bound by other imperatives, such as the precedents of his own court. He may focus on a literal distinction (speech versus conduct) that sometimes has no functional significance.

But when Black is done, there is more than an aggregate of his ethical premises—there is a doctrinally consistent, if perhaps analytically flawed, body of constitutional jurisprudence. In contrast, when one divorces Warren's opinions from their ethical premises, they evaporate. No overreaching doctrinal unity binds them; they are individual examples of beliefs leading to judgments. One may applaud the results, embrace the premises, and admire the instincts of the man, but one can never divorce Earl Warren's opinions from Earl Warren and treat them as anonymous contributions to constitutional literature.

Perhaps that is what Warren would have preferred. He never saw himself as the disembodied expositor of the Constitution, bound by the finite meaning of its language. He recognized that within every legal issue is an ethical judgment, and thus in that sense the separation of law and morals is fruitless. He was not deterred from asking basic questions about humanity and fairness by the purported authoritativeness of established legal sources. In many respects his opinions teach us to hold out for our instinctive beliefs in the whirling confusion of academic learning.

Warren kept his eye on the essentials of a case—the essential facts, the essential justice of the situation—and in so doing reminded us that when law gets encumbered with professional cant, it ceases to approximate reality and loses its reason for being. He tried to remember, in a profession where for every argument there is a counterargument, that a retreat to ethical principles is often a way out of the labyrinth. He raised ethics to a high judicial art, but he could not escape the burdens of he who chooses to proclaim for others what is fair and right and just. These burdens did not trouble Warren. He was an unusual judge.

ϫ

Earl Warren was one of the major figures in twentieth-century American history. He was one of the first beneficiaries of a political culture that deemphasized partisanship and machine organization and emphasized intangible personal qualities and effective use of the media. He dramatically altered the legal relationships between blacks

and whites in America and in so doing encountered and contributed
to a profound shift in the way Americans thought about race and
skin color as determinants of human worth. He and his liberal majorities on the Warren Court created a new role for the federal judiciary, that of champion of the rights of disadvantaged persons. In so
doing, he helped create a new set of public expectations about the
function of courts and contributed to the increasingly litigious character of contemporary American society.

Warren also personified many conventional mid-twentieth-
century American attitudes, and yet he showed a capacity to discard
conventional thinking as his career and the conditions of life in
America changed. He was, in the course of his career, a crusading
law enforcement official and the author of the *Miranda* warnings; a
zealous anti-Communist and an outspoken judicial foe of legislative
investigations of suspected Communists; a proponent of a state loyalty oath and a defender of academic freedom; a nativist with racial
prejudices against Orientals and the author of *Brown v. Board*. Warren was a patriot and an advocate of preparedness in World War I,
World War II, and Korea, but an opponent of the Vietnamese War.
He was an outspoken critic of the New Deal and a supporter of the
Great Society. He began his career in public life having close ties to
the Native Sons of the Golden West and ended it a devotee of world
peace through law. His first exposure to national politics was as a
supporter of Herbert Hoover; on his retirement he was called the
greatest Chief Justice of all time by Lyndon Johnson.

The history of Warren's public life becomes, in this vein, a history of a strand of twentieth-century American thought. Warren entered public service with a set of unexamined and unarticulated assumptions, based primarily on his instinctive Progressivism. Some
of these assumptions he held throughout his life, submerging them
for a time in his "conservative" law enforcement years, bringing them
gradually to the surface thereafter. Others, such as his nativism or
his distrust of federal power, he discarded. Many other Americans
likewise revised their attitudes on race, ethnicity, and federal-state
relations after the Second World War. Warren's "discovery" of the
significance of civil liberties in the late 1940s and fifties was a discovery many other Americans made. Few, however, had been supporters of Hiram Johnson, district attorneys, Red-baiters, and architects
of the Japanese relocation program before making that discovery. In
some respects Warren's changing stance on public issues was a tes-

tament to the ideological adjustments required of those who believed that twentieth-century American society should be marked by continuous progress.

Warren's contribution, however, was not as an ideologue. He was not a consummate Progressive and not a consummate liberal. He did not act from the perspective of a considered system of thought, but from his instincts for what was fair, honorable, politically feasible and sensible at the time. Like many public figures, he embodied attitudes rather than contributing to their intellectual development. Warren's greatest strengths and his most memorable qualities were intangibles: presence, timing, capacity for growth, persuasiveness, inner conviction, decency, persistence, reasonability. In possessing those qualities he functioned as a symbol for a large inarticulate body of the American public; he pursued Everyman's instinctive ideal of fairness and justice. If he was not a sophisticated or wholly consistent thinker, he was nonetheless a great man, not only for what he embodied but also for what he accomplished. In a public world of corruptible and self-serving actors, he set a standard of incorruptibility and humanity; in a society fraught with injustices, he sought to use the power of his offices to promote decency and justice. The end of his public career may be the end of a phase of American life.

# Appendix
## Opinions of Chief Justice Warren

As any regular reader of the United States Reports knows, Supreme Court opinions cannot easily be reduced to simple doctrinal categories. The classification scheme used in this Appendix has some necessarily arbitrary features, especially in those cases where an opinion embraced more than one doctrinal area (say the Fifth Amendment and the Due Process Clauses), and, by an instinctive judgment, was grouped in one or another doctrinal category.

Inveterate categorizers might be interested in the fact that appendix categories are not only arbitrary but change with time. If this list of opinions is compared with a similar list for Justice Holmes,* whose tenure on the Supreme Court began more than fifty years before Warren's, doctrinal categories that existed on the Holmes list will be seen to have disappeared and new ones to have taken their place. This is not primarily because the Constitution has changed, although there were several amendments between 1932, when Holmes left the Court, and 1969, when Warren retired. The changes in categories are primarily a result of judicial emphasis. A comparison of appendices for Holmes and Warren is a way of discerning the jurisprudential flavor of the Supreme Court at two different points in its history.

* See 44 Harv. L. Rev. 820 (1931). That list was prepared by Henry Hart and Felix Frankfurter.

## ADMINISTRATIVE LAW

United States v. An Article of Drug, 394 U.S. 784 (Apr. 28, 1969)

Thorpe v. Housing Authority, 393 U.S. 268 (Jan. 13, 1969)

King v. Smith, 392 U.S. 309 (June 17, 1968)

Federal Communications Commission v. Schreiber, 381 U.S. 279 (May 24, 1965)

Federal Trade Commission v. Colgate-Palmolive Co., 380 U.S. 374 (Apr. 5, 1965)

Udall v. Tallman, 380 U.S. 1 (Mar. 1, 1965)

Sperry v. Florida, 373 U.S. 379 (May 27, 1963)

New Jersey v. New York, Susquehanna & Western Railroad, 372 U.S. 1 (Feb. 18, 1963)

Gilbertsville Trucking Co. v. United States, 371 U.S. 115 (Dec. 3, 1962)

Civil Aeronautics Board v. Delta Air Lines, Inc., 367 U.S. 316 (June 12, 1961)

Brotherhood of Maintenance of Way Employees v. United States, 366 U.S. 169 (May 1, 1961)

Federal Power Commission v. Transcontinental Gas Pipe Line Corp., 365 U.S. 1 (Jan. 23, 1961)

American Trucking Associations v. United States, 364 U.S. 1 (June 27, 1960)

American Trucking Associations v. Frisco Transportation Co., 358 U.S. 133 (Dec. 15, 1958)

Schaffer Transportation Co. v. United States, 355 U.S. 83 (Dec. 9, 1957)

Peters v. Hobby, 349 U.S. 331 (June 6, 1955)

Federal Communications Commission v. American Broadcasting Co., 347 U.S. 284 (Apr. 5, 1954)


## ADMIRALTY

United States v. Isthmian Steamship Co., 359 U.S. 314 (Apr. 27, 1959)


## ANTITRUST AND TRADE REGULATION

Utah Public Service Commission v. El Paso Natural Gas Co., 395 U.S. 464 (June 16, 1969)

Federal Trade Commission v. Fred Meyer, Inc., 390 U.S. 341 (Mar. 18, 1968)

Kaplan v. Lehman Brothers, 389 U.S. 954 (Nov. 13, 1967) (dissenting opinion)

Federal Trade Commission v. Universal-Rundle Corp., 387 U.S. 244 (May 29, 1967)

Carnation Co. v. Pacific Westbound Conference, 383 U.S. 213 (Feb. 28, 1966)

United States v. Wise, 370 U.S. 405 (June 25, 1962)

Brown Shoe Co. v. United States, 370 U.S. 294 (June 25, 1962)

Federal Trade Commission v. Anheuser-Busch, Inc., 363 U.S. 536 (June 20, 1960)

United States v. Radio Corp. of America, 358 U.S. 334 (Feb. 24, 1959)

United States v. E. I. du Pont de Nemours & Co., 351 U.S. 377 (June 11, 1956) (dissenting opinion)

United States v. McKesson & Robbins, Inc., 351 U.S. 305 (June 11, 1956)

United States v. International Boxing Club, 348 U.S. 236 (Jan. 31, 1955)

United States v. Shubert, 348 U.S. 222 (Jan. 31, 1955)


BANKRUPTCY


Reading Co. v. Brown, 391 U.S. 471 (June 3, 1968) (dissenting opinion)

Bruning v. United States, 376 U.S. 358 (Mar. 23, 1964)

Kesler v. Department of Public Safety, 369 U.S. 153 (Mar. 26, 1962) (dissenting opinion)


CIVIL PROCEDURE


Will v. United States, 389 U.S. 90 (Nov. 13, 1967)

Walker v. City of Birmingham, 388 U.S. 307 (June 12, 1967) (dissenting opinion)

Amell v. United States, 384 U.S. 158 (May 16, 1966)

Hanna v. Plumer, 380 U.S. 460 (Apr. 26, 1965)

Communist Party of the United States v. Subversive Activities Control Board, 367 U.S. 1 (June 5, 1961) (dissenting opinion)

Wolfe v. North Carolina, 364 U.S. 177 (June 27, 1960) (dissenting opinion)

Dick v. New York Life Insurance Co., 359 U.S. 437 (May 18, 1959)

Hanson v. Denckla, 357 U.S. 235 (June 23, 1958)

Lawlor v. National Screen Service Corp., 349 U.S. 322 (June 6, 1955)
Lumbermen's Mutual Casualty Co. v. Elbert, 348 U.S. 48 (Dec. 6, 1954)
Partmar Corp. v. Paramount Pictures Theatres Corp., 347 U.S. 89 (Feb. 8, 1954) (dissenting opinion)

## COMMERCE CLAUSE

Shapiro v. Thompson, 394 U.S. 618 (Apr. 21, 1969) (dissenting opinion)
Halliburton Oil Well Cementing Co. v. Reilly, 373 U.S. 64 (May 13, 1963)

## CONGRESSIONAL POWERS

Powell v. McCormack, 395 U.S. 486 (June 16, 1969)
Yellin v. United States, 374 U.S. 109 (June 17, 1963)
Watkins v. United States, 354 U.S. 178 (June 17, 1957)
Bart v. United States, 349 U.S. 219 (May 23, 1955)
Emspak v. United States, 349 U.S. 190 (May 23, 1955)
Quinn v. United States, 349 U.S. 155 (May 23, 1955)

## COPYRIGHT AND TRADEMARK

Fleischmann Distilling Corp. v. Maier Brewing Co., 386 U.S. 714 (May 8, 1967)
Public Affairs Associates v. Rickover, 369 U.S. 111 (Mar. 5, 1962) (dissenting opinion)

## CRIMINAL LAW AND PROCEDURE

*Criminal Law (General)*

United States v. Nardello, 393 U.S. 286 (Jan. 13, 1969)
United States v. Johnson, 383 U.S. 169 (Feb. 24, 1966) (dissenting opinion)
Shuttlesworth v. City of Birmingham, 373 U.S. 262 (May 20, 1963)
United States v. Dege, 364 U.S. 51 (June 27, 1960) (dissenting opinion)
Gore v. United States, 357 U.S. 386 (June 30, 1958) (dissenting opinion)

Masciale v. United States, 356 U.S. 386 (May 19, 1958)
Sherman v. United States, 356 U.S. 369 (May 19, 1958)
Rathbun v. United States, 355 U.S. 109 (Dec. 9, 1957)
Benanti v. United States, 355 U.S. 96 (Dec. 9, 1957)
Prince v. United States, 352 U.S. 322 (Feb. 25, 1957)
Kinsella v. Krueger, 351 U.S. 470 (June 11, 1956) (dissenting opinion)


*Criminal Procedure (General)*

McCarthy v. United States, 394 U.S. 459 (Apr. 2, 1969)
Birnbaum v. United States, 394 U.S. 911 (Mar. 24, 1969) (dissenting opinion)
Kelly v. United States, 393 U.S. 963 (Nov. 25, 1968) (dissenting opinion)
Hoffa v. United States, 385 U.S. 293 (Dec. 12, 1966) (dissenting opinion)
Martin v. Texas, 382 U.S. 928 (Nov. 22, 1965) (dissenting opinion)
Fallen v. United States, 378 U.S. 139 (June 22, 1964)
Coppedge v. United States, 369 U.S. 438 (Apr. 30, 1962)
Piemonte v. United States, 367 U.S. 556 (June 17, 1961) (dissenting opinion)
Wyatt v. United States, 362 U.S. 525 (May 16, 1960) (dissenting opinion)
Smith v. United States, 360 U.S. 1 (June 8, 1959)
Brown v. United States, 359 U.S. 41 (Mar. 9, 1959) (dissenting opinion)
Carroll v. United States, 354 U.S. 394 (June 24, 1957)
Pollard v. United States, 352 U.S. 354 (Feb. 25, 1957) (dissenting opinion)
Mesarosh v. United States, 352 U.S. 1 (Nov. 5, 1956)
Parr v. United States, 351 U.S. 513 (June 11, 1956) (dissenting opinion)
Pereira v. United States, 347 U.S. 1 (Feb. 1, 1954)


*Due Process (Criminal)*

Bradford v. Michigan, 394 U.S. 1022 (May 5, 1969)
Burgett v. Texas, 389 U.S. 109 (Nov. 13, 1967) (concurring opinion)
Spencer v. Texas, 385 U.S. 554 (Jan. 26, 1967) (dissenting opinion)
Schmerber v. California, 384 U.S. 757 (June 20, 1966) (dissenting opinion)
Davis v. North Carolina, 384 U.S. 737 (June 20, 1966)
Wright v. Georgia, 372 U.S. 284 (May 20, 1963)
Hutcheson v. United States, 369 U.S. 599 (May 14, 1962) (dissenting opinion)
Garner v. Louisiana, 368 U.S. 157 (Dec. 11, 1961)
Culombe v. Connecticut, 367 U.S. 568 (June 19, 1961) (concurring opinion)
Blackburn v. Alabama, 361 U.S. 199 (Jan. 11, 1960)

Spano v. New York, 360 U.S. 264 (June 15, 1959)
Hoag v. United States, 356 U.S. 464 (May 19, 1958) (dissenting opinion)
Sweezy v. New Hampshire, 354 U.S. 234 (June 17, 1957)
Breithaupt v. Abram, 352 U.S. 432 (Feb. 25, 1957) (dissenting opinion)
Fikes v. Alabama, 352 U.S. 191 (Jan. 14, 1957)
Regan v. New York, 349 U.S. 58 (Apr. 25, 1955) (concurring opinion)
Chandler v. Fretag, 348 U.S. 3 (Nov. 8, 1954)


*Fifth Amendment Rights*

Jenkins v. Delaware, 395 U.S. 213 (June 2, 1969)
Haynes v. United States, 390 U.S. 85 (Jan. 29, 1968) (dissenting opinion)
Grosso v. United States, 390 U.S. 62 (Jan. 29, 1968) (dissenting opinion)
Marchetti v. United States, 390 U.S. 39 (Jan. 29, 1968) (dissenting opinion)
Johnson v. New Jersey, 384 U.S. 719 (June 20, 1966)
Miranda v. Arizona, 384 U.S. 436 (June 13, 1966)
Petite v. United States, 361 U.S. 529 (Feb. 23, 1960) (concurring opinion)
Mills v. Louisiana, 360 U.S. 230 (June 8, 1959) (dissenting opinion)
Knapp v. Schweitzer, 357 U.S. 271 (June 30, 1958) (dissenting opinion)


*Fourth Amendment Rights*

Wainwright v. City of New Orleans, 392 U.S. 598 (June 17, 1968) (dissenting opinion)
Sibron v. New York, 392 U.S. 40 (June 10, 1968)
Terry v. Ohio, 392 U.S. 1 (June 10, 1968)
Lewis v. United States, 385 U.S. 206 (Dec. 12, 1966)
Fahy v. Connecticut, 375 U.S. 85 (Dec. 2, 1963)
Lopez v. United States, 373 U.S. 427 (May 27, 1963) (concurring opinion)
Lanza v. New York, 370 U.S. 319 (June 4, 1962) (concurring opinion)


*Habeas Corpus*

Peyton v. Rowe, 391 U.S. 54 (May 20, 1968)
Townsend v. Sain, 372 U.S. 293 (Mar. 18, 1963)
Carbo v. United States, 364 U.S. 611 (Jan. 9, 1961) (dissenting opinion)
Parker v. Ellis, 362 U.S. 574 (May 16, 1960) (dissenting opinion)


*Sixth Amendment Rights*

Frank v. United States, 395 U.S. 147 (May 19, 1969) (dissenting opinion)
Washington v. Texas, 388 U.S. 14 (June 12, 1967)

Klopper v. North Carolina, 386 U.S. 213 (Mar. 13, 1967)
Estes v. Texas, 381 U.S. 532 (June 7, 1965) (concurring opinion)
Singer v. United States, 380 U.S. 24 (Mar. 1, 1965)


## DUE PROCESS (NONCRIMINAL)


Zemel v. Rusk, 381 U.S. 1 (May 3, 1965)
American Oil Co. v. Neill, 380 U.S. 451 (Apr. 26, 1965)
Gonzales v. United States, 364 U.S. 59 (June 27, 1960) (dissenting opinion)
Hannah v. Larche, 363 U.S. 420 (June 20, 1960)
Greene v. McElroy, 360 U.S. 474 (June 29, 1959)
Lerner v. Casey, 357 U.S. 468 (June 30, 1958) (dissenting opinion)
Beilan v. Board of Public Education, 357 U.S. 399 (June 30, 1958) (dissenting opinion)
Nelson v. City of New York, 352 U.S. 103 (Dec. 10, 1956)
Covey v. Town of Somers, 351 U.S. 141 (May 7, 1956)


## EQUAL PROTECTION CLAUSE


Williams v. Rhodes, 393 U.S. 23 (Oct. 15, 1968) (dissenting opinion)
Loving v. Virginia, 388 U.S. 1 (June 12, 1967)
Baxstrom v. Herold, 383 U.S. 107 (Feb. 22, 1966)
Lucas v. Forty-Fourth General Assembly, 377 U.S. 713 (June 15, 1964)
Roman v. Sincock, 377 U.S. 695 (June 15, 1964)
Davis v. Mann, 377 U.S. 678 (June 15, 1964)
Maryland Committee for Fair Representation v. Tawes, 377 U.S. 656 (June 15, 1964)
WMCA, Inc. v. Lomenzo, 377 U.S. 633 (June 15, 1964)
Reynolds v. Sims, 377 U.S. 533 (June 15, 1964)
Burns v. Ohio, 360 U.S. 252 (June 15, 1959)
Hernandez v. Texas, 347 U.S. 475 (May 3, 1954)


## FEDERAL CONTRACTS

Moseley v. Electronic & Missile Facilities, Inc., 374 U.S. 167 (June 17, 1963) (concurring opinion)

United States v. Mississippi Valley Generating Co., 364 U.S. 520 (Jan. 9, 1961)

## FEDERAL-STATE RELATIONS

Free v. Bland, 369 U.S. 663 (May 21, 1962)
Federal Land Bank v. Board of County Commissioners, 368 U.S. 146 (Dec. 11, 1961)
United States v. John Hancock Mutual Life Insurance Co., 364 U.S. 301 (Nov. 7, 1960)
United States v. Durham Lumber Co., 363 U.S. 522 (June 20, 1960)
Aquilino v. United States, 363 U.S. 509 (June 20, 1960)
Phillips Chemical Co. v. Dumas Independent School District, 361 U.S. 376 (Feb. 23, 1960)
McAllister v. Magnolia Petroleum Co., 357 U.S. 221 (June 23, 1958)
Pennsylvania v. Nelson, 350 U.S. 497 (Apr. 2, 1956)

## FIRST AMENDMENT

*Freedom of Association*

United States v. Robel, 389 U.S. 258 (Dec. 11, 1967)

*Freedom of Religion*

Flast v. Cohen, 392 U.S. 83 (June 10, 1968)
Gallagher v. Crown Kosher Super Market, Inc., 366 US. 617 (May 29, 1961)
Braunfeld v. Brown, 366 U.S. 599 (May 29, 1961)
McGowan v. Maryland, 366 U.S. 420 (May 29, 1961)

*Freedom of Speech*

Street v. New York, 394 U.S. 576 (Apr. 21, 1969) (dissenting opinion)
Gregory v. Chicago, 394 U.S. 111 (Mar. 10, 1969)
United States v. O'Brien, 391 U.S. 367 (May 27, 1968)
Bond v. Floyd, 385 U.S. 116 (Dec. 5, 1966)
Wood v. Georgia, 370 U.S. 375 (June 25, 1962)

Times Film Corp. v. City of Chicago, 365 U.S. 43 (Jan. 23, 1961) (dissenting opinion)
United States v. Harris, 347 U.S. 612 (June 7, 1954)

*Obscenity*

Tannenbaum v. New York, 388 U.S. 439 (June 12, 1967) (dissenting opinion)
Jacobs v. New York, 388 U.S. 431 (June 12, 1967) (dissenting opinion)
Jacobellis v. Ohio, 378 U.S. 184 (June 22, 1964) (dissenting opinion)
Roth v. United States, 354 U.S. 476 (June 24, 1957) (concurring opinion)
Kingsley Books, Inc. v. Brown, 354 U.S. 436 (June 24, 1957) (dissenting opinion)

## IMMIGRATION AND NATURALIZATION

Immigration & Naturalization Service v. Errico, 385 U.S. 214 (Dec. 12, 1966)
Foti v. Immigration & Naturalization Service, 375 U.S. 217 (Dec. 16, 1963)
Nishikawa v. Dulles, 356 U.S. 129 (Mar. 31, 1958)
Trop v. Dulles, 356 U.S. 86 (Mar. 31, 1958)
Perez v. Brownell, 356 U.S. 44 (Mar. 31, 1958)
Jay v. Boyd, 351 U.S. 345 (June 11, 1956) (dissenting opinion)
United States v. Zucca, 351 U.S. 91 (Apr. 30, 1956)
Barber v. Gonzales, 347 U.S. 637 (June 7, 1954)

## INDIAN RIGHTS

Poafpybitty v. Skelly Oil Co., 390 U.S. 365 (Mar. 18, 1968)
Squire v. Capoeman, 351 U.S. 1 (Apr. 23, 1956)

## LABOR LAW

NLRB v. Gissel Packing Co., 395 U.S. 575 (June 16, 1969)
NLRB v. Great Dane Trailers, 388 U.S. 26 (June 12, 1967)
International Union, UAW v. Scofield, 382 U.S. 205 (Dec. 7, 1965)
United States v. Brown, 381 U.S. 437 (June 7, 1965)

Fibreboard Paper Products Corp. v. NLRB, 379 U.S. 203 (Dec. 14, 1964)
Brotherhood of Locomotive Engineers v. Missouri-Kansas-Texas Railroad, 363 U.S. 528 (June 20, 1960)
Arnold v. Ben Kanowsky, Inc., 361 U.S. 388 (Feb. 23, 1960)
Mitchell v. Lublin, McGaughy & Associates, 358 U.S. 207 (Jan. 12, 1959)
NLRB v. United Steelworkers of America, 357 U.S. 357 (June 30, 1958) (dissenting opinion)
International Union, UAW v. Russell, 356 U.S. 634 (May 26, 1958) (dissenting opinion)
International Association of Machinists v. Gonzales, 356 U.S. 617 (May 26, 1958) (dissenting opinion)
Brotherhood of Railroad Trainmen v. Chicago River & Indiana Railroad, 353 U.S. 30 (Mar. 25, 1957)
San Diego Building Trades Council v. Garmon, 353 U.S. 26 (Mar. 25, 1957)
Amalgamated Meat Cutters Local 427 v. Fairlawn Meats, 353 U.S. 20 (Mar. 25, 1957)
Guss v. Utah Labor Relations Board, 353 U.S. 1 (Mar. 25, 1957)
NLRB v. Lion Oil Co., 352 U.S. 282 (Jan. 22, 1957)
Mitchell v. King Packing Co., 350 U.S. 247 (Jan. 30, 1956)
Steiner v. Mitchell, 350 U.S. 247 (Jan. 30, 1956)
NLRB v. Warren Co., 350 U.S. 107 (Dec. 12, 1955)
Amalgamated Clothing Workers v. Richman Brothers, 348 U.S. 511 (Apr. 4, 1955) (dissenting opinion)
Association of Westinghouse Salaried Employees v. Westinghouse Electric Corp., 348 U.S. 437 (Mar. 28, 1955) (concurring opinion)
United States v. Binghamton Construction Co., 347 U.S. 171 (Mar. 8, 1954)
Voris v. Eikel, 346 U.S. 328 (Nov. 9, 1953)


## SECURITIES REGULATION


Tcherepnin v. Knight, 389 U.S. 332 (Dec. 18, 1967)


## SEGREGATION


Drews v. Maryland, 381 U.S. 421 (June 1, 1965) (dissenting opinion)
Griffin v. Maryland, 378 U.S. 130 (June 22, 1964)
Lombard v. Louisiana, 373 U.S. 267 (May 20, 1963)
Peterson v. City of Greenville, 373 U.S. 244 (May 20, 1963)

Brown v. Board of Education, 349 U.S. 294 (May 31, 1955); 347 U.S. 483
(May 17, 1954)
Bolling v. Sharpe, 347 U.S. 497 (May 17, 1954)

## TAXATION

Commissioner v. Stidger, 386 U.S. 287 (Mar. 20, 1967)
Fribourg Navigation Co. v. Commissioner, 383 U.S. 272 (Mar. 7, 1966)
Enochs v. Williams Packing & Navigation Co., 370 U.S. 1 (May 28, 1962)
Hanover Bank v. Commissioner, 369 U.S. 672 (May 21, 1962)
Jarecki v. G. D. Searle & Co., 367 U.S. 303 (June 12, 1961)
James v. United States, 366 U.S. 213 (May 15, 1961)
United States v. Manufacturers National Bank, 363 U.S. 194 (June 13,
1960)
Flora v. United States, 362 U.S. 145 (Mar. 21, 1960); 357 U.S. 63 (June
16, 1958)
Fidelity-Philadelphia Trust Co. v. Smith, 356 U.S. 274 (Apr. 28, 1958)
General American Investors Co. v. Commissioner, 348 U.S. 434 (Mar. 28,
1955)
Commissioner v. Glenshaw Glass Co., 348 U.S. 426 (Mar. 38, 1955)

## TORTS

Becker v. Philco Corp., 389 U.S. 979 (Dec. 4, 1967) (dissenting opinion)
Curtis Publishing Co. v. Butts, 388 U.S. 130 (June 12, 1967) (concurring
opinion)
Pierson v. Ray, 386 U.S. 547 (Apr. 11, 1967)
United States v. Muniz, 374 U.S. 150 (June 17, 1963)
Shenker v. Baltimore & Ohio Railroad, 374 U.S. 1 (June 10, 1963)
Richards v. United States, 369 U.S. 1 (Feb. 26, 1962)
Howard v. Lyons, 360 U.S. 593 (June 29, 1959) (dissenting opinion)
Barr v. Matteo, 360 U.S. 564 (June 29, 1959) (dissenting opinion)

## VOTING RIGHTS

Kramer v. Union Free School District No. 15, 395 U.S. 621 (June 16,
1969)

McDonald v. Board of Election Commissioners, 394 U.S. 802 (Apr. 28, 1969)

Allen v. State Board of Elections, 393 U.S. 544 (Mar. 3, 1969)

South Carolina v. Katzenbach, 383 U.S. 301 (Mar. 7, 1966)

Harman v. Forssenius, 380 U.S. 528 (Apr. 27, 1965)

# Notes

## INTRODUCTION

1. H. Hirsch, *The Enigma of Felix Frankfurter* 11 (1981). See also J. Barber, *The Presidential Character* (1972); Alexander George, "Assessing Presidential Character," 26 *World Politics* 234 (1974).

## PART ONE

### CHAPTER 1—PREPARATION

1. The designation of Turner Street comes from an unpublished autobiography of Dr. Rebecca Lee Dorsey, a Los Angeles pediatrician, quoted in J. Weaver, *Warren: The Man, The Court, The Era* 357 (1967).
2. E. Warren, *The Memoirs of Earl Warren* 33 (1977).
3. Ibid. at 33, 43, 34.
4. Ibid. at 14.
5. Ibid. at 17.
6. Ibid. at 30–31.
7. Ibid. at 17.
8. Ibid. at 14–15.
9. Ibid. at 14, 16, 12.
10. See below, pp. 226–27.
11. An inscription dated November 29, 1866, on a Warren family Bible lists the birthdate of "Erik Matias Halvorsen Warren" as December 6, 1864. "Matias" became Methias in America.
12. Weaver, *supra*, note 1, at 19.
13. Ibid. at 16, 52, 187, 18, 27.

14. Ibid. at 9–52.

15. Ibid. at 32.

16. Ibid. at 35.

17. Ibid.

18. Ibid.

19. Ibid. at 42.

20. See *Los Angeles Times*, October 7, 1953, for the story, based on a television interview between Warren and Edward R. Murrow.

21. Warren, *supra*, note 2, at 43.

22. Ibid. at 32, 35.

23. Ibid. at 37.

24. Unnamed source (probably Horace M. Albright) to John D. Weaver, quoted in Weaver, *supra*, note 1, at 31.

25. Warren, *supra*, note 2, at 36.

26. Ibid.

27. Quoted in I. Stone, *Earl Warren: A Great American Story* 19 (1948). Stone's book, a campaign biography, is virtually an original source, since Warren provided much of the material.

28. Warren, *supra*, note 2, at 40, 46.

29. G. Mowry, *The California Progressives* 9, 13, 16, 17 (1951).

30. John R. Haynes, "Birth of Democracy in California," 1, unpublished manuscript, John Randolph and Dora Haynes Foundation, Los Angeles, California. See Mowry, *supra*, note 29, at 16.

31. Mowry, *supra*, note 29, at 99, 85.

32. Cf. White, "The Social Values of the Progressives," 70 *South Atlantic Quarterly* 62 (1971).

33. *California Weekly*, January 29, 1909, 2. The *California Weekly* became the *Outlook* in 1911.

34. Ibid.

35. Mowry, *supra*, note 29, at 113–14.

36. Warren, *supra*, note 2, at 38, 39, 171, 234.

37. Ibid. at 43, 44.

38. Ibid. at 45, 46, 47.

39. Ibid. at 48.

40. Ibid. at 50.

41. Ibid. at 52.

42. See below, pp. 193–94.

43. Warren, *supra*, note 2, at 52.

44. Ibid. at 55.

45. Ibid. at 55–56.

46. Ibid. at 61.

47. Quoted in *Oakland Tribune*, September 20, 1931. In Warren's memoirs Moley was remembered as simply calling Warren "the best district attorney in the United States." Warren, *supra*, note 2, at 118.

## CHAPTER 2—LAW ENFORCEMENT: THE DISTRICT ATTORNEY

1. E. Warren, *The Memoirs of Earl Warren* 61 (1977).

2. Ibid.

3. Ibid. at 62.

4. See, e.g., Merrell F. Small, "Letter Regarding Earl Warren's Court Appoint-

ment," in *Earl Warren: The Chief Justiceship* 4 (1977), Earl Warren Oral History Collection, Regional Oral History Office, The Bancroft Library, University of California at Berkeley. Hereafter cited as ROHO.

5. Warren, *supra*, note 1, at 68.
6. Ibid.
7. Ibid. at 67, 69.
8. Ibid. at 72.
9. Ibid. at 91.
10. Ibid.
11. Ibid. at 92.
12. Ibid. at 102.
13. Ibid. at 53, 54, 56, 60.
14. Discussions of Progressive attitudes toward the comparative virtues and faults of various branches of government can be found in Samuel Hays, *Conservation and the Gospel of Efficiency* (1959); Samuel Haber, *Efficiency and Uplift* (1964); and Robert Wiebe, *The Search for Order* (1967).
15. Warren, *supra*, note 1, at 98.
16. Ibid. at 94, 104.
17. Ibid. at 104.
18. Ibid. at 122.
19. Richard Rodda, "From The Capital Pressroom," in *Bee Perspectives Of The Warren Era* 12 (1976), ROHO.
20. Mary Shaw, "Perspectives of a Newspaperwoman," in 1 *Perspectives On the Alameda County District Attorney's Office* 11 (3 vols., 1972–74), ROHO.
21. J. Frank Coakley, "A Career in the Alameda County District Attorney's Office," in 3 *Perspectives*, *supra*, note 20, at 33.
22. Warren, *supra*, note 1, at 105.
23. Ibid. at 106.
24. Ibid. at 108.
25. Earl Warren, untitled interview, December, 1971, 22, ROHO.
26. Ibid. at 20.
27. Warren, *supra*, note 1, at 108.
28. Warren, *supra*, note 25, at 20.
29. Warren, *supra*, note 1, at 109.
30. Miriam Feingold, "The King-Ramsay-Conner Case: Labor, Radicalism, and the Law in California, 1936–1941," unpublished Ph.D. dissertation, University of Wisconsin-Madison 91 (1976). Hereafter cited as Feingold.
31. Warren, *supra*, note 25, at 40.
32. Earl Warren, untitled interview, April 17, 1972, 56, ROHO.
33. The practice of commenting on a defendant's failure to take the stand was declared unconstitutional by the Warren Court in 1965 in *Griffin v. California*, 380 U.S. 609. Warren, who declined to participate in that decision, said in 1971 that he thought the practice "was all right" when introduced in California in 1934 but that by 1965 he was "glad [it] was ruled unconstitutional." Warren, *supra*, note 25, at 67.
34. Warren, letter to registered voters, October, 1934, quoted in J. Weaver, *Warren: The Man, The Court, The Era* 52 (1967).
35. Warren, *supra*, note 1, at 110.
36. Ibid. at 117.
37. Powe, "Earl Warren: A Partial Dissent," 56 *No. Cal. L. Rev.* 408, 416 (1978).
38. See, e.g., Robert Holmes, "The King-Ramsay-Conner Frame-Up: Earl Warren's Murder Case," pamphlet, King-Ramsay-Conner Defense Committee, Septem-

ber, 1936, International Longshoremen's and Warehousemen's Union Library, San Francisco. Hereafter cited as ILWU Library. See also Feingold at 277–78.

39. Warren, *supra*, note 1, at 115.

40. The best source on the *Point Lobos* case is Feingold. My account of the background of the case relies heavily on Feingold; other sources are cited in subsequent notes.

41. Warren, *supra*, note 1, at 4.

42. See *Earl Warren's Bakersfield* 1 (1971), ROHO.

43. Lee Coe, "The Ship Murder: The Story of a Frame-Up," April, 1937, ILWU Library. See Feingold at 522–23.

44. Warren, *supra*, note 25, at 153.

45. The "Del Rio Letter" Murphy received from Wallace had been dictated to a third party. See *People v. King et al.* case files, Alameda County District Attorney's Office, County Courthouse, Oakland, California. Hereafter cited as ACDA files. See also Feingold at 207.

46. Statement of George Wallace, August 30, 1936, *People v. King et al.*, ACDA files.

47. *San Francisco Examiner*, August 29, 1936.

48. *San Francisco Chronicle*, September 1, 1936.

49. Howard E. Tupper, "Summary of Notes re: Frank Conner," September 3, 1936, ACDA files. See Feingold at 249.

50. Memorandum to Earl Warren, September 2, 1936, ACDA files. The apparent source of the memorandum was a letter (now lost?) from a New York City police official to Warren's office.

51. "Confidential Report," September 8, 1936, ACDA files.

52. "Notes on Benjamin Sakovitz," n.d., ACDA files; Warren, *supra*, note 1, at 116.

53. 4 *Transcript of People v. King et al.* 2181 (6 Vols.), ACDA files. *People v. King et al.* was an unreported case; the citation on appeal is 30 Cal. App.2d 185 (1938).

54. Feingold at 280.

55. 6 *Transcript of People v. King et al.*, *supra*, note 53, at 3854. See also Feingold at 280.

56. 6 *Transcript*, *supra*, note 53, at 3859.

57. Warren, *supra*, note 25, at 151.

58. Ralph E. Hoyt, Trial Notebook, n.d., ACDA files.

59. "Conference Notes," August 16, 1943, ibid.

60. *In re George Wallace*, 24 Cal.2d 933 (1944). Warren recalled that "after a lengthy hearing, Judge [Hartley] Shaw decided that the charge was not credible, and recommended that it be rejected. The Supreme Court [of California] affirmed his findings, and the case was ended." Warren, *supra*, note 1, at 116.

61. *San Francisco Chronicle*, October 16, 1940.

62. Culbert Olson to Earl Warren, October 17, 1940, ACDA files.

63. *San Francisco Examiner*, November 28, 1941.

64. *San Francisco Examiner*, December 3, 1941.

65. California Senate, 55th sess., *Report of the Joint Fact Finding Committee for Un-American Activities in California* 176 ff. (1943).

66. "Look At This Record," King-Ramsay-Conner Defense Committee, April, 1937, ILWU Library.

67. Warren, *supra*, note 1, at 113.

68. Ibid. at 116.

69. See generally Feingold at 1–17.

70. *Sacramento Bee*, October 22, 1934.

71. Commentators have severely criticized Warren's role in *Point Lobos*. See, e.g., L. Katcher, *Earl Warren: A Political Biography* 99 (1967) ("To many labor leaders and union men—and many liberals—there is still a stigma upon the case"); J. Pollack, *Earl Warren: The Judge Who Changed America* 57 (1979) ("Impartial legal students agreed that . . . there were certainly many dubious aspects to [the case's] handling which did not reflect well on Earl Warren"); Powe, *supra*, note 37, at 416 ("many people thought the defendants were innocent, despite the verdict").

72. *Spano v. New York*, 360 U.S. 315, 320–21 (1960).

73. 384 U.S. 436 (1966).

74. 388 U.S. 41 (1967).

75. 389 U.S. 347 (1967).

76. 384 U.S. 333 (1966).

77. 378 U.S. 368 (1964).

78. [Deering's] California Evidence Code, § 405.

79. 354 U.S. 449 (1957).

80. [Deering's] California Penal Code, § 859.

81. 318 U.S. 332 (1943).

82. 357 U.S. 433 (1958).

83. *Supra*, note 73, at 448, 457.

84. 277 U.S. 438 (1928).

85. 366 U.S. 717 (1961).

86. Warren, quoted in *San Francisco Chronicle*, September 1, 1936.

87. See *Los Angeles Times*, February 18, 1938.

88. Ibid.

89. See Weaver, *supra*, note 34, at 53.

90. This practice was changed in 1954. See J. Harris, *California Politics* 37–38, 67–72 (1961).

91. Warren, *supra*, note 1, at 123–24. For more on the California pension plans of the 1930s, see below, pp. 53–54.

92. Warren, *supra*, note 1, at 126.

93. Ibid. at 124–26.

94. Robert Coughlan, "California's Warren and Family," *Life*, April 24, 1944, 100, 106.


CHAPTER 3—LAW ENFORCEMENT: THE ATTORNEY GENERAL

1. E. Warren, *The Memoirs of Earl Warren* 129, 139 (1977).

2. Ibid. at 142.

3. Ibid. at 139, 143, 144.

4. Ibid. at 143.

5. Ibid. at 132.

6. *People v. Stralla*, 14 Cal.2d 617 (1939).

7. Warren, *supra*, note 1, at 133. He was relying on the Supreme Court case of *New Jersey v. City of New York*, 283 U.S. 473 (1931), where New Jersey was allowed to abate the dumping of garbage by New York City into New Jersey waters, sometimes at points over twenty miles from the New Jersey shoreline.

8. Warren, *supra*, note 1, at 142–43.

9. Quoted in Benno Schmidt, Jr., oral history (of his tenure as a law clerk with Warren) 96, Columbia University Oral History Research Office, Butler Library, Columbia University.

10. Warren, *supra*, note 1, at 155.

11. Ibid.

12. Ibid.

13. Many other Progressives did not: see O. Graham, *Encore for Reform: The Old Progressives and the New Deal* (1967).

14. See J. Voorhis, *Confessions of a Congressman* 17 (1947), for a capsule summary of "End Poverty in California." See also R. Burke, *Olson's New Deal for California* 3–5 (1953).

15. For Olson's early political history I have followed Burke, note 14, at 610.

16. Radio speech, August 26, 1938, Culbert L. Olson Papers, Bancroft Library, University of California at Berkeley. Hereafter cited as Olson papers.

17. Ibid. at 22.

18. On Samish's career, see A. Samish and B. Thomas, *The Secret Boss of California* (1971).

19. Carey McWilliams, "The Education of Earl Warren," *Nation*, October 12, 1974, 326.

20. See *Los Angeles Evening News*, June 17, 1938; Burke, *supra*, note 14, at 24.

21. Burke, *supra*, note 14, at 33.

22. Radio speech, November 2, 1938, Olson papers.

23. See Burke, *supra*, note 14, at 12–28.

24. William Schneiderman, "The Democratic Front in California," 17 *The Communist* 663 (1938); see Burke, *supra*, note 14, at 18–19.

25. "Platform of the California Republican Party—1938," quoted in Burke, *supra*, note 14, at 28; *Sacramento Bee*, November 3, 1938.

26. See Burke, *supra*, note 14, at 29–30.

27. Speech, October 26, 1938, Olson papers.

28. Burke, *supra*, note 14, at 36.

29. Warren, letter to Culbert Olson, January 7, 1939, quoted in *Oakland Tribune*, January 8, 1939.

30. See *San Francisco Chronicle*, February 24, 1939, for the prison board's action; see *San Francisco Chronicle*, October 18, 1939, for Olson's appeal to the Supreme Court.

31. Warren, *supra*, note 1, at 143, 144.

32. See Burke, *supra*, note 14, at 157.

33. Warren, *supra*, note 1, at 144.

34. Burke, *supra*, note 14, at 7–10.

35. *San Francisco Chronicle*, June 1, 1940.

36. Richard Graves, *Theoretician, Advocate, and Candidate in California State Government* 70, 72 (1973), in Earl Warren Oral History Collection, Regional Oral History Office, the Bancroft Library, University of California at Berkeley. Hereafter cited as ROHO.

37. William T. Sweigert, *Democrat, Friend, and Advisor to Earl Warren* 201 (in process), ROHO.

38. Warren, untitled interview, December, 1971, 199, ROHO.

39. *Los Angeles Examiner*, August 8, 1932.

40. See *San Francisco Chronicle*, December 19, 1941, for a contemporary account of Warren's challenge. Warren's language is from his *Memoirs*, *supra*, note 1, at 145.

41. Warren, *supra*, note 1, at 145.

42. Transcript, State Council of Defense Conference, August 28, 1942, Olson papers.

43. Warren, *supra*, note 1, at 145.

44. *San Francisco Chronicle*, February 28, 1942.

45. Warren, *supra*, note 38, at 242.

46. Earl Warren, Jr., "California Politics," in *The Governor's Family* 58 (1980), ROHO.

47. Max Radin, letter to Robert Gordon Sproul, February 26, 1936, Papers of Max Radin, Bancroft Library, University of California at Berkeley. Hereafter cited as Radin papers. Two good sources of biographical information on Radin are Albert Ehrenzweig, "Introduction," in Papers of Albert Ehrenzweig, University of California at Berkeley School of Law; and David M. Margolick, "The Nomination of Max Radin: People, Politics, and the California Supreme Court," unpublished manuscript, 1977. See also Margolick, "Earl Warren's Past: The Forgotten Blemish," *National Law Journal*, September 17, 1979, 15. Mr. Margolick's work has led me to some sources on the Radin nomination that might otherwise have escaped my attention, and I have profited from his observations on the Radin affair.

48. Burke, *supra*, note 14, at 211.

49. See, e.g., letter from Orrin K. McMurray to Culbert Olson, August 14, 1939, Radin papers; Roger Traynor to Olson, August 15, 1939, Radin papers; A. M. Kidd to Olson, August 15, 1939, Radin papers; P. O. McGovney to Olson, August 15, 1939, Radin papers. All four of those individuals were colleagues of Radin at Boalt Hall.

50. See Margolick, "The Nomination of Max Radin," *supra*, note 47, at 53n; Max Radin to Robert Gordon Sproul, February 26, 1935, Papers of Robert Gordon Sproul, Bancroft Library, University of California at Berkeley. Hereafter cited as Sproul papers.

51. See George D. Kidwell to Culbert Olson, July 15, 1939, Radin papers; Roger Traynor to Olson, *supra*, note 49.

52. Thurman Arnold to Culbert Olson, June 11, 1940, Radin papers; Jerome Frank to Culbert Olson, June 17, 1940, Radin papers; Culbert Olson to Max Radin, February 24, 1940, Radin papers.

53. *San Francisco Chronicle*, June 28, 1940.

54. This motive is suggested by the *San Francisco Chronicle*, August 9, 1939, and by Margolick, "The Nomination of Max Radin," *supra*, note 47, at 53n.

55. *Carter v. Commission on Qualifications of Judicial Appointments*, 14 Cal.2d 179 (1939).

56. See *Chronicle*, *supra*, note 53; Margolick, "The Nomination of Max Radin," *supra*, note 47, at 53n.

57. Radin to Robert Gordon Sproul, August 8, 1940, Sproul papers.

58. Jesse W. Carter, *California Supreme Court Justice Jesse W. Carter* 172 (1959), ROHO.

59. Radin, "Legislative Pardons: Another View," 27 *Calif. L. Rev.* 287, 397 (1938).

60. See Burke, *supra*, note 14, at 129–36.

61. Radin to Raymond Dunne, June 3, 1940, in files of the Commission on Judicial Qualifications, Administrative Office of the Courts, Library and Courts Building, San Francisco. Hereafter cited as Commission files.

62. Radin to Irving Neumiller, June 3, 1940, Commission files.

63. Radin to Philip Dunne, July 26, 1940, Radin papers.

64. Irving Neumiller to Max Radin, June 6, 1940, Commission files.

65. Warren to Gerald Hagar, June 28, 1940, Commission files.

66. Radin to Carey McWilliams, July 24, 1940, Radin papers. Warren's conduct in the Radin affair and in the *Point Lobos* case prompted McWilliams to launch an attack upon him in the *New Republic* in 1943, in which McWilliams called Warren, "among other things, a personification of smart reaction" and "an essentially grim and hard-boiled individual." McWilliams, "Warren of California," *New Republic*, October 18, 1943, 514. McWilliams had been appointed commissioner of immigration and housing by Olson; one of Warren's first actions as governor was to fire him in January, 1943.

67. Gerald Hagar to Grove J. Fink, July 2, 1940, Commission files.

68. *San Francisco News,* July 2, 1940.

69. *San Francisco Examiner,* July 8, 1940.

70. Warren to Gerald Hagar, July 8, 1940, Commission files.

71. Don Martin, statement before Committee of the Board of Governors of the State Bar of California, July 10, 1940, Commission files.

72. Max Radin, statement, ibid.

73. Report of Committee of the Board of Governors of the State Bar of California, July 19, 1940, Commission files.

74. Max Radin to Philip Dunne, June 26, 1940, Radin papers.

75. Resolution of the Board of Governors of the State Bar of California, July 19, 1940, Commission files.

76. J. Oscar Goldstein to Phil Gibson, July 19, 1940, Commission files.

77. Grove Fink to Warren, July 19, 1940, ibid.

78. Radin to Philip Dunne, *supra,* note 63.

79. The other vote against Radin was that of John T. Nourse, senior presiding judge of the California District Court of Appeals. Two days after the vote, Nourse was quoted in the *San Francisco Chronicle* as having said that Radin "was always found on the side of and defending people and groups with extreme views." *San Francisco Chronicle,* July 24, 1940. Chief Justice Phil S. Gibson apparently voted for Radin, although Warren did not disclose the commission's votes to the press. See the account given by the *San Francisco Chronicle,* July 23, 1940.

80. Radin to Carey McWilliams, *supra,* note 66.

81. Radin to Philip Dunne, *supra,* note 63; Max Radin to Norman Littell, August 5, 1940, Radin papers.

82. Robert Gordon Sproul, memorandum to "Special Problems" file, July 23, 1940, Sproul papers.

83. Radin to Robert Gordon Sproul, August 7, 1940, Radin papers.

84. Sproul, memorandum to "Special Problems" file, December 31, 1940, Sproul papers. See Margolick, "Earl Warren's Past," *supra,* note 47, at 15.

85. E. Warren, *supra,* note 1, at 40.

86. Ibid. at 118.

87. Ibid. at 40.

88. See M. Grodzins, *Americans Betrayed* 3–14 (1949); R. Daniels, *Concentration Camps USA* 1–7 (1971).

89. Daniels, *supra,* note 88, at 6–9; Warren, untitled interview, June 22, 1972, 302–3, ROHO.

90. *Los Angeles Times,* July 15, 1920. See Grodzins, *supra,* note 88, at 4–7; Daniels, *supra,* note 88, at 10, 15–21, 29–31.

91. *Porterfield v. Webb,* 263 U.S. 225 (1923).

92. Ibid. at 229. See also *Frick v. Webb,* 263 U.S. 326, 330 (1923) where Webb stated that the purpose of the statutes (known popularly as the Alien Land Law) was "to prevent ruinous competition by the Oriental farmer against the American farmer."

93. These figures are cited in L. Katcher, *Earl Warren: A Political Biography* 141 (1967).

94. Gordon Rogers, unpublished manuscript, quoted in J. Weaver, *Warren: The Man, The Court, The Era* 108 (1967).

95. Quoted in Katcher, *supra,* note 93, at 146.

96. Warren, quoted in Associated Press news release, January 30, 1942. See Grodzins, *supra,* note 88, at 94.

97. Warren, *supra,* note 1, at 145–148.

98. Quoted in Katcher, *supra,* note 93, at 135.

99. Proceedings, Conference of Sheriffs and District Attorneys Called by Attorney General Warren on the Subject of Alien Land Law Enforcement, February 2, 1942, 156–58, Bancroft Library, University of California at Berkeley.

100. Ibid. at 72.

101. Minutes, Meeting of California Joint Immigration Committee, February 7, 1942, 35–36, Bancroft Library, University of California at Berkeley.

102. See Wenig, "The California Attorney General's Office, the Judge Advocate General Corps, and the Japanese-American Relocation," in *The Japanese-American Relocation Reviewed: Decision and Exodus* 1–24 (1976), ROHO.

103. *Supra*, note 101, at 38.

104. Warren, testimony before the Select Committee Investigating National Defense Migration, U.S. Congress, House, 77th Cong., 2d sess., February 21, 23, 1942, Part 29, 11015.

105. *Supra*, note 99, at 176.

106. *Supra*, note 104, at 11015. See Grodzins, *supra*, note 88, at 98.

107. *Supra*, note 104, at 11012.

108. Bendesten, "The Story of Pacific Coast Japanese Evacuation," address to Commonwealth Club, San Francisco, May 20, 1942, quoted in Grodzins, *supra*, note 88, at 286.

109. *Supra*, note 101, at 36.

110. *San Francisco Chronicle*, February 9, 1942.

111. Lieutenant General John L. DeWitt to Secretary of War, "Final Recommendations . . . ," February 14, 1942, in Western Defense Command and Fourth Army, *Final Report, Japanese Evacuation from the West Coast* 33 (1943).

112. See Grodzins, *supra*, note 88, at 266–67.

113. Executive Order 9066, February 20, 1942, quoted in ibid. at 325.

114. See Daniels, *supra*, note 88, at 110–29.

115. *Los Angeles Times*, December 19, 1944.

116. *Proceedings of the Governors' Conference* 10 (1943).

117. *Supra*, note 111, at 34.

118. *San Francisco Chronicle*, April 14, 1943.

119. *Supra*, note 116, at 10.

120. See Grodzins, *supra*, note 88, at 92–93.

121. Attorney General's Opinion, No. 1-NS4083, February 8, 1942, Attorney General's Office, Sacramento, California.

122. Attorney General's Opinion, No. 1-NS4108, February 17, 1942, ibid.

123. *Supra*, note 115.

124. *Los Angeles Times*, December 20, 1944.

125. For an account of Congress's activities, see Grodzins, *supra*, note 88, at 325–48.

126. *Hirabayashi v. United States*, 320 U.S. 81 (1943); *Korematsu v. United States*, 323 U.S. 214 (1944).

127. Stone, Frankfurter, Douglas, and Black were with the majority in both *Hirabayashi* and *Korematsu*, Stone writing the opinion in the former case and Black in the latter. None of these justices subsequently suggested their decisions were misguided. Justice Tom Clark, who served with Warren on the Supreme Court from 1954 to 1966, said in 1966 that he had "made a mistake" in supporting the Japanese evacuation program, that "we picked up these people . . . our fellow citizens . . . and put them in concentration camps," and he was "amazed that the Supreme Court ever approved it." *San Diego Union*, July 10, 1966.

As for Lippmann, he conferred with Warren in February 8, 1942, about the Japanese "problem," and on February 12 wrote a syndicated column, "Today and

Tomorrow," in which he endorsed Warren's "no sabotage means well-organized sabotage" argument. See Grodzins, *supra*, note 88, at 99–100, 387.

128. McWilliams suggested that the House Select Committee on National Defense Migration hold hearings on the problem to give persons an opportunity to "blow off steam" and "blow down irresponsible rumors." He subsequently presented a plan for "selective" evacuation to the Committee on Immigration and Housing. *Supra*, note 104, Part 31, 11788–97. See Grodzins, *supra*, note 88, at 183–184.

129. See J. Pollack, *Earl Warren: The Judge Who Changed America* 364 (1979).

130. Warren, "The Bill of Rights and the Military," 37 *N.Y.U. L. Rev.* 181 (1962).

131. *Supra*, note 104, Part 29, 11018.

132. Katcher, *supra*, note 93, at 150.

133. Warren, untitled interview, June 22, 1972, 329, ROHO.

134. Warren, *supra*, note 1, at 146.

135. Ibid. at 149.

136. Ibid. at 147, 149.

137. Ibid. at 149.

138. Ibid. Italics added.

139. Ibid. at 156.

140. Warren, *supra*, note 38, at 197–99, 207–9.

141. "Politics on the West Coast," *Fortune*, March, 1940, 41, 140.

142. Katcher, *supra*, note 93, at 158.

143. Quoted in ibid.

144. William Sweigert, *supra*, note 37, at 247.

145. Quoted in Burke, *supra*, note 14, at 219.

146. McWilliams, "The Education of Earl Warren," *Nation*, October 12, 1974, 325.

147. See below, p. 87.

148. Olson, remarks during radio debate, quoted in *San Francisco News*, October 12, 1942; Olson, radio address, August 24, 1942, Olson papers.

149. Warren, address to Republican State Convention, September 17, 1942, quoted in Burke, *supra*, note 14, at 219.

150. Warren, *supra*, note 38, at 211.

151. Warren, *supra*, note 1, at 157. Italics added.


## CHAPTER 4—GOVERNING: OFFICEHOLDING AND POLICYMAKING

1. E. Warren, *The Memoirs of Earl Warren* 233, 237–39 (1977).

2. Ibid. at 234.

3. Ibid. at 238.

4. Ibid. at 27, 40.

5. Horace Albright to John Weaver, quoted in J. Weaver, *Warren: The Man, The Court, The Era* 35 (1967).

6. Warren, *supra*, note 1, at 65.

7. There is a discrepancy between Warren's account of Nina's employment at the time of their first meeting and two earlier accounts recorded in magazine articles on the Warren family. The articles reported that Nina had been employed as a stenographer at the Crane Plumbing Company in Oakland when she first met Warren. See Robert Coughlan, "California's Warren and Family," *Life*, April 24, 1944, 100, 105; Cameron Shipp, "The Golden Girls," *Redbook*, November, 1950, 36, 72.

8. Warren, *supra*, note 1, at 64.

9. Ibid. at 65, 64.

10. Warren, interview, June 22, 1972, 360, Earl Warren Oral History Collection,

Regional Oral History Office, the Bancroft Library, University of California at Berkeley. Hereafter cited as ROHO.

11. Warren, *supra*, note 1, at 345.

12. Horace Albright, quoted in Weaver, *supra*, note 5, at 41.

13. Weaver, "Happy Birthday, Earl Warren," *Los Angeles Times*, Magazine Section, March 8, 1970, 9.

14. McWilliams, "The Education of Earl Warren," *Nation*, October 12, 1974, 325.

15. Coughlan, *supra*, note 7, at 103.

16. Farnsworth Crowder, "The Governor's Never Too Busy," *Better Homes and Gardens*, August, 1947, 50, 51, 112, 121.

17. Shipp, *supra*, note 7, at 37, 38, 72, 73.

18. Sidney Shalett, "The Warrens: What A Family!," *Saturday Evening Post*, February 3, 1951, 17, 18, 74.

19. Al Stump, "They Live in the Limelight," *American Magazine*, June, 1953, 22, 99, 100, 102, 103, 105.

20. Coughlan, *supra*, note 7, at 105.

21. Quoted in Stump, *supra*, note 19, at 100.

22. Crowder, *supra*, note 16, at 120.

23. Ibid.

24. Warren, address to American Legion, August 16, 1943, quoted in *San Francisco Call-Bulletin*, August 17, 1943.

25. Coughlan, *supra*, note 7, at 105.

26. Warren, *supra*, note 1, at 173.

27. Ibid.

28. Nina Warren Brien, "Growing Up in the Warren Family," in *The Governor's Family* 26, 22 (1980), ROHO.

29. James Warren, "Recollections of the Eldest Warren Son," in ibid. at 53.

30. James Warren, quoted in Shalett, *supra*, note 18, at 74.

31. Earl Warren, Jr., "California Politics," in *supra*, note 28, at 5.

32. Warren, *supra*, note 1, at 245.

33. The use of the term "secretary" in these persons' titles was a then existing convention of the California governorship. None of the persons performed clerical or typing services.

34. Warren, *supra*, note 1, at 209.

35. William T. Sweigert, *Democrat, Friend, and Advisor to Earl Warren* 165 (in process), ROHO.

36. Warren, *supra*, note 1, at 209.

37. Sweigert, *supra*, note 35, at 83–84.

38. Interview with Judge William T. Sweigert, April 10, 1980, Carmel, California.

39. Sweigert, *supra*, note 35, at 190.

40. In 1947, for example, in an address to a legislative Joint Committee on Constitutional Revision, Warren said that "democratic government [should be made] even more responsive to . . . the needs of all, the needs of ordinary people . . ." Warren, "Constitutional Reform and the Development of State Government," in *The Public Papers of Chief Justice Earl Warren* 3, 9 (H. Christman, ed., 1959).

41. Warren, untitled interview, April 17, 1972, 94, ROHO.

42. Richard Graves, *Theoretician, Advocate, and Candidate in California State Government* 73 (1973), ROHO.

43. Sweigert, *supra*, note 38.

44. Warren, *supra*, note 1, at 209.

45. Sweigert, *supra*, note 35, at 235.

46. Ibid. at 251.

47. Warren, *supra*, note 1, at 210.
48. L. Katcher, *Earl Warren: A Political Biography* 114 (1967).
49. Walter P. Jones, "An Editor's Long Friendship with Earl Warren," in *Bee Perspectives of the Warren Era* 28 (1976), ROHO.
50. Scoggins, "Observations on California Affairs by Governor Earl Warren's Press Secretary," in *The Governor and the Public, the Press, and the Legislature* 31–36, 55 (1973), ROHO.
51. Warren, *supra*, note 1, at 210.
52. Warren, Jr., *supra*, note 31, at 53.
53. Warren, *supra*, note 1, at 211.
54. Warren, Jr., *supra*, note 31, at 53.
55. MacGregor, remarks in "Earl Warren—A Tribute," 58 *Calif. L. Rev.* 3, 36 (1970); Warren, Jr., *supra*, note 31, at 53.
56. Warren, *supra*, note 1, at 211.
57. MacGregor, *supra*, note 55, at 36.
58. Warren, *supra*, note 1, at 209.
59. MacGregor, *A Career in Public Service with Earl Warren* 6, 67 (1973), ROHO.
60. MacGregor, *supra*, note 55, at 33–34.
61. Ibid. at 35–36.
62. Quoted in Katcher, *supra*, note 48, at 56.
63. Quoted in ibid. at 217.
64. Quoted in ibid. at 155.
65. Warren, *supra*, note 1, at 214.
66. Warren, *supra*, note 41, at 96.
67. Quoted in Weaver, *supra*, note 5, at 195.
68. Ibid.
69. *Sacramento Bee*, December 21, 1969. Judge Sweigert confirmed the accuracy of Small's account in an interview in April, 1980, *supra*, note 38.
70. Merrell F. Small, *The Office of the Governor Under Earl Warren* 58, 136, 211, 153, 162, 157, 182 (1972), ROHO.
71. Warren, *supra*, note 1, at 137, 141, 152, 199, 200, 260.
72. Warren Olney III, *Law Enforcement and Judicial Administration in the Earl Warren Era* 137, 136, 173, 182 (1981), ROHO. Italics in original.
73. Warren, *supra*, note 1, at 275.
74. Benno Schmidt, Jr., oral history 275–76, Columbia University Oral History Research Office, Butler Library, Columbia University.
75. Warren, *supra*, note 1, at 277.
76. Ibid. at 277, 349.
77. Ibid. at 211.
78. Carey McWilliams, "Warren of California," *New Republic*, October 18, 1943, 514.
79. Warren, "What is Liberalism," *New York Times Magazine*, April 18, 1948, 10.
80. Quoted in Katcher, *supra*, note 48, at 171.
81. See, e.g., McWilliams, *supra*, note 14; Jones, *supra*, note 49; Small, *supra*, note 70. McWilliams' assessment represented a revision of comments he had made in 1947, when he called Warren's "occasional liberalism" the "opportunistic" response of "a very able conservative politician." McWilliams, "Earl Warren—A Likely Dark Horse," *Nation*, November 29, 1947, 583.
82. See G. Mowry, *The California Progressives* 296–99 (1951).
83. See J. Stevenson, *The Undiminished Man* 25 (1980).
84. For a text of the letter see *Los Angeles Times*, July 24, 1938; Stevenson, *supra*, note 83, at 166.

85. See O. Graham, *An Encore for Reform: The Old Progressives and the New Deal* 103–26 (1967).
86. Warren, *supra*, note 79, at 10–11.
87. Warren, *supra*, note 1, at 171, 176, 231–32.
88. Mowry, *supra*, note 82, at 145–50.
89. Warren, *supra*, note 1, at 171, 231–32.
90. Ibid. at 178.
91. Ibid. at 185.
92. Ibid. at 193.
93. Ibid. at 195.
94. Ibid. at 190.
95. Ibid. at 191.
96. Ibid. at 191–93.
97. Ibid. at 200–201.
98. Quoted in *Los Angeles Times*, June 3, 1947.
99. Quoted in ibid., May 11, 1949.
100. See ibid., May 12, 1949.
101. See ibid., March 6, 1952.
102. Warren, address, January 7, 1951, quoted in Weaver, *supra*, note 5, at 174.
103. Lewis, "Earl Warren," in 4 L. Friedman and F. Israel, *The Justices of the United States Supreme Court* 2721, 2728 (4 vols., 1949).
104. Warren, *supra*, note 1, at 14.
105. Ibid. at 187.
106. Ibid. at 188.
107. Sweigert, *supra*, note 38. See also Merrell Small, "Governor Earl Warren Was Far Ahead of His Time In Proposing Prepaid Medical Care," *Sacramento Bee*, December 7, 1969.
108. See W. Sweigert, "Earl Warren: Biography and Record," in Irving Stone papers, Powell Library, University of California at Los Angeles.
109. See Small, *supra*, note 107.
110. The issue of regulated versus unregulated competition was one on which Progressives throughout the nation were deeply divided. See Graham, *supra*, note 85, at 9–14.
111. See, e.g., California Assembly Committee on Public Health, "Majority Report," *Assembly Journal*, 56th sess., April 5, 1945, 1601–5. Among the epithets used to describe Warren's health plan were "revolutionary," "alien," "unamerican," "collectivist," and "state socialist."
112. For example, in an address to a legislative committee charged with revising the California Constitution in 1947, Warren spoke of "protect[ion] . . . against domination [of individuals] by the state," and urged that California "avoid centralized bureaucracy which . . . we deplore." Quoted in Christman, *supra*, note 40, at 7, 10.
113. McWilliams, *supra*, note 81, at 583.
114. Warren, *supra*, note 1, at 187–89.
115. Address to Legislative Joint Interim Committee on Constitutional Revision, quoted in Christman, *supra*, note 40, at 7.
116. G. Stewart et al., *The Year of the Oath* 11, 10 (1950).
117. Warren, *supra*, note 1, at 218, 223.
118. See D. Gardner, *The California Oath Controversy* 85–86 (1967).
119. Warren, *supra*, note 1, at 218.
120. The original text of the oath is recorded in Minutes of the Regents of the University of California, Regular Session, March 25, 1949, The Bancroft Library,

University of California at Berkeley. See also Gardner, *supra*, note 118, at 21–26; Stewart, *supra*, note 116, at 28.

121. Gardner, *supra*, note 118, at 14.

122. See memorandum, June 16, 1949, Robert Gordon Sproul papers, Bancroft Library, University of California at Berkeley.

123. Ibid.

124. 19 *University of California Faculty Bulletin* 46 (July, 1949); see also Gardner, *supra*, note 118, at 30.

125. See *San Francisco Chronicle*, June 13, 1949.

126. My account of the details of the oath controversy relies on Stewart, *supra*, note 116, and Gardner, *supra*, note 118, in addition to the primary sources cited.

127. Minutes of the Regents Special Committee on Communist Activities, June 24, 1949, Bancroft Library, University of California at Berkeley. Hereafter cited as Minutes . . . Bancroft, with sessions and dates.

128. See Stewart, *supra*, note 116, at 31.

129. Minutes, Executive Session, September 23, 1949, Bancroft.

130. See Minutes, *supra*, note 127; see also Gardner, *supra*, note 118, at 42.

131. Newlan, "Resolution," in Minutes, Executive Session, February 24, 1950, Bancroft.

132. See 19 *University of California Faculty Bulletin* 62 (December, 1949); Gardner, *supra*, note 118, at 82.

133. Frank L. Kidner to David P. Gardner, May 7, 1965; see Gardner, *supra*, note 118, at 107.

134. Minutes, Executive Session, February 24, 1950, Bancroft.

135. *San Francisco Chronicle*, February 25, 1950.

136. Ibid., March 1, 1950.

137. Report of the Special Committee on Conference with the Regents, Northern Section, March 7, 1950, 13–14, Bancroft Library, University of California at Berkeley. See also Gardner, *supra*, note 118, at 111–12.

138. Minutes, Regular Session, March 31, 1950, Bancroft.

139. Minutes, Regular Session, April 21, 1950, Bancroft.

140. Ibid.

141. Minutes, Regular Session, August 25, 1950, Bancroft.

142. *Oakland Tribune*, September 22, 1950; see also Gardner, *supra*, note 118, at 217–19.

143. Edward Tolman to Benjamin Fine, November 5, 1951, Group for Academic Freedom Files, Bancroft Library, University of California at Berkeley. See also Gardner, *supra*, note 118, at 427.

144. See *Oakland Tribune*, September 22, 1950.

145. Thomas Kuchel to Olaf Lindberg, October 13, 1950, quoted in *Berkeley Daily Gazette*, October 13, 1950. Lindberg was the Comptroller of the University of California.

146. See Gardner, *supra*, note 118, at 218; J. Pollack, *Earl Warren: The Judge Who Changed America* 131 (1979).

147. See Weaver, *supra*, note 5, at 285–86; Katcher, *supra*, note 48, at 252–54.

148. Warren, *supra*, note 1, at 221–22.

149. See Gardner, *supra*, note 118, at 233–34.

150. Minutes, Regular Session, October 19, 1951, Bancroft.

151. Quoted in Katcher, *supra*, note 48, at 252.

152. See *San Francisco Chronicle*, April 8, 1951.

153. Warren, *supra*, note 1, at 218.

154. Ibid.

155. Cal. Const. Art 9, § 9.
156. Cal. Const. Art. 20, § 3.
157. Warren, *supra*, note 1, at 221.
158. *Sweezy v. New Hampshire*, 354 U.S. 234, 250 (1957).
159. Max Radin wrote an article on the loyalty oath controversy in March and April, 1950, in which he said that "the integrity of the University . . . requires an atmosphere where there is no sense of fear or constraint, no shadow of a Kommissar," and that "when men engaged in intellectual pursuits . . . are told that each year they may not go about their tasks until they have made a public protestation of their loyalty, the atmosphere is poisoned and full intellectual freedom ends." Radin also singled out Warren for praise in voting against the oath. Radin, "The Loyalty Oath at The University of California," 36 *Amer. Assoc. of University Professors Bulletin* 237, 244–45 (Summer, 1950). Radin died on June 22, 1950, at the age of seventy. He would not have signed the oath had he still been a member of the Berkeley faculty, but he had retired two years earlier.
160. See *Oakland Tribune*, September 22, 1950.
161. *Sweezy v. New Hampshire*, *supra*, note 158.
162. E.g., *Elfbrandt v. Russell*, 384 U.S. 11 (1966); *Keyishian v. New York Board of Regents*, 385 U.S. 580 (1967).

## CHAPTER 5—GOVERNING: NATIONAL POLITICS AND THE SUPREME COURT NOMINATION

1. E. Warren, *The Memoirs of Chief Justice Warren* 5 (1977).
2. Ibid. at 238.
3. *Oakland Tribune*, May 28, 1936.
4. *Los Angeles Times*, September 13, 1936.
5. Warren, *supra*, note 1, at 122.
6. Ibid. at 123.
7. Quoted in L. Katcher, *Earl Warren: A Political Biography* 202–3 (1967).
8. *Los Angeles Times*, June 28, 1944.
9. Warren, *supra*, note 1, at 241.
10. See, e.g., *New York Times*, April 18, 1944, where Warren criticized the New Deal's emphasis on federal rather than state relief programs.
11. See Walter P. Jones, "An Editor's Long Friendship with Earl Warren," in *Bee Perspectives of the Warren Era* 18 (1976), Earl Warren Oral History Collection, Regional Oral History Office, the Bancroft Library, University of California at Berkeley. Hereafter cited as ROHO. Warren's remarks to Jones are quoted below, pp. 135.
12. This language is from Warren's declaration of his candidacy for governor in 1942, quoted in *Oakland Tribune*, April 10, 1942.
13. Quoted in Frank Taylor, "California's Bigger Headache," *Saturday Evening Post*, August 14, 1948, 21.
14. Warren, *supra*, note 1, at 240.
15. Ibid. at 241.
16. Stone, "Earl Warren's Friend and Biographer," in *The Warrens: Four Personal Views* 18 (1976), ROHO.
17. Warren, *supra*, note 1, at 241.
18. Ibid. at 242.
19. Jones, *supra*, note 11.

20. William T. Sweigert, *Democrat, Friend, and Advisor to Earl Warren* 184 (in process), ROHO.

21. Ibid. at 185, 186.

22. Warren, *supra*, note 1, at 244.

23. Quoted in J. Weaver, *Warren: The Man, The Court, The Era* 158 (1967); *Los Angeles Times*, November 5, 1948.

24. Quoted in Katcher, *supra*, note 7, at 234.

25. Quoted in *Los Angeles Times*, June 26, 1948.

26. Merrell F. Small, *The Office of the Governor Under Earl Warren* 84 (1972), ROHO.

27. Warren, *supra*, note 1, at 249.

28. Quoted in Katcher, *supra*, note 7, at 216–17.

29. J. Gunther, *Inside U.S.A.* 18 (1947).

30. Warren, *supra*, note 1, at 249.

31. Quoted in Katcher, *supra*, note 7, at 240.

32. *Los Angeles Times*, September 17, 1949.

33. Ibid., September 21, 1949.

34. Warren, *supra*, note 1, at 248.

35. Quoted in Katcher, *supra*, note 7, at 262.

36. Warren, *supra*, note 1, at 249.

37. See Katcher, *supra*, note 7, at 251–52.

38. Warren, *supra*, note 1, at 253.

39. *Time Magazine*, June 21, 1948, 23.

40. D. Eisenhower, *Mandate for Change: The White House Years* 228 (1963).

41. Warren, *supra*, note 1, at 252.

42. See R. Nixon, *Six Crises* 321–23 (1962).

43. A caucus of Republican governors, including Warren, met at the convention and voted to support the idea of a "fairplay" resolution. See *San Francisco Chronicle*, July 4, 1952.

44. In an oral interview for the Regional Oral History Office's Earl Warren Project, Herbert Brownell stated that he believed that "the outcome of the convention was going to depend upon [the 'fair play' resolution]," but that Warren sponsored the resolution because otherwise the nomination would have gone to Taft. Brownell, interview with Amelia R. Fry, October 29, 1974, in *Earl Warren: The Chief Justiceship* 43–44 (1977), ROHO. Most other accounts of the convention maintain that Taft could at best hope for a deadlock on the first ballot. See sources cited in L. Henderson, "Earl Warren and California Politics," unpublished Ph.D. dissertation, University of California at Berkeley, 381–97 (1965).

45. The Douglas campaign story is set forth in Katcher, *supra*, note 7, at 26–62.

46. Quoted in *Los Angeles Times*, November 16, 1951.

47. Warren, *supra*, note 1, at 252, 254.

48. See J. Pollack, *Earl Warren: The Judge Who Changed America* 214–15 (1979).

49. Benno Schmidt, Jr., oral history 71, Columbia University Oral History Research Office, Butler Library, Columbia University.

50. *San Francisco Chronicle*, September 27, 1952.

51. Quoted in *Los Angeles Times*, November 1, 1952.

52. Weaver, *supra*, note 23, at 185.

53. Warren, *supra*, note 1, at 259–60.

54. Ibid. at 260–61.

55. See Katcher, *supra*, note 7, at 270.

56. Warren's and Brownell's accounts of the Brownell-Warren meeting in May, 1953 and of Warren's subsequent consideration of the solicitor generalship offer are remarkably similar. Warren's account is in Warren, *supra*, note 1, at 268–69; Brow-

nell's in "Earl Warren's Appointment to the Supreme Court: A Retrospective Memorandum," in *Brownell, supra*, note 44, at 4–5.

57. Brownell, *supra*, note 56, at 4, 5.
58. Warren, *supra*, note 1, at 268.
59. Ibid. at 269.
60. Notably Justice Felix Frankfurter. See below, pp. 177–78.
61. Brownell, *supra*, note 44, at 53, 54, 62.
62. Brownell, *supra*, note 56, at 15.
63. See *Los Angeles Times*, September 6, 1953; *Nation*, August 15, 1953, 153.
64. Brownell, *supra*, note 56, at 7; Brownell, *supra*, note 44, at 57–59.
65. Brownell, *supra*, note 56, at 8.
66. Eisenhower, *supra*, note 40.
67. Brownell, *supra*, note 56, at 9, 6; Eisenhower, *supra*, note 40, at 228–29.
68. Brownell, *supra*, note 56, at 10.
69. Warren, interview, June 22, 1972, 378–79, ROHO; Warren, *supra*, note 1, at 270.
70. Brownell, *supra*, note 56, at 65–66.
71. Merrell Small, letter to Amelia R. Fry, November 16, 1972, in *Earl Warren: The Chief Justiceship, supra*, note 44, at 1, 4.
72. Ibid. at 3.
73. Ibid.
74. Merrell Small, "Warren Held His Ground When Top Bench Opened," *Sacramento Bee*, June 7, 1970.
75. Weaver, *supra*, note 23, at 193.
76. Warren, *supra*, note 1, at 270.
77. Brownell, *supra*, note 56, at 65.
78. See Small, *supra*, note 71, at 3.
79. Ibid.
80. Brownell, *supra*, note 56, at 54.
81. Ibid at 60.
82. Ibid at 61.
83. See *Los Angeles Times*, October 1, 1953.
84. Brownell, *supra*, note 56, at 70–71.
85. Earl Warren to Herbert Brownell, November 8, 1945, in Earl Warren papers, California State Archives, Sacramento.


PART TWO

CHAPTER 6—THE CRUCIBLE OF *BROWN V. BOARD OF EDUCATION*

1. E. Warren, *The Memoirs of Earl Warren* 275–77 (1977).
2. Ibid. at 277.
3. Ibid. at 280.
4. *Voris v. Eikel*, 346 U.S. 328 (1953).
5. Warren, *supra*, note 1, at 272.
6. 347 U.S. 483 (1954).
7. See R. Kluger, *Simple Justice* (1976); W. Harbaugh, *Lawyer's Lawyer* 507, 516–17 (1973); White, "The Supreme Court's Public and the Public's Supreme Court," 52 *Va. Q.* 370, 385–86 (1976); Ulmer, "Earl Warren and the *Brown* Decision," 32 *Journal of Politics* 689 (1971). See also Harbaugh, Book Review, 62 *Va. L. Rev.* 1311, 1314–16 (1976). Of these sources Kluger's is the most extensive. Except where suc-

ceeding notes suggest a discrepancy between my findings and Kluger's, I have followed his account.

8. See William Byran Rumford, *Legislator for Fair Employment, Fair Housing, and Public Health* 34 (1973) in Earl Warren Oral History Collection, Regional Oral History Office, the Bancroft Library, University of California at Berkeley. Hereafter cited as ROHO.

9. Earl Warren, Jr., "California Politics," in *The Governor's Family* 37 (1980), ROHO.

10. Earl Warren, untitled interview, June 22, 1972, 281, ROHO; see also Warren, *supra*, note 1, at 4.

11. Rumford, *supra*, note 8, at 35.

12. 163 U.S. 537 (1896).

13. Earl Warren, interview with Richard Kluger, June 16, 1971, quoted in Kluger, *supra*, note 7, at 683.

14. Diary, Harold H. Burton papers, Manuscripts Division, Library of Congress (hereafter cited as Burton papers); see also Ulmer, *supra*, note 7, at 694.

15. See Lewis, "Earl Warren," in 4 L. Friedman and F. Israel, *The Justices of the United States Supreme Court* 2721, 2725 (4 vols., 1969).

16. See Burton papers, relied upon by Kluger, *supra*, note 7, at 611–12. Kluger also cites Justice Robert Jackson's conference notes in the Jackson papers, currently in the possession of Professor Philip Kurland of the University of Chicago Law School.

17. Frankfurter, memorandum to Justices, January 15, 1954, Felix Frankfurter papers, Manuscripts Division, Library of Congress. See generally Kluger, *supra*, note 7, at 684–87.

18. No sources are authoritative on this point. Warren told Richard Kluger, in a June 16, 1971 interview, that the vote was taken in late February and also indicated that February was the month in his memoirs. Warren, *supra*, note 1, at 285. But Warren also told *Ebony* magazine, in an interview published in May, 1974, that the vote took place in late March, (Slater, "1954 Revisited," 29 *Ebony* 116, 126 (1954)) and told Jack Harrison Pollack, who was preparing a biography of Warren, that the justices were "ready to vote" in late February and "took a vote early in March." Quoted in J. Pollack, *Earl Warren: The Judge Who Changed America* 174 (1979). Because of his strong feelings about confidentiality in the Court's proceedings and his interest in minimizing his role in the *Brown* decision, Warren chose to reveal very little information about the inner history of the case, and that which he did reveal must be used with care. Warren said, in his memoirs, for example, that "on the first vote [which he claimed to be in February] we unanimously agreed that the 'separate but equal' doctrine had no place in public education." (Warren, *supra*, note 1, at 285). But evidence has surfaced that Justice Reed had prepared a dissent in February, which he subsequently withdrew. See Kluger, *supra*, note 7, at 691–93.

19. See Frankfurter, memorandum to Justices, *supra*, note 17; Warren, *supra*, note 1, at 286–87.

20. Warren, memorandum to Justices, May 7, 1954, Burton papers, quoted in Ulmer, *supra*, note 7, at 698, and in Kluger, *supra*, note 7, at 696.

21. See Kluger, *supra*, note 7, at 696–98; diary entry for May 12, 1954, Burton papers.

22. George Mickum, interview with Richard Kluger, October 20, 1974, quoted in Kluger, *supra*, note 7, at 698.

23. See Warren, *supra*, note 1, at 286.

24. Felix Frankfurter to Learned Hand, July 21, 1954, Felix Frankfurter papers, Manuscripts Division, Library of Congress.

25. Felix Frankfurter to Learned Hand, October 12, 1957, Learned Hand papers, in possession of Professor Gerald Gunther, Stanford Law School. Reprinted with permission.

26. Frankfurter to Hand, *supra*, note 24.

27. Diary, Burton papers; see Ulmer, *supra*, note 7, at 699–700.

28. *New York Times*, March 4, 1954.

29. 358 U.S. 1 (1958). See discussion below, pp. 183–84.

30. See, e.g., Frankfurter to Hand, *supra*, note 25.

31. 347 U.S. at 493.

32. Ibid. at 494.

## CHAPTER 7—REACTING TO FELIX FRANKFURTER

1. The best sources on Frankfurter's pre-Court career are H. Hirsch, *The Enigma of Felix Frankfurter* (1981) and J. Lash, ed., *From The Diaries of Felix Frankfurter* (1975). See also L. Baker, *Felix Frankfurter* (1969) and G. White, *The American Judicial Tradition* 325–28 (1976), and sources cited therein. Frankfurter's role as Brandeis' paid agent, which was a closely guarded secret, is revealed in Murphy, "Elements of Extrajudicial Strategy," 69 *Geo. L. J.* 101, 111–13 (1980).

2. Felix Frankfurter to Oliver Wendell Holmes, September 6, 1913, Oliver Wendell Holmes papers, Harvard Law School Library, Cambridge, Massachusetts.

3. Felix Frankfurter to Julian Mack, September 25, 1927, Felix Frankfurter papers, Manuscripts Division, Library of Congress.

4. Hirsch, *supra*, note 1, at 132.

5. Felix Frankfurter to Marion Denman Frankfurter, June 25, 1929, Felix Frankfurter papers, *supra*, note 3. See Hirsch, *supra*, note 1, at 62.

6. M. Freedman, ed., *Roosevelt and Frankfurter: Their Correspondence, 1928–1945* 27 (1967).

7. See H. Ickes, 2 *The Secret Diary of Harold Ickes* 539–40 (2 vols., 1953); Lash, *supra*, note 1, at 64; Hirsch, *supra*, note 1, at 124–25.

8. Hirsch, *supra*, note 1, at 177.

9. See ibid. at 174–91.

10. See Lash, *supra*, note 1, at 35–36.

11. See Hirsch, *supra*, note 1, at 43, 98; Lash, *supra*, note 1, at 76–79.

12. Of this group Frankfurter was close only to Roberts, although his relationships with Burton, Clark, and Reed were cordial and his exchanges with Murphy were marked by good humor. See Hirsch, *supra*, note 1, at 155–71.

13. See R. Kluger, *Simple Justice* 664 (1976).

14. 349 U.S. 294 (1955).

15. Felix Frankfurter to Charles C. Burlingham, February 15, 1954, Felix Frankfurter papers, Manuscripts Division, Library of Congress. Subsequent citations refer to the Library of Congress collection unless otherwise indicated. Schwartz, "Felix Frankfurter and Earl Warren: A Study of a Deteriorating Relationship," 1980 S.Ct. Rev. 115, an article which came to my attention after this chapter had been written, is a helpful analysis of Frankfurter and Warren's relationship which, while concentrating primarily on Frankfurter's reactions to Warren, corroborates many of this chapter's findings.

16. Frankfurter to Robert Jackson, April 15, 1954, ibid.

17. Frankfurter to Burlingham, May 4, 1954, ibid.

18. Frankfurter to Burlingham, May 28, 1954, ibid.

19. Warren to Frankfurter, August 6, 1954, ibid.

20. Frankfurter to Warren, May 1, 1954, ibid.

21. Frankfurter to Warren, June 5, 1954, ibid.

22. Frankfurter to Warren, August 20, 1954, ibid.

23. Frankfurter to Warren, December 7, 1954, ibid.

24. Frankfurter to Warren, March 26, 1955, ibid.

25. Frankfurter to Warren, June 5, 1954, ibid.
26. Frankfurter to Warren, March 26, 1955, ibid.
27. Hand to Frankfurter, June 5, 1954, ibid.
28. Hand to Frankfurter, September 15, 1955, ibid.
29. Hand to Frankfurter, January 1, 1956, ibid.
30. Hand to Frankfurter, October 25, 1956, ibid.
31. Hand to Frankfurter, July 21, 1959, ibid.
32. Hand to Frankfurter, October 25, 1956, ibid.
33. Hand to Frankfurter, February 26, 1956, ibid.
34. Frankfurter to Hand, January 1, 1956, Learned Hand papers, in possession of Gerald Gunther, Stanford Law School. Reprinted by permission.
35. Jacobs, "The Warren Court—After Three Terms," 10 *Western Pol. Q.* 937, 940, 942, 946, 953, 954 (1956).
36. Ibid. at 937.
37. Frankfurter to Hand, June 30, 1957, Learned Hand papers, in possession of Gerald Gunther, Stanford Law School. Reprinted by permission.
38. Frankfurter to Harlan, September 2, 1958, Frankfurter papers.
39. Kurland to Frankfurter, July 18, 1960, ibid.
40. Frankfurter to Bickel, March 18, 1963, ibid.
41. Frankfurter to Bickel, June 8, 1964, ibid.
42. Frankfurter to Kurland, October 16, 1964, ibid.
43. Warren to Frankfurter, December 11, 1958, ibid.
44. 357 U.S. 549 (1958).
45. Ibid. at 559.
46. See Brennan, "Chief Justice Warren," 88 *Harv. L. Rev.* 1, 2 (1974).
47. 354 U.S. 178 (1957). For a fuller discussion of the case's significance for Warren's jurisprudence, see below, pp. 243–49.
48. Warren to Frankfurter, May 22, 1957, Frankfurter papers.
49. Frankfurter to Hand, *supra*, note 37.
50. 358 U.S. 1 (1958).
51. William T. Sweigert, *Democrat, Friend, and Advisor to Earl Warren* 218 (in process), Earl Warren Oral History Collection, Regional Oral History Office, the Bancroft Library, University of California at Berkeley.
52. E. Warren, *The Memoirs of Earl Warren* 298–99 (1977).
53. 356 U.S. 86 (1958).
54. See Lewis, "Earl Warren," in 4 L. Friedman and F. Israel, *The Justices of the United States Supreme Court* 2730 (1969).
55. See Rodell, "It Is The Warren Court," *New York Times Magazine*, March 13, 1966, 93–94.
56. Warren, "Mr. Justice Harlan, As Seen by a Colleague," 85 *Harv. L. Rev.* 369, 370 (1971).
57. Quoted in H. Black, Jr., *My Father: A Remembrance* 229 (1975).
58. Warren, "A Tribute to Hugo L. Black," 85 *Harv. L. Rev.* 1, 2 (1971).
59. Frankfurter to Hand, July 4, 1961, Learned Hand papers, Harvard Law School Library.
59a. 369 U.S. 186 (1962).
60. 376 U.S. 254 (1964).
61. See Benno Schmidt, Jr., oral history 69–70, Oral History Research Office, Butler Library, Columbia University. Schmidt's language was "The Chief told us—and I'm sure it's true—that he refused to even look at the stuff, that he delegated to Justice Brennan the task of reviewing the stuff . . . ."
62. See below, pp. 258–59.

63. 369 U.S. 186 (1962).
64. 329 U.S. 549 (1946).
65. See Ely, "The Chief," 88 *Harv. L. Rev.* 11, 12 (1974).


## CHAPTER 8—THE WARREN COMMISSION

1. U.S. Congress, House, Select Committee on Assassinations, *Final Report of the Select Committee on Assassinations*, 96th Cong., 1st sess., 1979, 256. Hereafter cited as *Final Report*.
2. Ibid. at 261, 256.
3. L. Johnson, *The Vantage Point* 26 (1971).
4. E.g., Norman Redlich, Warren Commission staff member, in testimony before the House Select Committee on Assassinations, November 8, 1977, cited in U.S. Congress, House, Select Committee on Assassinations, *Appendix to Hearings Before the Select Committee*, 96th Cong., 1st sess., 1979, 23. Hereafter cited as *Appendix*.
5. E. Warren, *The Memoirs of Earl Warren* 356 (1977).
6. Ibid. at 358.
7. Johnson, *supra*, note 3, at 27.
8. Melvin Eisenberg, memorandum to files, February 17, 1964, quoted in *Appendix*, *supra*, note 4, at 27.
9. W. David Slawson, testimony before the Select Committee, November 15, 1977, cited in ibid. at 12, 13.
10. Warren, *supra*, note 5, at 362, 367.
11. Ibid. at 364, 365, 367.
12. Ibid., at 365, 363–64.
13. Ibid. at 362.
14. Johnson, *supra*, note 3, at 26.
15. *Final Report*, *supra*, note 1, at 4.
16. Gerald R. Ford, testimony before the Select Committee, September 21, 1978, cited in *Appendix*, *supra*, note 4, at 16.
17. Warren Commission executive session transcript, December 5, 1963, National Archives.
18. Ibid.
19. Warren Commission executive session transcript, December 6, 1963, National Archives.
20. Warren Commission executive session transcript, December 16, 1963, National Archives.
21. G. Ford, *Portrait of the Assassin* 14 (1965).
22. Warren Commission executive session transcript, January 22, 1964, National Archives.
23. Warren Commission executive session transcript, December 16, 1963, National Archives.
24. Warren Commission executive session transcript, January 27, 1964, National Archives.
25. See *Appendix*, *supra*, note 4, at 53. See generally U.S. Congress, Senate, Select Committee to Study Governmental Operations, *The Investigation of the Assassination of President John F. Kennedy: Performance of the Intelligence Agencies*, 94th Cong., 2d sess., 1976. Hereafter cited as *Performance*.
26. *Performance* at 46 (remarks of J. Edgar Hoover).
27. Ibid. at 85.
28. See generally *Appendix*, *supra*, note 4, at 57–71.

29. J. Lee Rankin, deposition before Select Committee, August 17, 1978, cited in *Appendix, supra,* note 4, at 49.

30. Burt W. Griffin, testimony before Select Committee, November 17, 1977, cited in ibid. at 43.

31. Wesley Liebeler, testimony before Select Committee, November 15, 1977, cited in ibid. at 42.

32. W. David Slawson, *supra,* note 9, cited in ibid. at 42.

33. *Watkins v. United States,* 354 U.S. 178, 195 (1957).

34. Warren, *supra,* note 5, at 371.

35. See *Appendix, supra,* note 4, at 92, 93.

36. Arlen Specter, testimony before Select Committee, November 8, 1977, quoted in ibid. at 102.

37. Warren, *supra,* note 5, at 371.

38. Rankin, *supra,* note 29, cited in *Appendix, supra,* note 4, at 379.

39. The texts of Specter's memoranda are set forth in *Appendix, supra,* note 4, at 92–93.

40. Specter, *supra,* note 36, cited in ibid. at 92.

41. See Warren Commission executive session transcripts, *supra,* notes 20 and 22.

42. President's Commission on the Assassination of President Kennedy, *Report* 19 (1964). Hereafter cited as *Warren Commission Report.*

43. *Final Report, supra,* note 1, at 3.

44. Ibid.

45. Specter, *supra,* note 36, cited in *Appendix, supra,* note 4, at 92.

46. Warren Commission executive session transcript, April 30, 1964, National Archives.

47. Quoted in J. Pollack, *Earl Warren: The Judge Who Changed America* 239–40 (1979).

48. Redlich, *supra,* note 4, cited in *Appendix, supra,* note 4, at 113.

49. Norman Redlich to J. Lee Rankin, February 28, 1964, cited in *Appendix, supra,* note 4, at 126–27.

50. Warren Commission executive session transcript, *supra,* note 20.

51. Warren Commission executive session transcript, *supra,* note 22.

52. See testimony of Specter, *supra,* note 36, cited in *Appendix, supra,* note 4, at 85.

53. Earl Warren, Jr., "California Politics," in *The Governor's Family* 25 (1980), Earl Warren Oral History Collection, Regional Oral History Office, The Bancroft Library, University of California at Berkeley. Hereafter cited as ROHO.

54. The text of Warren's remarks is set forth in Pollack, *supra,* note 47, at 224.

55. See, e.g., *Time Inc. v. Hill,* 385 U.S. 374 (1967).

56. Compare Warren Commission executive session transcript, January 21, 1964, National Archives, with Warren Commission executive session transcript, April 30, 1964, *supra,* note 46.

57. Warren, *supra,* note 5, at 372.

58. Rankin, *supra,* note 29, cited in *Appendix, supra,* note 4, at 348.

59. See ibid. at 101.

60. Warren Commission executive session transcript, January 21, 1964, *supra,* note 56.

61. Howard P. Willens, testimony before Select Committee, November 17, 1977, cited in *Appendix, supra,* note 4, at 333.

62. See Warren, "Oral History Project," 28, National Archives.

63. Ibid.

64. *Supra,* note 60.

65. Slawson, *supra,* note 9, cited in *Appendix, supra,* note 4, at 28.

66. Willens, *supra,* note 61, cited in ibid. at 384.

67. Specter, *supra*, note 36, cited in ibid. at 101.
68. See ibid.
69. See comments by Wesley Liebeler and Howard Willens in ibid. at 88, 326.
70. John Sherman Cooper, testimony before Select Committee, September 21, 1978, cited in ibid. at 25.
71. *Warren Commission Report*, *supra*, note 42, at 19.
72. See *Appendix*, *supra*, note 4, at 25.
73. *Warren Commission Report*, *supra*, note 42, at 21–22. Italics added. See Rankin, *supra*, note 29, cited in *Appendix*, *supra*, note 4, at 355–56.
74. Rankin, *supra*, note 29, cited in ibid. at 356.
75. Slawson, *supra*, note 9, cited in ibid. at 197.
76. The language here is Arlen Specter's. See ibid. at 101.
77. See R. Kluger, *Simple Justice* 674 (1976).
78. For a summary of criticism of the Warren Commission, see J. David, *The Weight of the Evidence* (1968).
79. See Warren Commission executive session transcript, January 27, 1964, National Archives.
80. See Warren Commission, executive session transcript, December 16, 1963, National Archives.
81. See David, *supra*, note 78.
82. Earl Warren, Jr., *supra*, note 53, at 7.
83. Benno Schmidt, Jr., oral history 57, Columbia University Oral History Research Office, Butler Library, Columbia University.


PART THREE

CHAPTER 9—WARREN'S THEORY OF JUDGING

1. Lewis, "Earl Warren," in 4 L. Friedman & F. Israel, *The Justices of the United States Supreme Court* 2721, 2723–24 (4 vols., 1969).
2. This chapter frequently employs the terms "ethical imperatives" and "ethical structure of the Constitution." The terms are not intended to convey a precise philosophical meaning. They refer to a conception of the Constitution as embodying certain fundamental ethical principles—such as fairness, equality, privacy, and dignity—that are not necessarily enumerated in the Constitution's literal text but may be extrapolated from the text and the history of American civilization. For variations on this theme, see C. Black, *Structure and Relationship in Constitutional Law* (1969); Grey, "Do We Have an Unwritten Constitution?," 27 *Stan. L. Rev.* 703 (1975).
3. While Warren's theory of adjudication consistently produced "one right answer," it should not be seen as anticipating the "rights thesis" of Ronald Dworkin, see R. Dworkin, *Taking Rights Seriously* 82–90 (2d ed., 1978). Warren did not reason from orthodox doctrinal principles to explanations and "best possible justifications." He brought his own ethical values to the reasoning process and used them in his search for ethical imperatives, sometimes ignoring orthodox doctrinal principles in his reasoning. Although this process may seem to be result-oriented, I shall argue that such a characterization of Warren is hasty and incomplete.
4. See G. White, *Patterns of American Legal Thought* 99–153 (1978); G. White, *The American Judicial Tradition* 291–94, 317–25 (1976). Hereafter cited as *American Judicial Tradition*.
5. 198 U.S. 45 (1905).
6. Ibid. at 76.

7. See Fuller, "Reason and Fiat in Case Law," 59 *Harv. L. Rev.* 376 (1946).
8. See H. Hart and A. Sacks, *The Legal Process* 161 (tent. ed., 1958).
9. See Wechsler, "Toward Neutral Principles of Constitutional Law," 73 *Harv. L. Rev.* 1 (1959).
10. Kurland, Book Review, 87 *Yale L. J.* 225, 233 (1977). Emphasis in original.
11. E. Warren, *The Memoirs of Earl Warren* 332–33 (1977.)
12. See, e.g., Chamberlain, "Reminiscences About the Alameda County District Attorney's Office," in 2 *Perspectives on the Alameda County District Attorney's Office* 24–25 (3 vols., 1972–74, Earl Warren Oral History Collection, Regional Oral History Office, The Bancroft Library, University of California at Berkeley.

A manifestation of Warren's efforts to maintain principles of ethics in law enforcement was his practice of face-to-face confrontations with persons he was about to prosecute. After having prepared his case, but before actually proceeding to trial, Warren would call in his target for an interview, reveal to him the decision to prosecute, and offer him a "compromise" in which the target agreed to cease his operations so as to avoid public censure. The talks did not amount to any effort on Warren's part to play favorites; he offered the compromises to anyone and was not moved by the overtures of close acquaintances. See Warren, *supra*, note 11, at 92–103, 131–33 (1977).

13. Warren to Robert Kenny, July 20, 1938, reprinted in *Los Angeles Times*, July 24, 1938.
14. Warren, "The Law and the Future," *Fortune*, November, 1955, 106, 224.
15. Ibid. at 107.
16. Ibid. at 106.
17. Ibid. at 226.
18. Ibid.
19. Ibid.
20. Ibid.
21. Ibid. at 224.
22. Ibid. at 226.
23. Ibid.
24. Warren, address, Louis Marshall Award Dinner at the Jewish Theological Seminary in America, November 11, 1962, 1, 8–9.
25. Ibid. at 11.
26. *Poe v. Ullman*, 367 U.S. 497 (1961).
27. *Griswold v. Connecticut*, 381 U.S. 479, 486 (1965).
28. 347 U.S. 497 (1954).
29. 347 U.S. 483 (1954).
30. Hutchinson, "Unanimity and Desegregation: Decisionmaking in the Supreme Court, 1948–1958," 68 *Geo. L. Rev.* 1, 44–50 (1979).
31. Warren, Memorandum on the District of Columbia Case (May 7, 1954), Harold Burton papers, Manuscript Divison, Library of Congress.
32. Ibid.
33. 347 U.S. at 493.
34. See, e.g., *Lincoln Fed. Labor Union v. Northwestern Iron & Metal Co.*, 335 U.S. 525, 537 (1949) (Black); *A.F.L. v. American Sash & Door Co.*, 335 U.S. 538, 542 (1949) (Frankfurter).
35. 347 U.S. at 499–500.
36. Ibid. at 500.
37. See Hutchinson, *supra*, note 30, at 44–47.
38. Benno Schmidt, Jr., oral history 84, Columbia University Oral History Research Office, Butler Library, Columbia University.

39. Ibid. at 35.
40. Warren, *supra*, note 14, at 106.
41. See, e.g., Lewis, *supra*, note 1, at 2725; Greenberg, "Earl Warren—A Tribute," 58 *Calif. L. Rev.* 3, 30 (1970).
42. See *Brown v. Board of Education*, 347 U.S. 483 (1954).
43. See *Reynolds v. Sims*, 377 U.S. 533 (1964).
44. See *Escobedo v. Illinois*, 378 U.S. 478 (1964).
45. See *Marchetti v. United States*, 390 U.S. 39, 77 (1968) (dissent).
46. See *Ginzburg v. United States*, 383 U.S. 463 (1966); *Roth v. United States*, 354 U.S. 476, 494 (1957).
47. Warren, *supra*, note 24, at 10.
48. See Schmidt, *supra*, note 38, at 33–34.
49. Ibid. at 26.

## CHAPTER 10—LESSONS IN CIVICS

1. See E. Warren, *A Republic, If You Can Keep It* (1972).
2. 356 U.S. 44 (1958).
3. 356 U.S. 86 (1958).
4. Ibid. at 92.
5. 356 U.S. at 64.
6. Ibid.
7. *Fletcher v. Peck*, 10 U.S. (6 Cranch) 87, 139 (1810).
8. 356 U.S. at 102.
9. Ibid. at 101.
10. Ibid.
11. Ibid.
12. Ibid. at 103.
13. Ibid. at 104.
14. See, e.g., A. Bickel, *The Supreme Court and the Idea of Progress* (1970); Kurland, Book Review, 87 *Yale L.J.* 225 (1977); Wechsler, "Toward Neutral Principles of Constitutional Law," 73 *Harv. L. Rev.* 1 (1959).
15. Warren, *supra*, note 1, at 83–85.
16. 377 U.S. 533 (1964).
17. 357 U.S. 916 (1958).
18. Ibid. at 916–17.
19. 369 U.S. 186 (1962).
20. E. Warren, *The Memoirs of Earl Warren* 310 (1977).
21. Ibid. at 308.
22. 377 U.S. at 555.
23. See ibid. at 593–608 (Harlan, J., dissenting).
24. Ibid. at 555.
25. See *Lucas v. Forty-Fourth Gen. Assembly of Colo.*, 377 U.S. 713 (1964).
26. 377 U.S. at 566.
27. See, e.g., *Colegrove v. Green*, 328 U.S. 549 (1946).
28. 377 U.S. at 580.
29. Ibid. at 624–25 (Harlan, J., dissenting).
30. See Warren, *supra*, note 20, at 310.
31. Ibid.
32. See Hexter, Book Review, 16 *Hist. & Theory* 306, 331–37 (1977).
33. 384 U.S. 436 (1966).

34. Lewis, "A Talk with Warren on Crime, the Court, the Country," *New York Times Magazine*, October 19, 1969, 35.
35. 347 U.S. 442 (1954).
36. 349 U.S. 331 (1955).
37. 354 U.S. 178 (1957).
38. Ibid. at 195.
39. Ibid. at 185.
40. Ibid. at 198.
41. Ibid. at 197.
42. Ibid. at 200.
43. Ibid. at 205.
44. Ibid.
45. Ibid. at 205–6.
46. Ibid. at 198–99.
47. Ibid. at 215.
48. Ibid. at 195.
49. Ibid.
50. Ibid. at 215.
51. 354 U.S. 234 (1957).
52. Ibid. at 236–37.
53. Ibid. at 245.
54. Ibid. at 250.
55. Ibid. at 251.
56. 360 U.S. 474 (1959).
57. See ibid. at 508.
58. Ibid. at 492.
59. Ibid. at 496.
60. Ibid. at 506–7.
61. See ibid. at 512–13 (Clark, J., dissenting).
62. See ibid. at 515.
63. Ibid. at 497.
64. Ibid. at 509 (Harlan, J., concurring).
65. See *Barenblatt v. United States*, 360 U.S. 109, 134 (1959).
66. See *Uphaus v. Wyman*, 360 U.S. 72, 82 (1959).
67. See *Barenblatt v. United States*, 360 U.S. at 134.
68. See *Albertson v. Subversive Activities Control Bd.*, 382 U.S. 70 (1965).
69. See *Aptheker v. Secretary of State*, 378 U.S. 500 (1964).
70. See *United States v. Brown*, 381 U.S. 437 (1965).
71. See *United States v. Robel*, 389 U.S. 258 (1967).
72. See *United States v. Brown*, 381 U.S. at 456–62.
73. Quoted in Warren, *supra*, note 20, at 5.
74. Ibid. at 322–25.
75. Ibid. at 5–6.
76. Ibid. at 5.
77. Ibid. at 6.

## CHAPTER 11—"A RIGHT TO MAINTAIN A DECENT SOCIETY"

1. M. Landsberg, *A Conversation With Chief Justice Earl Warren* 6 (1969).
2. See Benno Schmidt, Jr., oral history 96, Columbia University Oral History Research Office, Butler Library, Columbia University.

3. E. Warren, *A Republic, If You Can Keep It* 159–60 (1972).
4. Ibid. at 160.
5. Schmidt, *supra*, note 2, at 70.
6. *Roth v. United States*, 354 U.S. 476, 494 (1957).
7. Ibid. at 496.
8. Warren, *supra*, note 3, at 160.
9. 386 U.S. 767 (1967).
10. See Schmidt, *supra*, note 2, at 69–70.
11. Quoted in Landsberg, *supra*, note 1.
12. Warren, *supra*, note 3, at 159–60.
13. Ibid. at 159.
14. Quoted in Landsberg, *supra*, note 1.
15. Warren, *supra*, note 3, at 159.
16. Quoted in Landsberg, *supra*, note 1, at 6–7.
17. See *Newsweek*, May 11, 1964, 28.
18. 354 U.S. 436 (1957).
19. 354 U.S. 476 (1957).
20. 354 U.S. at 445–46 (dissent).
21. 354 U.S. at 495–96 (dissent).
22. Ibid. at 495.
23. Ibid.
24. Ibid.
25. Ibid. at 494.
26. Ibid. at 495.
27. 365 U.S. 43 (1961).
28. Ibid. at 55 (dissent).
29. See ibid. at 69.
30. Ibid. at 51.
31. *New York Times*, December 8, 1965.
32. 354 U.S. at 489.
33. See Kalven, "The Metaphysics of the Law of Obscenity," 1960 *Sup. Ct. Rev.* 1, 43.
34. 378 U.S. 184 (1964).
35. Ibid. at 200, 202–3 (dissent).
36. See ibid. at 201.
37. 383 U.S. 463 (1966).
38. Ibid. at 474.
39. Justice Brennan's opinion in a companion case again emphasized that obscene material must be utterly without redeeming social importance. See *A Book Named "John Cleland's Memoirs of a Woman of Pleasure" v. Attorney Gen.*, 393 U.S. 413, 418–19 (1966).
40. See 378 U.S. at 195.
41. 394 U.S. 557 (1969).
42. Ibid. at 568.

## CHAPTER 12—LAW ENFORCEMENT FROM THE OTHER SIDE

1. In 1926, in a campaign for district attorney of Alameda County, Warren reportedly boasted about his conviction record as a prosecutor and was subsequently embarrassed about the incident. See Miriam Feingold, "The King-Ramsay-Conner Case: Labor, Radicalism, and the Law in California, 1936–1941," unpublished Ph.D. dis-

sertation, University of Wisconsin-Madison 789 (1976). See also *Oakland Tribune*, December 28, 1968.

2. Quoted in *Oakland Tribune*, September 20, 1931.

3. Quoted in Lewis, "A Talk With Warren on Crime, the Court, the Country," *New York Times Magazine*, October 19, 1969, 124.

4. Warren, "The Law and the Future," *Fortune*, November, 1955, 106, 226.

5. Quoted in *Oakland Tribune*, December 28, 1968.

6. Jester, "Reminiscences of an Inspector in the District Attorney's Office," in 2 *Perspectives on the Alameda County District Attorney's Office* 65–66 (3 vols., 1972–74), Earl Warren Oral History Collection, Regional Oral History Office, The Bancroft Library, University of California at Berkeley. Hereafter cited as ROHO.

7. Quoted in Lewis, *supra*, note 3, at 124.

8. Ibid. at 35.

9. 347 U.S. 128 (1954).

10. 347 U.S. 442 (1954).

11. 347 U.S. 522 (1954).

12. Quoted in Lewis, *supra*, note 3, at 125.

13. See 347 U.S. at 137–38.

14. 338 U.S. 25 (1949).

15. Quoted in Lewis, *supra*, note 3, at 125.

16. Ibid.

17. See Benno Schmidt, Jr., oral history 129–30, Columbia University Oral History Research Office, Butler Library, Columbia University.

18. See *Mapp v. Ohio*, 367 U.S. 643 (1961).

19. See *Cicenia v. Lagay*, 357 U.S. 504, 511 (1958); *Crooker v. California*, 357 U.S. 433, 441 (1958).

20. *Crooker v. California*, 357 U.S. at 444.

21. 360 U.S. 315 (1959).

22. Ibid. at 323.

23. Ibid. at 320–21.

24. 384 U.S. 436 (1966).

25. 378 U.S. 478 (1964).

26. Quoted in Lewis, *supra*, note 3, at 35.

27. Ibid. Italics in original.

28. 384 U.S. at 445.

29. Ibid.

30. Ibid. See ibid. at 445–58.

31. Ibid. at 455.

32. Ibid. at 450.

33. Ibid. at 457.

34. Ibid. at 460.

35. See *Malloy v. Hagen*, 378 U.S. 1 (1964); *Ziang Sung Wan v. United States*, 266 U.S. 1 (1924); *Bram v. United States*, 168 U.S. 532 (1897).

36. 384 U.S. at 510.

37. Ibid. at 512 n. 9.

38. See ibid.

39. Ibid. at 467–68.

40. Ibid. at 469.

41. Ibid. at 470.

42. Ibid. at 473.

43. Ibid. at 475.

44. Ibid. at 476.

45. Ibid.
46. Ibid. at 477.
47. See ibid. at 492.
48. See ibid. at 494.
49. See ibid. at 496.
50. See ibid. at 498.
51. Quoted in Lewis, *supra*, note 3, at 126.
52. See *Miranda v. Arizona*, 384 U.S. at 523 (Harlan, J., dissenting). In dissent Justice Harlan addressed the "ironic untimeliness" of the new rules imposed by *Miranda*, see ibid., and argued that legislative reform was preferable to the Court's changing of the constitutional standards on confessions.
53. See Coakley, "A Career in the Alameda County District Attorney's Office," in 3 *Perspectives on the Alameda County District Attorney's Office* 53–54 (3 vols., 1972–74), ROHO.
54. 384 U.S. at 542.
55. 367 U.S. 643 (1961).
56. 381 U.S. 618 (1965).
57. Ibid. at 636.
58. See *Desist v. United States*, 394 U.S. 244, 258 (1969) (Harlan, J., dissenting).
59. 384 U.S. 719 (1966).
60. Ibid. at 726–27.
61. Ibid. at 731.
62. Ibid. at 733.
63. See *Desist v. United States*, 394 U.S. 244 (1969); *Stovall v. Denno*, 388 U.S. 293 (1967).
64. *Jenkins v. Delaware*, 395 U.S. 213, 219 (1969).
65. 395 U.S. 213 (1969).
66. Ibid. at 218.
67. Ibid. at 219, 221.
68. Ibid. at 221. In dissent, Harlan pointed out that a defendant in a post-*Miranda* retrial had not yet been convicted; thus, he argued, there was no " 'conviction . . . validly obtained' " which might be " 'needlessly aborted' " by giving the defendant the benefit of the *Miranda* rules. Ibid. at 223.
69. Quoted in Lewis, *supra*, note 3, at 126.
70. Miriam Feingold, who interviewed Warren in 1971 and 1972 for the California Regional Oral History Office, referred to his "selective memory" on "the more unpleasant aspects of his earlier career." Feingold, *supra*, note 1, at 790.
71. 392 U.S. 1 (1968).
72. Ibid. at 14.
73. Ibid. at 12.
74. Ibid. at 24–25.
75. See ibid. at 30–31.
76. See *Breithaupt v. Abram*, 352 U.S. 432, 440 (1957) (dissent).
77. 392 U.S. at 38–39.
78. Warren, *supra*, note 4, at 226.


## CHAPTER 13—WARREN THE ECONOMIC THEORIST: LABOR AND ANTITRUST

1. See generally St. Antoine, "Judicial Valour and the Warren Court's Labor Decisions," 67 *Mich. L. Rev.* 317 (1968).

2. See *United States v. Hutcheson,* 312 U.S. 219 (1941); *Allen Bradley Co. v. International Brotherhood of Electrical Workers,* 325 U.S. 797 (1945).
3. The Portal to Portal Act, 61 Stat. 84 (1947).
4. The Railway Labor Act, 44 Stat. 477 (1926).
5. The Taft-Hartley Act, 61 Stat. 146 (1947).
6. *Steinger v. Mitchell,* 350 U.S. 247 (1956); *Mitchell v. King,* 350 U.S. 260 (1956).
7. 348 U.S. 511 (1955).
8. Ibid. at 525 (dissent).
9. 353 U.S. 1 (1957).
10. Ibid. at 11.
11. 353 U.S. 20 (1957).
12. Ibid. at 23.
13. Ibid. at 24.
14. 353 U.S. 26 (1957).
15. Ibid. at 16.
16. Ibid. at 11.
17. Landrum-Griffin Act, 73 Stat. 519 (1959). See 29 U.S.C. § 164(c).
18. 356 U.S. 617 (1958).
19. Ibid. at 620.
20. 356 U.S. 634 (1958).
21. Ibid. at 631.
22. Ibid. at 628.
23. Ibid. at 623.
24. Ibid. at 650.
25. Ibid. at 651.
26. Ibid. at 653.
27. 359 U.S. 236 (1959).
28. Ibid. at 241–43.
29. 379 U.S. 203 (1964).
30. Ibid. at 207.
31. Ibid. at 225.
32. Ibid. at 224.
33. Ibid. at 209.
34. Ibid. at 215.
35. Ibid. at 205.
36. Ibid. at 215.
37. Ibid. at 218.
38. Ibid. at 225.
39. Ibid. at 214.
40. 388 U.S. 26 (1967).
40a. Ibid. at 33–34.
41. Ibid. at 35. Italics in original.
42. Ibid. at 38, 35 (Harlan, J., dissenting).
43. 395 U.S. 575 (1969).
44. Ibid. at 580.
45. Ibid. at 581.
46. Ibid. at 582.
47. Ibid. at 589.
48. Ibid. at 600.
49. Ibid. at 604.
50. Ibid. at 589.
51. Ibid. at 617.

52. Ibid. at 618.
53. Ibid. at 619, 620.
54. See *supra*, note 2.
55. See *United Mine Workers v. Pennington*, 381 U.S. 657 (1965); *Amalgamated Meat Cutters & Butchers v. Jewel Tea Co.*, 381 U.S. 676 (1965).
56. See Kauper, "The 'Warren Court' and the Antitrust Laws," 67 *Mich. L. Rev.* 325, 337 (1969); Arnold, "The Supreme Court and the Antitrust Laws 1953–1967," 34 *A.B.A. Antitrust L. J.* 2, 7 (1967).
57. 351 U.S. 305 (1956).
58. 50 Stat. 693 (1937).
59. 66 Stat. 632 (1952).
60. 351 U.S. at 314.
61. Ibid. at 317 (Harlan, J., dissenting).
62. Ibid. at 316 (Harlan, J., dissenting).
63. 351 U.S. 377 (1956).
64. Ibid. at 400.
65. Ibid. at 418.
66. Ibid. at 417.
67. Ibid. at 424.
68. Ibid. at 426.
69. *Federal Trade Commission v. Anheuser-Busch*, 363 U.S. 536 (1959).
70. Ibid. at 543.
71. *Federal Trade Commission v. Colgate Palmolive Co.*, 380 U.S. 374 (1965).
72. *Federal Trade Commission v. Fred Meyer, Inc.*, 390 U.S. 341 (1968).
73. Ibid. at 349.
74. Ibid. at 352.
75. Ibid. at 361, 360.
76. *Kaplan v. Lehman Brothers*, 389 U.S. 954, 957, 958 (1967) (dissent).
77. 395 U.S. 464 (1969).
78. *Cascade Natural Gas Corp. v. El Paso Natural Gas Co.*, 386 U.S. 129 (1967).
79. 395 U.S. at 477 (Harlan, J., dissenting).
80. Ibid. at 471.
81. 15 U.S.C. § 18 (1950).
82. 370 U.S. 294 (1962).
83. Kauper, *supra*, note 56, at 329.
84. 370 U.S. at 319.
85. Ibid. at 320. Italics in original.
86. Ibid. at 322.
87. Ibid. at 344.
88. Ibid.
89. Ibid. at 346.
90. 374 U.S. 321 (1963).
91. 377 U.S. 271 (1964).
92. 384 U.S. 546 (1966).
93. Ibid. at 549.
94. 384 U.S. 270.
95. Ibid. at 275.

PART FOUR

CHAPTER 14—THE LAST YEARS

1. 388 U.S. 1 (1967).
2. 383 U.S. 663 (1966).
3. 392 U.S. 409 (1968).
4. *Escobedo v. Illinois*, 378 U.S. 478 (1964); *Miranda v. Arizona*, 384 U.S. 436 (1966).
5. Rodell, "It Is The Earl Warren Court," *New York Times Magazine*, March 13, 1966, 30, 96, 97.
6. For a collection of critical and favorable articles on the Warren Court, see L. Levy, ed., *The Supreme Court Under Earl Warren* (1972).
7. See Mersky and Blanstein, "Rating Supreme Court Justices," 58 *A.B.A. J.* 1183 (1972); Fleming and Owens, "The Great, Near Great, and Not so Great: Ranking of the U.S. Supreme Court Justices," unpublished manuscript, quoted in Powe, "Earl Warren: A Partial Dissent," 56 *No. Cal. L. Rev.* 408 (1978).
8. See J. Pollack, *Earl Warren: The Judge Who Changed America* 266 (1979).
9. Earl Warren to Lyndon Johnson, April 2, 1968, quoted in ibid. at 274.
10. See Pollack, *supra*, note 8, at 275.
11. Earl Warren to Lyndon Johnson, June 13, 1968, quoted in *New York Times*, June 27, 1968.
12. *New York Daily News*, June 22, 1968.
13. *New York Times*, June 27, 1968.
14. Quoted in *Washington Post*, July 6, 1968.
15. *New York Times*, June 30, 1968; *New Republic*, July 20, 1968.
16. *Wall Street Journal*, June 14, 1968.
17. Robert Griffin to Robert Shogan, January 26, 1970, quoted in R. Shogan, *A Question of Judgment* 154 (1972).
18. Abe Fortas, quoted in Rodell, "The Complexities of Mr. Justice Fortas," *New York Times Magazine*, July 28, 1968, 67; Shogan, *supra*, note 17, at 119.
19. Quoted in *Washington Post*, July 6, 1968.
20. Quoted in Shogan, *supra*, note 17, at 176.
21. John Hughes (press aide to Senator Griffin) to Robert Shogan, quoted in Shogan, *supra*, note 17, at 178.
22. U.S. Congress, Senate, Judiciary Committee, *Nominations of Abe Fortas and Homer Thornberry*, 90th Cong., 2d. sess., 1968, 1286–1304.
23. This same theme was to surface in the incident, the revelation of which eventually forced Fortas to resign from the Court. *Life* magazine, in its May 9, 1969 issue, reported that Fortas had accepted a $20,000 fee from a foundation established by Louis Wolfson, a financier who had recently been convicted of securities violations. The article suggested that when Wolfson's charges were pending, he had claimed that Fortas would render him assistance. No evidence was found to document this suggestion, and Fortas returned the $20,000 fee eleven months after it was offered, but the resultant public reaction to the ethical implications of Fortas' conduct eventually resulted in his resigning from the Court on May 14, 1969. The fee that Fortas had accepted was the first payment of a series that were to last the duration of Fortas' life and then extend to his widow. When evidence of the arrangement between Fortas and Wolfson came into the hands of the Justice Department, Attorney General John Mitchell secretly informed Warren, who was shocked and outraged by Fortas' conduct. See Shogan, *supra*, note 17, at 191–261.
24. Quoted in Pollack, *supra*, note 8, at 282.

25. *New York Times*, December 5, 1968.
26. 395 U.S. 486 (1969).
27. Ibid. at 549.
28. Ibid. at 550.
29. See Sandalow, "Comments on Powell v. McCormack," 17 *U.C.L.A. L. Rev.* 172 (1969).
30. This claim is made by B. Woodward and S. Armstrong, *The Brethren* 11 (1979).
31. See Benno Schmidt, Jr., oral history 74, Columbia University Oral History Research Office, Butler Library, Columbia University.
32. E. Warren, *A Republic, If You Can Keep It* (1972).
33. Powe, *supra*, note 7, at 409.
34. E. Warren, *The Memoirs of Earl Warren* 176, 213, 220, 233 (1977).
35. Ibid. at 249, 248, 250, 252.
36. Ibid. at 285, 313, 316.
37. Powe, *supra*, note 7, at 409.
38. Warren, *supra*, note 34, at 320, 371, 113.
39. Powe, *supra*, note 7, at 419.
40. Totenberg, " 'Miniature' High Court Is Planned," *National Observer*, November 11, 1972.
41. Friendly's remarks were quoted in the *New York Times*, November 14, 1972; Stewart's in the *Harvard Law Record*, November 21, 1972.
42. Freund, "Why We Need the National Court of Appeals," 59 *A.B.A. J.* 247 (1973).
43. In his memoirs Warren discussed his feud with the American Bar Association in some detail. Among other things, the ABA falsely announced that Warren had been dropped from its membership for nonpayment of dues. See Warren, *supra*, note 34, at 321–31.
44. Totenberg, "Warren Mobilizes Drive to Block Plan for Miniature Supreme Court," *National Observer*, December 16, 1972.
45. The address was reprinted in 59 *A.B.A. J.* 725 (1973).
46. 59 *A.B.A. J.* 725, 726 (1973).
47. Ibid. at 726, 727.
48. Ibid. at 728, 729.
49. Ibid. at 728.
50. Ibid. at 730.
51. Ibid.
52. Warren, *supra*, note 34, at 321.
53. 58 *A.B.A. J.* 721, 723, 724 (1973).
54. Warren, address, De Paul University College of Law, December, 1973, quoted in Pollack, *supra*, note 8, at 318; Warren, letter to Dean E. McHenry, May 30, 1974, quoted in Pollack, *supra*, note 8, at 323.
55. Quoted in Pollack, *supra*, note 8, at 321.
56. *United States v. Nixon*, 418 U.S. 683 (1974).


CHAPTER 15—FROM PROGRESSIVE TO LIBERAL

1. The intellectual history of Progressivism is extensive, although surprisingly little work has been done in the past decade. From an earlier portrait of Progressivism that identified the movement as a precursor of the New Deal and an example of the American tradition of recurrent liberalism, exemplified in such works as Henry Steele Commager, *The American Mind* (1950), and Arthur Link, *American Epoch* (1955),

historians moved to a posture that emphasized discontinuities between Progressivism and subsequent reform movements. With this emphasis on discontinuity came
efforts to construct explanatory models of Progressivism, the most influential of
which was Richard Hofstadter's "status revolution" model, put forth in *The Age of
Reform* (1955). Hofstadter argued that Progressivism was a protest by "displaced"
upper-middle-class native Americans against the potential loss of their prestige and
power in a rapidly industrializing culture where "new wealth" was becoming increasingly influential.

Hofstadter's synthesis, and other comparable efforts to generalize about Progressivism, such as George Mowry, *The Era of Theodore Roosevelt* (1958), stimulated a flurry
of activity among historians in the 1950s and sixties that sought to demonstrate the
diversity and complexity of Progressive thought. Such studies as David Noble's *The
Paradox of Progressive Thought* (1958), Daniel Levine's *Varieties of Reform Thought* (1964),
and Richard Abrams' *Conservatism in a Progressive Era* (1964) exemplified this trend.
Other historians sought to show that Progressive reforms were not "liberal" in the
conventional sense of that term, since the economic interests being singled out for
attack by reformers often benefited from the reforms. Robert Wiebe's *Businessmen
and Reform* (1962), and Gabriel Kolko's *The Triumph of Conservatism* (1963) were in
this vein, with Kolko claiming that Progressivism was from first to last a movement
designed to further the interests of established economic elites.

By the 1960s most historians had retreated into a posture that permitted only
modest generalizations about Progressivism. Hofstadter noted in 1963 that "one cannot . . . ignore the possibility that Progressivism could mean something different
in the countryside from what it meant in the city, that it might have different
principles in the Northeast as opposed to the South and West, [or] that businessmen
who favored some Progressive measures might have different hopes from those of
professional and middle class Progressives." (R. Hofstadter, *The Progressive Movement, 1900–1915* 3 (1963).) Otis Graham, in his 1967 study, *An Encore for Reform*,
spoke of "increased . . . respect for the need for qualification, for careful attention
to subdivisions and subgroups, for accounts that allow for complicated and overlap-
ping line of dissent." (O. Graham, *An Encore for Reform: The Old Progressives and the
New Deal* 234 (1967).)

That posture still seems to be dominant. But with the sharp divergence between
the attitudes of belief held by Progressives and those currently prevalent in America, the time appears ripe for a reemphasis on the common features and values of
Progressivism. One may note, in this vein, that all of the distinctively "Progressive"
attitudes of belief I have identified—morality, patriotism, and progress—have ceased
to have a common meaning in contemporary American educated thought. Moreover, the idea of affirmative government has passed from a commonplace policy
assumption to an issue of sharp controversy.

I made a preliminary step toward a reemphasis of the common features of Progressivism in "The Social Values of the Progressives: Some New Perspectives," 70
*South Atlantic Quarterly* 62 (1971). That effort can be seen as (unconsciously) responsive to themes suggested by David Thelan in his "Social Tensions and the Origins
of Progressivism," 56 *J. Am. Hist.* 323, 341 (1969), where Thelan called for a search
for what made Progressives "seek a common cause." Such a search has not seemed
to consume historians in the 1970s and eighties. Peter Filene despaired of a satisfactory characterization of Progressivism in his "Obituary for the Progressive Movement," 22 *Am. Q.* 20 (1970), and perhaps his despair has been contagious. Close to
a decade of inactivity would itself seem to provide a reason for a reassessment of
Progressivism.

2. W. White, *The Old Order Changeth* 31 (1970).

3. E. Warren, *A Republic If You Can Keep It* 2 (1972).
4. For a fuller discussion of these two economic theories, known respectively as the "New Nationalism" and the "New Freedom," see S. Fine, *Laissez-Faire and the General Welfare State* (1957) and A. Schlesinger, *The Crisis of the Old Order* (1957).
5. Warren, *supra*, note 3, at 159.
6. Ibid. at 2, 5–6.
7. E.g., *Barsky v. Board of Regents*, 347 U.S. 442 (1954).
8. Warren, *supra*, note 3, at 6.
9. Lewis, "A Talk With Warren About Crime, the Court, the Country," *New York Times Magazine*, October 19, 1969, 35.
10. E. Warren, *The Memoirs of Earl Warren* 122 (1977).
11. Warren, *supra*, note 3, at 83.
12. See G. Mowry, *The California Progressives* 92–103 (1950).
13. See Graham, *supra*, note 1.
14. Warren, "What Is Liberalism," *New York Times Magazine*, April 18, 1948, 10, 11.
15. Ibid.
16. See Pound, "Liberty of Contract," 8 *Colum. L. Rev.* 605 (1908). For an analysis of the relationship between Progressivism and sociological jurisprudence, see G. White, *Patterns of American Legal Thought* 99–135 (1978).
17. Earl Warren, letter to Robert Kenny, July 20, 1938. The letter is quoted in full in J. Stevenson, *The Undiminished Man* 166–67 (1980).
18. E. Warren, *supra*, note 10, at 6.
19. Some discretionary admissions policies, however, such as the imposition of preferential quotas based on race, have been held to violate the Equal Protection Clause. See *University of California Regents v. Bakke*, 438 U.S. 265, 320 (1978).
20. Ibid.

## CHAPTER 16—ETHICS AND ACTIVISM

1. See A. Paul, *Conservative Crisis and the Rule of Law* (1960); G. White, *The American Judicial Tradition*, 178–99 (1976).
2. See, e.g., *Osborn v. Bank of United States*, 22 U.S. (9 Wheat.) 737, 866 (1824) (Marshall, C.J.) ("Courts are the mere instruments of the law, and can will nothing"); Brewer, "The Nation's Safeguard," in *Proceedings of the N.Y. State Bar Ass'n* 46 (1894) ("The Courts . . . make no laws, they establish no policy, they never enter into the domain of public action").
3. 198 U.S. 45, 76 (1905).
4. Warren to Felix Frankfurter, August 6, 1954, Felix Frankfurter papers, Manuscripts Division, Library of Congress.
5. 358 U.S. 1 (1958).
6. Frankfurter to William J. Brennan, April 10, 1957, Frankfurter papers.
7. Harlan, "Mr. Justice Black—Remarks of a Colleague," 81 *Harv. L. Rev.* 1–2 (1967).
8. E. Warren, *The Memoirs of Earl Warren* 6 (1977).
9. For a discussion of Traynor, see White, *supra*, note 1, at 292–316.
10. See generally J. Howard, *Mr. Justice Murphy: A Political Biography* (1968).
11. Lewis, "Earl Warren," in 4 L. Friedman and F. Israel, eds., *The Justices of the United States Supreme Court* 2721, 2726 (4 vols., 1969).
12. See, e.g., L. Fuller, *The Morality of Law* (1964).
13. See Wechsler, "Toward Neutral Principles of Constitutional Law," 73 *Harv. L. Rev.* 1 (1959).

14. 390 U.S. 39 (1968).

15. Warren did note, in his dissent in *Grosso v. United States*, 390 U.S. 62 (1968), that no "protected . . . rights" were at stake because "[t]he occupation of gambling can in no sense be called a 'protected' activity." 390 U.S. at 80.

16. Cf. E. Corwin, *John Marshall and the Constitution* 116 (1919) ("Marshall's 'original bias,' to quote Story's own words, '. . . was to general principles and comprehensive views, rather than to technical or recondite learning' ").

# Index